CHEMICAL EVOLUTION: ORIGIN OF LIFE

Studies in
Chemical Evolution and the Origin of Life

Series Editor: Cyril Ponnamperuma

Published Volumes and Volumes in Preparation

Prebiological Self Organization of Matter
Cyril Ponnamperuma and Frederick R. Eirich (Eds.)

Trieste Conferences on
Chemical Evolution and the Origin of Life

Chemical Evolution: Origin of Life
Proceedings of Trieste Conference October 1992
Cyril Ponnamperuma and Julian Chela-Flores (Eds.)

Chemical Evolution: Self Organization of the
Macromolecules of Life
Proceedings of Trieste Conference October 1993
Julian Chela-Flores et al.

CHEMICAL EVOLUTION: ORIGIN OF LIFE

Edited by

Cyril Ponnamperuma
Laboratory of Chemical Evolution
University of Maryland at College Park
College Park, Maryland, USA

and

Julian Chela-Flores
International Centre for Theoretical Physics
Trieste, Italy
and
International Institute for Advanced Studies
Caracas, Venezuela

Proceedings of
The Trieste Conference on Chemical Evolution and the Origin of Life, 26–30 October 1992

A. DEEPAK Publishing **1993**
A Division of Science and Technology Corporation
Hampton, Virginia USA

A. DEEPAK Publishing
A Division of Science and Technology Corporation
101 Research Drive
Hampton, Virginia 23666–1340 USA

Library of Congress Cataloging-in-Publication Data

Trieste Conference on Chemical Evolution and the Origin of Life (1992)
Chemical evolution—origin of life : proceedings of the Trieste Conference on Chemical Evolution and the Origin of Life, 26–30 October 1992 / edited by Cyril Ponnamperuma and Julian Chela-Flores.
p. cm.
Includes bibliographical references.
ISBN 0-937194-31-X : $48.00
1. Molecular evolution—Congresses. 2. Life—Origin—Congresses. I. Ponnamperuma, Cyril, 1923– . II. Chela-Flores, Julian. III. Title.
QH325.T74 1992
577—dc20 93-34768
CIP

Printed in the United States of America

CONTENTS

DISCUSSION SESSIONS

PREFACE

Modern scientific thinking leads to the conclusion that the emergence of life is a necessary consequence of the evolutionary process in the universe. This interdisciplinary book consists of the proceedings of the Conference on Chemical Evolution and the Origin of Life, held in Trieste, Italy, 26–30 October 1992. The purpose of the Conference was to provide a forum to explore the many aspects of cosmic evolution from the chemical history of the universe, the processes of prebiological chemistry, self-organization, the earliest evidence for life, and the origin of chirality to the search for extraterrestrial intelligence.

These subjects were covered by 15 speakers from France, Germany, Pakistan, the Russian Federation, Sri Lanka, Sweden, UK, USA, and Venezuela. In the discussion sessions 15 short contributions were presented by participants from the Czech Republic, France, Germany, Hungary, India, Israel, Mexico, People's Republic of China, Spain, UK, USA, and Yugoslavia. Funds were provided by the International Centre for Science, the International Centre for Theoretical Physics, the European Community, and UNESCO.

This event brought together an international group of 66 scientists, of whom 28 were speakers at the plenary sessions and discussion periods. Altogether 34 countries were represented, including 17 from the Third World.

This volume will be of interest not only to the students of chemical evolution, but also to a broad spectrum of researchers and students in the fields of biochemistry, biophysics, and other related fields. It is hoped that the record of this presentation will stimulate many students in the basic sciences to be involved in the exciting subject of chemical evolution.

C. Ponnamperuma
University of Maryland, USA

J. Chela-Flores
ICTP, Italy, and IDEA, Venezuela

PHOTOGRAPH OF PARTICIPANTS

Participants: 1) I. Draganic, 2) K. Jamil, 3) V.J.A. Novak, 4) A. Negron-Mendoza, 5) W. Wang, 6) J. Chela-Flores, 7) R. Weerasooriya, 8) A. Salam, 9) C. Ponnamperuma, 10) H. Baltscheffsky, 11) N. Rahman, 12) G.O. Arrhenius, 13) C.-X. Wang, 14) B. Mathew Graham, 15) M.K. Rahman, 16) I. Arif, 17) B. Dorjsuren, 18) A. Popescu, 19) B. Heinz, 20) M.S. Chadha, 21) J.M. Diaz, 22) S. Dhanuskodi, 23) A.R.J. Figureau, 24) P. Aich, 25) T. Yohannes, 26) Y.Y.J. Yan, 27) N.Y.K. Baryeh, 28) G.H. Shehu, 29) K. Danforth (Conference Secretary), 30) K.N. Nedev, 31) A.J. MacDermott, 32) Observer, 33) P. Schuster, 34) L. Peliti, 35) R.K. Khanna, 36) V. Avetisov, 37) C.M. Gallo Lopez-Aliaga, 38) O.R. Olguin, 39) J.C. Cecchini, 40) P.A.S. Silva, 41) P. Fernandes Machado, 42) C. Cosmovici, 43) J. Heidmann, 44) W. Hebel, 45) F.L. Babani, 46) L.-F. Luo, 47) V. Lopac, 48) R. Weiner, 49) M. A. Khan, 50) L. Barron, 51) G. Gilat, 52) Observer, 53) G. Lenaz, 54) P. Boya, 55) L.J. Boya, 56) A. Garay, 57) T.O. Diener, 58) L. Keszthelyi, 59) Observer, 60) M. Rizzotti, 61) Observer, 62) J.W. Schopf, 63) A. Brack, 64) Observer, 65) S. Chen, 66) Observer

THE ORIGIN, EVOLUTION, AND DISTRIBUTION OF LIFE IN THE UNIVERSE

Cyril Ponnamperuma
Laboratory of Chemical Evolution
Department of Chemistry and Biochemistry
University of Maryland
College Park, Maryland 20742

ABSTRACT

According to the Oparin-Haldane Hypothesis of Chemical Evolution, the formation of molecules of biological significance was a necessary preamble to the emergence of life on earth. The analysis of carbonaceous chondrites has revealed the presence of these same molecules in the early samples of the solar system. Radio astronomers have discovered a vast array of organic molecules in the interstellar medium. We are thus led to the inescapable conclusion that life must be commonplace in the Cosmos. Laboratory experiments help us to retrace the path of chemical evolution on earth. Planetary probes search for evidence of life in our solar system. Radio signals from outer space may reveal the presence of our distant neighbors.

1. INTRODUCTION

In an authoritative document, the National Academy of Sciences has set down the search for extraterrestrial life as the prime goal of space biology.

> "It is not since Darwin and, before him, Copernicus, that science has had the opportunity for so great an impact on man's understanding of man. The scientific question at stake in exobiology is the most exciting, the most challenging, and profound issue, not only of this century, but of the entire naturalistic movement that has characterized the history of western thought for over three hundred years. If there is life on Mars and if we can demonstrate its independent origin, then we shall have an enlightening answer to the question of the improbability and uniqueness in the origin of life. Arising twice in a single planetary system, it must surely occur abundantly elsewhere in the staggering number of comparable planetary systems."

CHEMICAL EVOLUTION: ORIGIN OF LIFE
C. Ponnamperuma and J. Chela-Flores (Eds.)

ISBN 0-937194-31-X

2. EVIDENCE FOR LIFE IN THE UNIVERSE

Modern astronomy by its exhaustive study of galactic, stellar and planetary evolution has come to the inescapable conclusion that life in the universe must be of common occurrence. On the basis of our sampling of galaxy populations to the limit attainable by present-day telescopes, we can readily compte that there are more than 10^{20} stars in the universe. Each one of these can maintain the photochemical reactions that are the basis of plant and animal life. Let us impose a number of restrictions in considering the stars that can support life. Suppose that because of doubling, clustering, and secondary collisions, only one star in a thousand has a planetary system, has a planet at the right distance from a star to provide the water and the warmth that protoplasm requires. In our own planetary system, we have two such planets. Further, let us suppose that only one out of a thousand of those stars with planets at the right distance has a planet large enough to hold an atmosphere. In our planetary system at least seven of the nine can do that. Suppose a further restriction is made, and we suggest that the right chemical composition for life to arise occurs only once in a thousand times. Assuming all these four restrictions, we come to the conclusion that there are at least a hundred million possibilities for the existence of life. This is a conservative estimate made by Harlow Shapley.

The astronomer, Su-Shu Huang, has been less rigorous in his requirements for the existence of life in the universe. In considering the time scales of biological and stellar evolution, the habitable zones of a star, dynamic and other considerations, he has come to the astonishing conclusion that at least five percent of the stars in the universe must support life. This means that there are at least a hundred billion, billion sites for the existence of life.

This conclusion which astronomers have reached by the rigorous analysis of scientific data was already prophetically foretold by Giovanni Bruno in the sixteenth century:

> "Sky, universe, all-embracing ether, and immeasurable space alive with movement ... all these are of one nature. In space there are countless constellations, suns and planets; we see only the suns because they give light; the planets remain invisible, for they are small and dark. There are also numberless earths circling around their suns, no worse and no less than this globe of ours. For no reasonable mind can assume that heavenly bodies which may be far more magnificent than ours would not bear upon them creatures similar or even superior to those upon our human Earth."

3. THE SEARCH

In our search for the existence of extraterrestrial life, three possible approaches present themselves to us.

Firstly, we have the experimental attack on the problem. Here life is considered an inevitable consequence of the evolution of matter. Since the laws of chemistry and physics are universal laws, the retracing, in the laboratory, of the path by which life appeared on earth would give strong support to our belief in its existence elsewhere in the universe.

Secondly, the landing of instruments or man somewhere in the universe. With our present knowledge, this attempt would undoubtedly be restricted to our own planetary system.

A final method is **via** radio contact with civilizations in outer space. This presupposes the existence of intelligent beings in space with a technology as advanced or even greater than our own.

4. THE EXPERIMENTAL APPROACH

Laboratory experiments on earth can reveal which materials and conditions available in the universe might give rise to chemical components and structural attributes of life was we know it. As we noted earlier, the retracing of the pattern by which life appeared on earth would give strong support to the theory of its existence elsewhere in the universe. This is the study of chemical evolution.

The Darwinian theory of evolution has postulated the unity of the earth's entire biosphere. According to Darwin, the higher forms of life evolved from the lower over a very extended period in the life of this planet. Fossil analysis has shown that the oldest known forms of life may be about three billion years old. Geochemical data tells us that the earth is about four and one-half billion years old. Life, indeed, had a beginning on this planet. A question immediately arises as to the history of our own planet between its birth four and one-half billion years ago and the emergence of life. This idea was uppermost in the mind of the physicist Tyndall, when in 1871, he wrote in his "Fragments of Science of Unscientific People,"

> "Darwin placed at the root of life a primordial germ, from which he conceived that the amazing richness and variety of the life now upon the earth's surface might have deduced. If this hypothesis were true, it would not be final. The human imagination would infallibly look beyond the germ and, however hopeless the attempt, would enquire into the history of its genesis.... A desire immediately arises to connect the present life of our planet with the past. We wish to know something of

our remotest ancestry... Does life belong to what we know as matter, or is it an independent principle inserted into matter at some suitable epoch, when the physical conditions became such as to permit of the development of life?"

The consideration of biological evolution thus leads us logically to another form of evolution, namely, chemical evolution.

Recent biochemical discoveries have underlined the remarkable unity of living matter. In all living organisms, from the smallest microbe to the largest mammal, there are two basic molecules. Their interaction appears to result in that unique property of matter which is generally described by the word "life." these two molecules are the nucleic acids and protein. While each one of these molecules is complex in form, the units comprising them are few in number. The nucleic acid molecule consists of nucleotides strung together like beads along a chain. The nucleotides, in turn, are made up of a purine or pyrimidine base, a sugar, ad a phosphate. In the protein molecule, twenty amino acids link up with one another to give the macromolecule. A study of the composition of living matter thus leads us to the inescapable conclusion that all living organisms must have had some common chemical ancestry. A form of evolution purely chemical in nature must of necessity have preceded biological evolution.

Chemical evolution may be considered to have taken place in three stages. From inorganic chemistry to organic chemistry, and from organic chemistry to biological chemistry. The first stage of chemical evolution perhaps began with the very origin of matter. In a series of cataclysmic reactions during the birth of a star, the elements of the periodic table must have been formed. Almost 15 billion years later, when the solar system was being formed, the highly reactive elements which occur in living organisms, probably existed in combination with hydrogen -- carbon as methane, nitrogen as ammonia, and oxygen as water. Four-a-half billion years ago, when the planet earth was being born from the primitive dust cloud. the rudimentary molecules, which were the forerunners of the complex biological polymers of today, were perhaps already in existence. Within this framework, life appears to be a special property of matter, a property which arose at a particular period in the existence of our planet and which resulted from its orderly development.

The idea of life arising from non-life, or the theory of spontaneous generation, had been accepted for centuries. One had only to accept the evidence of the senses, thought the ancients: worms from mud, maggots from decaying meat, and mice from old linen. Aristotle had propounded the doctrine of spontaneous generation in his *Metaphysics*. He had traced the generation of fireflies to morning dew and the birth of mice to moist soil. His teaching was accepted by the long line of Western thinkers who had turned to him as the final authority in matters metaphysical and physical. Newton, Harvey, Descartes, van Helmont, all accepted the idea of spontaneous generation without serious question. Even the English Jesuit, John Tuberville Needham, could subscribe to this view, for Genesis tells not that God created plants and animals directly but that he bade the earth and waters to bring them forth. The world's literature is full of allusions to this popular belief in spontaneous generation. Virgil in his *Georgics* tells us how a swarm of bees arose form the carcass of a calf. Recall *Anthony and Cleopatra*, Act II, Scene VII, where Lepidus tells Mark Antony, "Your serpent of Egypt is bred ... now of your mud by the operation of your sun - so is your crocodile."

The great impetus, however, to the experimental study of the origin of life began with the Russian pointing out, "that there was no fundamental difference between a living organism and brute matter. The complex combination of manifestations and properties so characteristic of life must have arisen in the process of the evolution of matter." According to Oparin,

> "... at first there were the simple solutions of organic substances whose behaviour was governed by the properties of their component atoms and the arrangement of these atoms in the molecular structure. But gradually, as a result of growth, and increasing complexity of the molecules, new properties have come into being and a new colloidal chemical order was imposed on the more simple organic chemical relations. These newer properties were determined by the spatial arrangement and mutual relationship of the molecules. In this process biological orderliness already comes into prominence."

Independently of Oparin, Haldane in 1928, had speculated on the early conditions suitable of the emergence of terrestrial life.

> "When ultraviolet light acts on a mixture of water; carbon dioxide and ammonia, a variety of organic substances are made, including sugars and apparently some of the materials from which proteins are built up. Before the origin of life they must have accumulated until the primitive oceans reached the constituency of a hot dilute soup."

A starting point for any experimental consideration of the origin of life must turn round the question of the cosmic distribution of elements. Astronomical spectroscopy reveals that with surprising uniformity the most abundant elements in our galaxy are, in the order of rank; hydrogen, helium, oxygen, nitrogen and carbon. Hydrogen, oxygen, nitrogen and carbon are indeed the basic constituents of living systems. The table on the composition of the sun illustrates the distribution of these elements very clearly (Table 1). In the presence of hydrogen, carbon will be in the form of methane, oxygen as water, and nitrogen as ammonia. It is this atmosphere of water vapor, methane, ammonia, and small amounts of hydrogen which will be considered in this discussion as the primitive atmosphere of the earth.

TABLE 1. Composition of the Sun

	Percent
Hydrogen	87.0
Helium	12.9
Oxygen	0.025
Nitrogen	0.02
Carbon	0.01
Magnesium	0.003
Silicon	0.002
Iron	0.001
Sulfur	0.001
Others	0.038

The energies available for the synthesis of organic compounds under primitive earth conditions are ultraviolet light from the sun, electric discharges, ionizing radiation, and heat. It is evident that sunlight was the principal source of energy. Photochemical reactions would have taken place in the upper atmosphere, and the products transferred by convection. Next in importance as a source of energy are electric discharges such as lighting and corona discharges from pointed objects. These occur close to the earth's surface, and hence, would more efficiently deposit the reaction products in the primitive oceans. A certain amount of energy was also available from the disintegration of uranium, thorium, and potassium 40. While some of this energy may have been expended on the solid material such as rocks, a certain proportion of it was available in the oceans and the atmosphere. Heat from volcanoes may also have been effective in primordial organic synthesis. In comparison to the energy from the sun and electric discharges, this was, perhaps, not too widely distributed and its effect may have been only local on the sides of volcanoes for example. Most of these forms of energy have been used in the laboratory for the synthesis of organic molecules. Simulation experiments have been devised to study the effect of ionizing radiation, electric discharges, heat and ultraviolet light on the assumed early atmosphere of the earth. The analysis of the end products has often yielded, very surprisingly, the very compounds which we consider today as important for living systems.

In the experiments in our own laboratory, we have adopted the simple working hypothesis that the molecules which are fundamental now, were fundamental at the time of the origin of life. We are analyzing "the primordial soup" described by Haldane. The various forms of energy which are thought to have been present in the primitive earth have been used by us in a series of experiments. In the experiments with methane, ammonia, and water, electron irradiation was used as convenient source of ionizing radiation simulating the K40 on the primitive earth. The results of this investigation clearly establish adenine as a product of the irradiation of methane, ammonia, and water. It is the single largest non-volatile compound produced. The apparent preference for adenine synthesis may be related to adenine's multiple roles in biological systems. Not only is it a constituent of both the nucleic acids DNA and RNA, but it is also a unit of many important co-factors. In these and other experiments, most of the molecules necessary for life have been synthesized.

5. LIFE ON MARS

Our effort to land an instrument or eventually a scientist astronaut on the neighboring planet is primarily directed to the planet Mars. The possibility of life on Mars has often been raised. The "canal-like structures" on Mars and the seasonal wave of darkening across the planet have led many to believe that there must be some form of life on Mars. Some have suggested the existence of highly intelligent beings, who by incredible feats of engineering, have saved for themselves the depleting water supply on the planet, by building mammoth canals crisscrossing the planet. All these speculations have fired the imagination of the planetary scientist and have made him determined to find out the answer to the question, "Is there life on Mars?"

When we leave speculation aside and consider the actual conditions that exist on Mars today, we must very likely exclude the existence of advanced forms of life. However, the physical conditions are such that low forms of life, such as microorganisms, could survive on the planet. The atmosphere of Mars is made up of a small amount of carbon dioxide and a trace of water. Oxygen has not been detected, and if it exists at all, it must be in very low concentration. There must be a high ultraviolet flux which reaches the surface of Mars, since the Martian atmosphere does not appear to have a built in protection from ultraviolet light as our own earth. With this high incidence of ultra-violet radiation, Martian organisms would have to protect themselves by burrowing into the surface or, they may have involved a mechanism compatible with the existence of a high ultraviolet flux.

Polarization data have shown that the polar caps of Mars consist of solid carbon dioxide mixed with ice. They wax and wane with the seasons. They are probably extremely thin, and perhaps not more than a few centimeters in thickness. In the summer, the pole cap recedes to about thirty-five kilometers per day. As one pole cap recedes, the other is under a cloud. A dark band has been observed to follow the receding pole cap. It is this dark band observed to follow the receding pole cap. It is this dark band which has led many to speculate on the existence of vegetation. The wave of darkening proceeds from pole to equator at the rate of about thirty-five kilometers per day during the spring and summer. It seems attractive to interpret this wave of darkening as a gradual growth of vegetation across the planet. The average temperature of Mars is considerably lower than the average temperature of the earth. However, the extremes may not be incompatible with life.

The atmospheric pressure on Mars was long disputed. The recent data confirms the low estimate that it is only one hundredth of that of the earth. While this low pressure may not in itself be a factor which affects the survival of microorganisms, it might have an effect on the availability of water. The amount of water present on Mars is about one-thousandth of that found in the earth's atmosphere. This again does not preclude the existence of micro-environments in which above average accumulations of water occur.

A very sketchy survey of the physical parameters of Mars indicate to us that, although the conditions are rigorous as compared to the earth, they are within the range in which microorganisms can survive. Indeed, laboratory experiments in which these conditions have been simulated have shown that some earth microorganisms can survive and even multiply under such conditions. Furthermore, if we consider planetary evolution, on account of the smallness of the planet Mars, the processes of chemical evolution may have proceeded very rapidly. Life may have evolved and disappeared. Visitors to Mars may be greeted by relics or fossils of a once thriving biosphere.

To celebrate the 200th anniversary of the birth of the United States, two spacecraft -- Viking I and Viking II -- were landed on Mars in July 1976. On board were a series of experiments -- a camera to take photographs of Mars, a weather station, devices for chemical analysis and of the greatest interest to all of us, a life detection package.

The photographs were magnificent, and revealed a most stunning landscape of the red planet. The mass spectrometer on the other hand gave us the astonishing result that the total amount of carbon on the planet was less than five parts per billion. The life detection experiments performed well. There were some exciting preliminary signals However, in the absence of organic matter, we had to interpret these as a result of the surface of Mars mimicking Biology. The Viking mission did not give nay evidence of life on Mars.

6. INTELLIGENT LIFE

According to our calculations of the distribution of life in the universe, it is reasonable to assume that intelligent life must have evolved in a large number of sites. An estimate of the distribution of intelligent life in the universe, made by Professor Carl Sagan of Cornell University, puts the figure, at a million in our own galaxy. The distance between these civilizations may be as large as thousand light years. The separation between civilization seems to be almost infinite by human standards.

> "There is one race of men; one race of gods; both have breath of life from a single mother. But sundered power holds us divided, so that one is nothing, while for the other the brazen sky is established their sure citadel forever"

wrote Pindar in the sixth *Nemean Ode*.

Among the methods available for interstellar communication, nuclear particle radiation and electromagnetic radiation present themselves as possible methods. Nuclear particle radiation can carry information at the speed of light while known useful nuclear particles possess far more energy than photons, their information content is very similar to that of electromagnetic radiation. Electromagnetic radiation may travel at the velocity of light with little interference to distances of thousand light years or more.

If electromagnetic communication is chosen as the best methods of solving the problem, the next question centers around the best wave length at which to transmit a message. The first theoretical discussion of this was made in 1959, by Cocconi and Morrison who suggested that of the hydrogen line at twenty-one centimeters. The sub-harmonics and harmonics of the hydrogen line frequency have also been suggested as possible candidates. One search for intelligent signals has already been made on the hydrogen line frequency. The hydrogen line at twenty-one centimeters has been considered to be the one that intelligent beings somewhere in the universe might use. This is presumably because frequencies of lower wave length will be jammed by natural radio noise. According to Professor Frank Drake who directed Project Ozma, this monumental search for intelligent beings in the universe, may be compared to meeting a friend in New York City, without making arrangements in advance about a meeting place. One does not just wander the streets looking everywhere. Instead, the most likely places to look are the places that are already familiar -- Grand Central Station, for instance, or Times Square. There are similar places in every city with which most people are acquainted. This is what the radio astronomer is in search of -- a Grand Central Station of the galaxy -- some special frequency about which everyone in the Milky Way would know. This indeed is the emanation of the hydrogen line.

Perhaps the best place to seek for intelligent signals with our present state of knowledge seems to be to concentrate on the nearest stars. the two nearest for consideration are Tau Ceti in the constellation of Cetus, and the star, Epsilon Eridani, in the constellation of Eridanus. Both of these are about ten light years away.

In the search conducted at the Green Bank Observatory, each day at three in the morning the telescope was pointed to Tau Ceti. When Tau Ceti set beyond the mountains, the telescope was swung to Epsilon Eridani, and the search carried on. A signal was picked up twice; the same signal each time. The discovery was exciting, but moving the antenna showed that these were actually signals from the earth and not from space. At the end of two months endeavor, no evidence of signals from space could be found on the records. There were the first steps of the endeavor at the National Astronomical Observatory in the Project Ozma, named after the Princess Os, who lived in the mythical land of Oz. It is clear that one should not be disappointed or discouraged because the first attempts of Project Ozma to find another civilization have failed. It will take many long years with larger radio-telescopes and more sensitive receivers to have a good chance of success. Let us assume that after years of futile listening, we receive a peculiar series of pulses and spaces from - Eridani. The message is repeated every twenty-two hours and fifty-three minutes, apparently the length of their day. The pulses occur at separations which are integral multiples of a minimum separation. Writing ones for pulses, and filling the blanks with the appropriate numbers of zeros, we get a binary series. It consists of 1271 ones and zeros. 1271 is the products of two prime numbers, 31 x 41. This strongly suggests that we arrange the message in a 31 x 41 array (Figure

1). When we do so, putting blanks for zeros and a dot for each pulse, we get the nonrandom pattern. Apparently we are in touch with a race of erect bipeds who reproduce sexually; there is even a suggestion that they might be mammals. The crude circle and column of dots at the left suggest their sun and planetary systems. The figure is pointing to the fourth planet, evidently their home. The way line over the third planet indicates that it is covered with water and the fish-like form shows that there is marine life there. The bipeds know this so they must have space travel.

7. CONCLUSION

There is no reason to doubt that we shall rediscover, one by one, the path of chemical evolution. We may even reproduce the intermediate steps in the laboratory. Looking back upon the biochemical understanding gained during the span of one human generation, we have the right to be quite optimistic. In contrast to unconscious nature which had to spend billions of years, conscious nature has a purpose and know the outcome.

Before the dawn of the twenty-first century, further manned exploration of the solar system may tell us whether we are alone in our solar system. Radio telescopes scanning the distant galaxies for intelligible messages may reveal the presence of our more remote neighbors. Laboratory experiments will endeavor to retrace the path of chemical evolution and may support our belief in the existence of extraterrestrial life.

To conclude with Harlow Sharpley:

> "The new discoveries and developments contribute to the unfolding of a magnificent universe, ... with our confreres on distant planets; with our fellow animals and plants of land, air and sea; with the rocks and waters of all planetary crusts and the photons and atoms that make up the stars ... with all these we are associated in an existence and an evolution ... and as groping philosophers and scientists we are thankful for the mysteries that still lie beyond our grasp."

Figure 1. Extraterrestrial message.

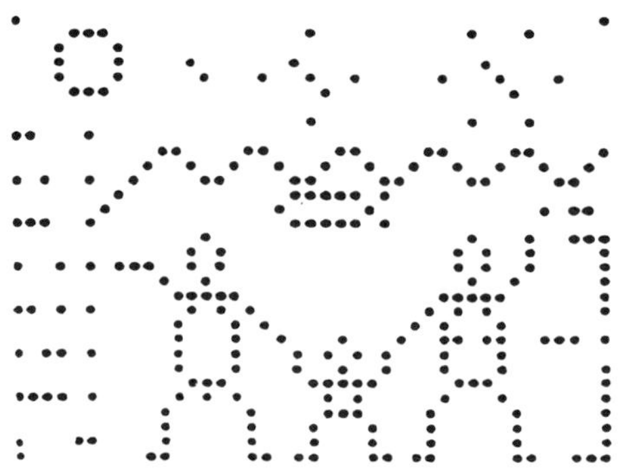

REFERENCES

A Review of Space Research. Washington, DC: National Academy of Sciences National Research Council Publication No. 1079, 1962, chapt. 9.

Bruno, G. *Dialoghi Italiani*, edited by G. Aquileccha, 3rd ed., 1958.

Cocconi, G., and P. Morrison. *Nature* 1894, 844, 1959.

Darwin, C. *Life and Letters 3*, Notes of the Royal Society, London, 14, No. 1, 1959.

Drake, F. *Intelligent Life in Space.* New York: Macmillan, 1964.

Haldane, J. B. S. The Origin of Life. *Rationalist Annu.* 148:3-10, 1928.

Huang, S.-S. *American Scientist* 47:397, 1959.

Oparin, A. I. *Origin of Life.* New York: Dover, 1938.

Ponnamperuma, C. Experimental Studies on the Origin of Life. *J. Br. Interplanetary Soc.* 42:397-400, 1989.

Shapley, H. *Of Stars and Men.* Boston, MA: Beacon, 1958.

Tyndall, J. *Fragments of Science for Unscientific People.* London: Longmans, Green, 1871.

**This paper was published by The American Physiological Society in *The Physiologist*, 35, 5, 273-78, Oct. 1992.

CHEMICAL ORIGIN AND EARLY EVOLUTION OF BIOLOGICAL ENERGY CONVERSION

Herrick Baltscheffsky
Department of Biochemistry, Arrhenius Laboratories for Natural Sciences, Stockholm University, S-106 91 Stockholm, Sweden

ABSTRACT

This presentation starts with a brief general introduction of some essentials in biological energy conversion and some questions about its molecular evolution in connection with the origin of life. Here we consider only the kind of life on earth of which something may be said to be known, i.e. life based on the interplay between information and function at the level of nucleic acids and proteins, as this would seem to have been a prerequisite for the evolution of all so far known organisms. Attention will then be focussed on the main current alternatives for primordial energy conversion reactions of potential significance for the origin of life. Among those energy-rich compounds which have been shown to exist also in living cells, inorganic pyrophosphate (PPi) appears to be a most suitable candidate for having been involved in energy conversion already at the pre-nucleotide level of chemical evolution. In addition, geochemical characteristics of PPi which could be connected with its apparent role in chemical evolution will be considered, as will some new aspects of the inorganic pyrophosphatases (PPases). Finally, a sequence of anastrophes, which may well have occurred and have been of particular significance in the early evolution of biological energy conversion, and which may possibly still be traced among existing enzymes, will be briefly discussed.

1. INTRODUCTION

1.1 SOME ESSENTIALS IN BIOLOGICAL ENERGY CONVERSION

When life on earth appeared, some four billion years ago, the cosmic evolutionary process, on this planet, left its chemical and entered its biological phase. So did also the molecular evolution of energy conversion.

Just as the origin of life problems may be approached both from the preceding chemical evolution side and from the subsequent biological evolution side, so may also the problems connected with the origin and early evolution of biological energy conversion.

ISBN 0-937194-31-X

The coupling between energy liberating and energy requiring chemical reactions is a basic property of life, as is the capacity of every living cell to exchange energy and matter with its environment. The two main kinds of useful energy taken up are light (photic) energy and chemical energy, the latter becoming metabolically liberated from the compounds in which it resides.

In photosynthetic organisms, containing a suitable light absorbing pigment, usually chlorophyll, light energy is transformed to metabolically useful chemiosmotic and/or chemical energy. In photosynthesis with chlorophyll, light initiates series of oxidation-reduction reactions, several being quite similar to those occurring in both anaerobic and aerobic cellular respiration, which, at the "electron transport level", are coupled to chemiosmosis and phosphorylation. These major, membrane bound bioenergetic reactions are also linked to the transport of various compounds through cell membranes. In addition, various well-known fermentation, glycolytic and similar reactions in the cytoplasm, such as the formation of the cellular "energy currency" ATP (adenosine triphosphate) over 1,3-phospho-glyceric acid or phosphoenolpyruvate, as well as the formation of other energy-rich phosphate or thioester compounds, are of importance in biological energy conversion, at the "substrate level".

It may be pointed out at this stage, that whereas closed systems progress towards maximum chaos, or entropy, living systems seem to do the opposite. This apparent contradiction may also be expressed as a tendency towards an increase in free energy in living systems, in contrast to one towards minimum free energy in closed systems, such as those treated by classical thermodynamics. However, as has long been known, this increase in energy and order of living cells is at the expense of a greater amount of energy obtained from the environment.

1.2 SOME QUESTIONS ABOUT THE EARLY EVOLUTION OF ENERGY CONVERSION

Was the original energy conversion process, which led to early living systems driven by light or by energy-rich chemical compounds, or both? Were the earliest energy-rich compounds, which converted energy on the primitive earth, of condensed phosphate or of thioester nature, or both? Did the earliest energy conversion processes on the way to life occur at membranes or other solid matter, or in solution? Was chemiosmosis through membranes, involving protons and other ions, an early energy conversion process, perhaps regulating pH, osmotic pressure and ionic environment? Some of these questions will be discussed below in considerable detail.

In general, I adhere to the view that a functioning energy conversion system was a necessary prerequisite for the emergence of nucleic acid type molecules. It is still a very open question whether any such energy conversion system became incorporated essentially unchanged into early living cells and, somehow, produced by their genetic machinery, or whether major changes in energy conversion occurred in connection with the origin of life.

2. CHEMICAL ORIGINS

2.1 EARLY "WORLDS"

In the 1980s, when it had been firmly established that RNA was capable of functioning as a "ribozyme", i.e. could catalyze some of its own reactions, Gilbert (1986) created the concept of an "RNA world", pointing out that, if RNA enzymes can catalyze the synthesis of a new RNA molecule from precursors and an RNA template, then there is no early need for protein enzymes.

Recently, additional early "worlds" have been suggested. Three of them are of interest in connection with early energy conversion, already at the pre-nucleotide level, and in addition, one of these would appear to be of potential relevance for early information transfer, one for phosphorylation and one for oxidation-reduction. These three are called the "thioester world" (de Duve, 1991), the "pyrophosphate world" or "PPi world" (H. Baltscheffsky and M. Baltscheffsky, 1993) and the "iron-sulphur world" (Wächtershäuser, 1992). With the introduction of a "PPi world" it became practical to suggest also an "ATP world" (H. Baltscheffsky and M. Baltscheffsky, 1993), for two main reasons:
1) with PPi as the favoured candidate of ours for a central role in energy conversion at the pre-nucleotide level of molecular evolution, there exists an intriguing possibility that a pathway, still traceable in protein structures, exists for an evolution from energy conversion with PPi to energy conversion with ATP, and 2) both the heteropolymeric RNA with its informational and functional capabilities, and the monomeric ATP with its pivotal position in the energy conversion of all known living cells, may well have been nucleotides of major importance in an early nucleotide world.

All these "worlds" may easily be collected in the scheme given in Fig. 1, which shows their relative positioning in time and some interrelationships.

As Fig. 1 indicates, the three "worlds" which were suggested to have existed before nucleotides entered into the molecular evolutionary process may have occurred more or less simultaneously on the primitive earth and may each have been of energetic significance for the origin of life. The thioesters may, in addition, also have been instrumental in early information transfer in connection with the formation of oligopeptides, as has been pointed out by de Duve (1991), with reference to the earlier discovery by Lipmann (1971) of cellular pantetheine-dependent peptide synthesis and his suggestion that it may have preceded the RNA-dependent mechanism. PPi

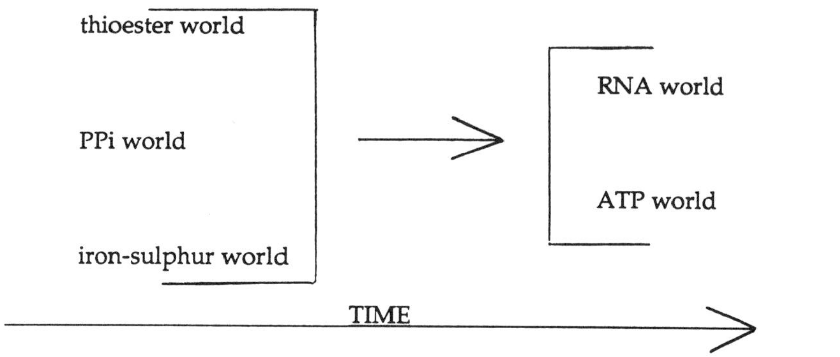

Figure 1. Earlier suggested non-nucleotide and nucleotide "worlds". Further details are given in the text and the references cited above.

may be assumed to have functioned not only as an energy donor, but also as a phosphorylating agent, in agreement with its known functions in various living organisms (M. Baltscheffsky and H. Baltscheffsky, 1992). And iron-sulphur compounds have long been known to participate in a multitude of biological oxidation-reduction reactions. A central concept in the iron-sulphur world panorama is the suggested oxidative formation of pyrite (FeS_2) from hydrogen sulphide and ferrous ions or iron sulphide (FeS), a reaction claimed to be a most plausible source of reducing power for the early evolution of life (Wächtershäuser, 1992). Pyrite mineralization has been demonstrated in living cells (Mann et al., 1990), and pyrite formation has recently been implicated to occur in a number of redox experiments of potential geochemical and early biochemical significance (Blöchl et al., 1992).

2.2 PPi IN MAGMA AND IN A MINERAL

The demonstration by Yamagata et al. (1991) that appreciable amounts of PPi are obtained both from hot volcanic gas and in model experiments simulating magmatic conditions could well constitute a breakthrough in our evaluation of the possibilities for the existence and continuous supply of PPi on the early earth. The earlier finding of a PPi containing mineral (Rouse et al., 1988), canaphite ($CaNa_2P_2O_7x4H_2O$), serves to underline the stability of PPi as well as the fact that a sufficiently high energy of activation may actually prevent a thermodynamically favourable reaction, as in this case the hydrolysis of PPi, from occurring at any significant rate.

As is shown by both these findings, there are now good indications supporting earlier assumptions that PPi may have been around during chemical evolution on the primitive earth.

3. ON PPi AND BIOLOGICAL EVOLUTION

3.1 FORMATION AND UTILISATION OF PPi

Our discovery of light-induced formation of PPi, as a first alternative to the formation of ATP, in bacterial photophosphorylation (H. Baltscheffsky et al., 1966; H. Baltscheffsky and von Stedingk, 1966), and the subsequent demonstration that PPi in the same system could serve as a useful alternative to ATP as energy donor (M. Baltscheffsky et al., 1967; M. Baltscheffsky, 1967), led us to propose that PPi may have been the original phosphorylated product of photophosphorylation , later substituted in evolution by ATP (H. Baltscheffsky, 1967). As PPi and PPases have recently been reviewed also from the evolutionary point of view (M. Baltscheffsky and H. Baltscheffsky, 1992), it may suffice here to emphasize the above mentioned observations bearing on the existence and possible roles of PPi on the primitive earth and to focus attention on the question: did PPi metabolism evolve to ATP metabolism and can this evolution still be traced in the structures of proteins?

Such a suggested stepwise evolution may be visualised as:

PPi metabolism → → → → ATP metabolism

In essence, this indicated multistep evolution may be supposed to have encompassed steps leading from specificity in an enzyme, or enzymes, for PPi, over a broader specificity, for PPi and ATP, and finally to ATP specificity. Although PPi still functions as energy and phosphate donor in living cells, an early rise of the more complex and more versatile ATP to its central position in contemporary biological energy conversion can be easily understood. ATP can do what PPi can, and in addition, it can pyrophosphorylate and, what is of well known importance in cellular biosynthesis, it can adenylate.

3.2 A "PRIMITIVE" ACTIVE SITE OF PPASES ?

We have recently drawn attention to the possible evolutionary significance of the fact that soluble cellular PPases appear to have active sites with several charged amino acid side groups, a number of the negatively charged groups being involved in binding 3-4 magnesium ions, which are of importance for the binding and the reaction of the negatively charged substrate, Pi or PPi (H. Baltscheffsky and M. Baltscheffsky, 1993).

To be more specific, suitably charged surfaces, with metals attached, could well be imagined to catalyse hydrolysis (and even synthesis) of PPi. Thus, based on the known three-dimensional structure of existing PPases and their active sites (Kuranova, 1988; Vihinen et al., 1992), experimental attempts to extrapolate backwards towards early chemical catalysis of the hydrolysis of PPi should now be possible, for example with suitable metal-binding oligopeptides.

3.3 SUBUNIT STRUCTURES OF PPASES

The great majority of the known PPases consist of two or more identical subunits, three, four and six have also been reported. Six identical subunits have been found in several bacterial PPases, including the soluble PPase from the thermoacidophilic archaebacterium *Thermoplasma acidophilum* (Richter and Schäfer, 1992a), which shows a high degree of deduced amino acid sequence similarity to other bacterial PPases (Richter and Schäfer, 1992b). The other known archaebacterial PPase, isolated from *Methanotrix soehngenii*, appears to be composed of subunits with molecular masses of 35 and 33 kDa, in an $\alpha_2\beta_2$ oligomer (Jetten et al., 1992).

The two varieties of oligomeric structure in the presently known archaebacterial PPases could be of more than passing interest in connection with the early evolution of energy conversion in biological membranes.We have earlier drawn attention to the six identical subunits of bacterial PPases in relation to the fact that the three α and three β subunits of the F_1ATP synthases (F_1ATPases) appear to have a common origin (M. Baltscheffsky and H. Baltscheffsky, 1992). The regulatory α- and the catalytic β-subunit of the F_1ATPases seem to be in close structural-functional cooperation during photosynthetic and respiratory formation of ATP, and the conformational change mechanism for this ATP formation from Pi and ADP would seem to require two or three αβ oligomers (Cross, 1992).

Both these arrangements related to F_1ATPases and their apparent evolution and energy conserving mechanism, that is, six identical subunits and more than one non-identical subunit pair, respectively, are thus present in archaebacterial PPases. Does this indicate the existence of still unknown evolutionary and mechanistic relationships between PPases and ATPases? Attention has earlier been drawn to some similarities between these two groups of enzymes (H. Baltscheffsky et al., 1987, Sakai-Nore et al., 1992). An additional similarity is given in Fig. 2. which compares amino acid sequences of yeast mitochondrial PPase (Lundin et al., 1991), the integrally membrane bound and proton pumping *Arabidopsis thaliana* vacuolar PPase (Sarafian et al., 1992) and two β-subunits of F_1ATPases, in a well conserved region of the active site of the yeast (*Saccharomyces cerevisiae*) mitochondrial PPase and several cytoplasmic PPases.

Although the vacuolar PPase is very different from all the other sequenced PPases, being an integrally membrane bound proton pump, its amino acid sequence segment shown in Fig. 2 shows some similarity both to the mitochondrial PPase and to the ATPase β-subunit segments. Also when

-110CDVALKGDNDPLDCCE IGSDVLEMGS-	PPase, mitochondrial (*S. c.*)
-341AE LG IY PAV DPLDS TSRIMNPNIV GS-	ATPase β-subunit (bovine)
-387AS LG IY PAV DPLDS TSRQLDPLVVGQ-	- " - (*E. coli*)
-430TS NAYS PVQDVADSCRTGAAANVIFG-	PPase, vacuolar (*A. thaliana*)

Figure 2. Comparison of two PPase and two ATPase β-subunit segments, chosen to include in a central position the amino acid sequence -DPLD- of yeast mitochondrial PPase. This sequence is known to line its active site and to be in a strongly conserved region of soluble PPases.

taken together, all the similarities so far found between PPases and ATPase β-subunits are far too weak to allow any conclusions about a possible common origin for these two groups. More sequence and other structural information will be needed before the speculations in this direction can become either substantiated or disproved. In particular, an amino acid sequence from an integrally membrane bound photobacterial PPase would be important to obtain in this connection. specially because of the demonstrated immunological cross-reactivity between the proton pumping and energy coupling bacterial enzyme and the proton-pumping vacuolar PPase from *A. thaliana* (Nore et al.,1991).

4. FROM PPi TO ATP?

4.1 ANASTROPHES

An anastrophe is the opposite to a catastrophe, just as, in metabolism, anabolism is building up and catabolism is breaking down, the word anastrophe having been suggested to fill a language gap by describing momentous changes in a constructive, rather than destructive, direction (H. Baltscheffsky, 1977a; 1977b; 1992). It has not been considered necessary to define very specifically the word anastrophe, but it can cover both an event and its consequences, just as the word catastrophe. Anastrophes would appear to be particularly pronounced in open, living systems. In fact, biological evolution may be considered to consist of series of anastrophes. A typical example would be a site specific mutation leading to a protein with advantageous properties as compared with the original protein (H. Baltscheffsky, 1977a).

4.2 DID ANASTROPHIC EVENTS LEAD FROM PPi TO ATP?

In the evolutionary process leading to the first organisms from which all present life has evolved, three different proposals for energy conversion at the pre-nucleotide level have been discussed above. The "PPi world" is the one which may have been a very direct predecessor of the "ATP world", where ATP constitutes the only major, direct chemical link between the energy liberating and the energy requiring reactions of living cells, that is, the world in which we live today. It goes without saying, that we here focus attention on bioenergetics rather than, for example, replication.

The few similarities found so far between proteins involved in the metabolism of PPi and proteins involved the metabolism of ATP, do not suffice to signal an evolutionary pathway from the former to the latter. However, when more structural information on PPi and ATP metabolising proteins becomes available, I expect it to become possible to support or refute the hypothesis that the partial similarity of the energy rich molecules PPi and ATP may well have given rise to a stepwise evolution from proteins with PPi specificity to proteins with ATP specificity. So, we can still only ask the question, whether a sequence of anastrophes may have led from a postulated use of PPi in energy conversion at the pre-nucleotide level to the use of the comparatively advanced and versatile ATP, at the nucleotide level of the molecular evolution of life.

We may summarize the main reasons why PPi has emerged as a particularly attractive compound and an early "PPi world" as a useful concept in our current considerations about the possible origin and early evolution of biological energy conversion. As was discussed above, recent geochemical findings are in line with requirements for the production, stability and thus presence of PPi, during chemical evolution on an early earth, preceding the origin of life. The inorganic nature of PPi allows its possible operation both in an inorganic and in an organic, "prebiotic soup" containing world. The partial, close similarity between PPi and ATP stimulates ideas on a stepwise molecular evolution from an energy metabolism with PPi to one with ATP. Luckily, many kinds of living cells still show examples of both, with PPi in different ways being involved in useful action as what has sometimes been called, due to its lower content of free energy, a "poor man´s ATP". Further progress relating to the possible roles of PPi in early energy conversion may be expected from continued investigations both on the chemical and on the biological side of the molecular evolutionary process.

REFERENCES

Baltscheffsky, H. (1967): Inorganic Pyrophosphate and the Evolution of Biological Energy Transformation. Acta Chem. Scand. 21, 1973-1974.

Baltscheffsky, H. (1977a): Livet ett led i en kemisk utveckling. Forskning och framsteg (7), 6-11.

Baltscheffsky, H. (1977b): Livets uppkomst - liv i universum. Fauna och flora 72 (2/3), 53-58.

Baltscheffsky, H. (1992): Hur började livet på jorden? Fauna och flora 87 (4/5), 31-36.

Baltscheffsky, H., Alauddin, M., Falk, G. and Lundin, M. (1987): Similarities Between Soluble Inorganic Pyrophosphatase from Yeast and some Nucleotide-Binding Polypeptides. Acta Chem. Scand. B41, 106-107.

Baltscheffsky, H. and Baltscheffsky, M. (1993): Molecular Origin and Evolution of Early Biological Energy Conversion. In: S. Bengtson (Ed.), Early Life on Earth. Nobel Symposium No 84. Columbia University Press (in press).

Baltscheffsky, H. and von Stedingk, L.-V. (1966): Bacterial photophosphorylation in the absence of added nucleotide. A second intermediate stage of energy transfer in light-induced formation of ATP. Biochem. Biophys. Res. Commun. 22, 722-728.

Baltscheffsky, H, von Stedingk, L.-V., Heldt, H.-W. and Klingenberg, M. (1966): Inorganic pyrophosphate: Formation in bacterial photophosphorylation. Science 153, 1120-1122.

Baltscheffsky, M. (1967): Inorganic pyrophosphate and ATP as energy donors in chromatophores from *Rhodospirillum rubrum*. Nature 216, 241-243.

Baltscheffsky, M. and Baltscheffsky, H. (1992): Inorganic pyrophosphate and inorganic pyrophosphatases. In L. Ernster (Ed.), Molecular Mechanisms in Bioenergetics, Elsevier, Amsterdam, 331-348.

Baltscheffsky, M., Baltscheffsky, H. and von Stedingk, L.-V. (1967): Light-Induced Energy Conversion and the Inorganic Pyrophosphatase Reaction in Chromatophores from *Rhodospirillum rubrum*.. Brookhaven Symp. Biol. 19, 246-257.

Blöchl, E., Keller, M., Wächtershäuser, G. and Stetter, K.O. (1992): Reactions depending on iron sulfide and linking geochemistry with biochemistry. Proc. Natl. Acad. Sci. USA 89, 8117-8120.

Cross, R.L. (1992): The reaction mechanism of F_0F_1-ATP synthases. In L. Ernster (Ed.), Molecular Mechanisms in Bioenergetics, Elsevier, Amsterdam, 317-330.

de Duve, C. (1991): Blueprint for a Cell: The Nature and Origin of Life. Patterson, Burlington, N.C. 275 pp.

Gilbert, W. (1986): The RNA world. Nature 318, 618.

Kuranova, J.P. (1988): Three-dimensional structure of yeast inorganic pyrophosphatase and its active site. Biokhimiya 53, 1821-1827.

Jetten, M.S.M., Fluit, T. J., Stams, A. J. M. and Zehnder, A. J. B. (1992): A fluoride - insensitive inorganic pyrophosphatase isolated from *Methanothrix soehngenii*. Arch. Microbiol. 157, 284-289.

Lundin, M., Baltscheffsky, H. and Ronne, H. (1991): Yeast *PPA2* gene encodes a mitochondrial inorganic pyrophosphatase that is essential for mitochondrial function J. Biol. Chem. 266, 12168-12172.

Mann, S., Sparks, N.H.C., Frankel, R.B., Bazylinski, D.A. and Jannasch, H.W. (1990): Biomineralization of ferrimagnetic greigite (Fe_3S_4) and iron pyrite (FeS_2) in a magnetotactic bacterium. Nature 343, 258-261.

Nore, B.F., Sakai-Nore, Y., Maeshima, M., Baltscheffsky, M. and Nyrén, P. (1991): Immunological cross-reactivity between proton-pumping inorganic pyrophosphatases of widely phylogenic separated species. Biochem. Biophys. Res. Commun. 181, 962-967.

Richter, O.-M.H. and Schäfer, G. (1992a): Purification and enzymic characterization of the cytoplasmic pyrophosphatase from the thermoacidophilic archaebacterium *Thermoplasma acidophilum*. Eur. J. Biochem. 209, 343-349.

Richter, O.-M.H. and Schäfer, G. (1992b): Cloning and sequencing of the gene for the cytoplasmic inorganic pyrophosphatase from the thermoacidophilic archaebacterium *Thermoplasma acidophilum*. Eur. J. Biochem. 209, 351-355.

Sakai-Nore, Y., Nyrén, P. and Strid, Å. (1992): Amino acid sequence similarities between proton pumping proteins. In: N. Murata (Ed.), Research in Photosynthesis, Vol. II. Kluwer, Dordrecht, 693-696.

Sarafian, V., Kim, Y., Poole, R.J. and Rea, P.A. (1992): Molecular cloning and sequence of cDNA encoding the pyrophosphate-energized vacuolar membrane proton pump of Arabidopsis thaliana Proc. Natl. Acad. Sci. USA 89, 1775-1779.

Rose, R.C., Peacor, D.R. and Freed, R.L. (1988): Pyrophosphate groups in the structure of canaphite, $CaNa_2P_2O_7x4H_2O$: the first occurence of a condensed phosphate as a mineral. Am. Mineral. 73, 168-171.

Vihinen, M., Lundin, M. and Baltscheffsky, H. (1992): Computer modeling of two inorganic pyrophosphatases. Biochem. Biophys. Res. Commun. 186, 122-128.

Wächtershäuser, G. (1992): Groundworks for an Evolutionary Biochemistry: The Iron-Sulphur World. Prog. Biophys. molec. Biol. 58, 85-201.

Yamagata, Y., Watanabe, H., Saitoh, M. and Namba, T. (1991): Volcanic production of polyphosphates and its relevance to prebiotic evolution. Nature, 352, 516-519.

PHOSPHATE IN MODELS FOR CHEMICAL EVOLUTION[1]

G. Arrhenius, B. Gedulin and S. Mojzsis
NASA Exobiology Center
Scripps Institution of Oceanography
La Jolla, CA 92093-0220, U.S.A.

ABSTRACT

Phosphate ion is unique as a link between living systems and the inorganic world. Phosphate minerals for this reason occupy an important place in tracing the earliest life forms and in the reconstruction of biomolecular evolution in the Archean. The discovery of paths to phosphorylation of simple organic compounds poses a challenge since an aqueous environment does only in special cases allow organic esterification and dehydration reactions.

Recent work has revealed natural mechanisms for solid state condensation of orthophosphate minerals to reactive oligophosphates, and for concentration of ortho- and oligophosphate ions from dilute solution into common hydroxide minerals. Furthermore, model experiments have been carried out showing efficient condensation of simple aldehyde phosphates in the hydroxide mineral interlayer to form hexose sugar phosphates, which may be considered as models for precursor components of modern RNA. Finally, ordered complexes between these minerals and nucleoside phosphates suggest a model for molecular recognition and memory function at a primitive level.

Because of the high solubility of magnesium salts relative to corresponding calcium compounds, a high magnesium/calcium ratio is maintained in the ocean and throughout the hydrosphere. This high relative magnesium content has a controlling effect on the phosphate minerals that may form; one consequence is the formation of the protonated magnesium-calcium and calcium phosphates referred to above, potentially yielding condensed phosphates. Another geochemically important effect is the exclusion of basic calcium phosphate, apatite, except at high pH (above 8.5) at natural magnesium/calcium ratios in the source solution. This chemical control can be broken by the physiological interaction of organisms, which are capable, even at high magnesium content in the surroundings, to selectively exclude this ion and to preferably incorporate calcium by intracellular mechanisms, and thus to deposit apatite out of equilibrium with the surrounding medium. Another technique, employed by some microorganisms, is to secrete enough ammonia into the interstitial microenvironment to eliminate magnesium as insoluble magnesium ammonium phosphate hydrate (struvite).

In deposits known or inferred to have formed from solutions with pH less than 8.5 (and with ammonia concentration less than isomolar with magnesium) authigenic sedi-

[1] Support for this research from NSF by grant EAR 8916467 and from NASA by grants NAGW 1031 and NAGW 2881 (NSCORT Exobiology) is gratefully acknowledged.

ISBN 0-937194-31-X

mentary apatite thus appears as a reliable indicator of live organisms, whose presence can be established even in the absence of identifiable microfossil remains.

Abundant apatite of suggestive microbial origin has been found in the 2.5 b.y. Brockman banded iron-formation (Hamersley Group, Western Australia), and closely similar, commonly occurring chemical fossils have been identified in the oldest metasedimentary rocks known on Earth, the Isua formation (Southwestern Greenland) with an age of 3.8 b.y. The presence of life is consequently seen as far back as a sedimentary geological record exists, and the time period available for evolution of cellular life on Earth appears to be substantially less than 1 b.y.

1. THE CHOICE OF PHOSPHATE

At the base of the present geochemical studies is the concept that minerals selectively concentrated and ordered organic molecules from dilute solution, and catalyzed their growth to larger, functional units. The speculation that such processes gave rise to organic life goes back to Descartes or beyond, and was expressed in its modern form by Bernal (1949) and Goldschmidt (1952). Cairns-Smith (1982) has placed this hypothesis in a new context by suggesting that the hybrid structures of minerals and organic molecules actually constituted the first life forms

The unique and central role of phosphate in biochemistry and therefore probably also in biopoesis has been somewhat of a puzzle to organic chemists, who rarely would use phosphate ion as an efficient reactant. Likely reasons for nature's choice of phosphate have been identified and discussed by Westheimer (1987), who pointed out the selective advantage conferred by the relative stability of the phosphate ester bond against nucleophilic attack, thus safeguarding the integrity of the sequences in the DNA-RNA polymer. In addition, the negatively charged phosphate envelope protects RNA and DNA from nucleophilic attack on the interior moieties, and from its loss by diffusion through the cell membrane.

The experiments on hybridization between minerals and organic structures point to another effect to be added to Westheimer's list: the efficient extraction of the negatively charged phosphate- and phosphate ester ions from dilute solution into catalytically active anion exchange minerals. The resulting spontaneous and highly selective synthesis of sugar phosphates, similar to the monomeric components of RNA, has opened new avenues to modeling biopoesis and the early evolution of a genetic code (Pitsch *et al.*, 1993; Eschenmoser and Dobler, 1992).

After Wöhler's demonstration in 1828 that urea, an organic compound, can be synthesized from inorganic components, and Berzelius' 1837 postulation of enzymes as peptide-related molecules, much of the thinking about the transition from the inorganic to the organic world became focused on amino acids, the fundamental components of polypeptides and proteins. A notable change took place, following the discovery by teams led by Cech (Kruger *et al.*, 1982) and Altman (Guerrier-Takada *et al.*, 1983) of catalytically active RNA, "ribozymes," demonstrating that RNA can in principle both serve as carrier of information and undertake the role of catalyst, previously thought to be reserved for peptides. Recent work (Joyce, 1989; Beaudry and Joyce, 1992; Szostak, 1992) has further demonstrated experimentally that synthetic ribozymes may undergo darwinian evolution. All of this has contributed to the concept of the early organic environment as an "RNA world" (Gilbert, 1986). In geochemical terms this translates as a phosphate-aldehyde-cyanide world.

2. SURFACE ACTIVE MINERALS AS PRIMORDIAL ENZYMES

The three source components of RNA, and presumably of its evolutionary precursors: phosphate, cyanide and carbonate, are all anionic. Several hydrous minerals have structures where excess positive charge on the framework leads to efficient attraction and bonding of anions and negatively charged molecular complexes. Among such minerals, the double layer hydroxides are found to be exceptionally efficient in sorbing high charge density anions from dilute (micromolar) aqueous solution into strongly concentrated form (~ 7 molar). In this concentrated state, and in contact with the catalytically active metal hydroxide sheets, the sorbed components, including simple aldehyde phosphates, can undergo reactions that otherwise would be practically excluded in a natural setting (Fig. 2.2). In this respect, the double layer hydroxide minerals can be compared to the enzymes and ribozymes that are responsible for this function in evolved forms of life.

In contrast to the homogeneous glycolaldehyde phosphate reaction, which in the presence of formaldehyde gives products shown in Fig. 2.1.b, the mineral-induced form of this reaction (Fig. 2.2(1-3)) offers striking geochemical advantages. It utilizes dilute solutions of the reactant phosphate, concentrating it by internal sorption. It proceeds rapidly under mildly alkaline conditions, and selectively produces the altrose form of hexose phosphate, rather than allose, which dominates the hexose products in the homogeneous reaction. Altrose-, in contrast to allose phosphate, optimizes base pairing in hexose-derived polynucleic acid analogs[2], similar to homo-DNA (Fig. 2.3). It should be emphasized, however, that the hexose phosphates produced in the mineral induced reaction, as well as in the homogeneous Eschenmoser reaction, are phosphate triesters in contrast to the phosphate diesters that form the backbone in RNA and DNA.

The spontaneous, selective synthesis of sugar phosphates gives support to the notion that commonly occurring hydrous crystalline phases may have played a role in a transition from well-organized but inflexible "mineral life" to purely organic systems with similar capability for self assembly and replication but with unlimited freedom for mutation.

3. THE STRUCTURE AND FUNCTION OF DOUBLE LAYER HYDROXIDE MINERALS

The properties of these common but inconspicuous minerals (Table 3.1; page 11) have been explored by Bernal (1959), Allmann (1970), Brindley (1976), Taylor (1982), Miyata (1983), Drits (1985) and others, whose results have been seminal for our own

[2] Fischer, R., A. Helg, A. Giger, C. Leumann, 1991: In preparation.

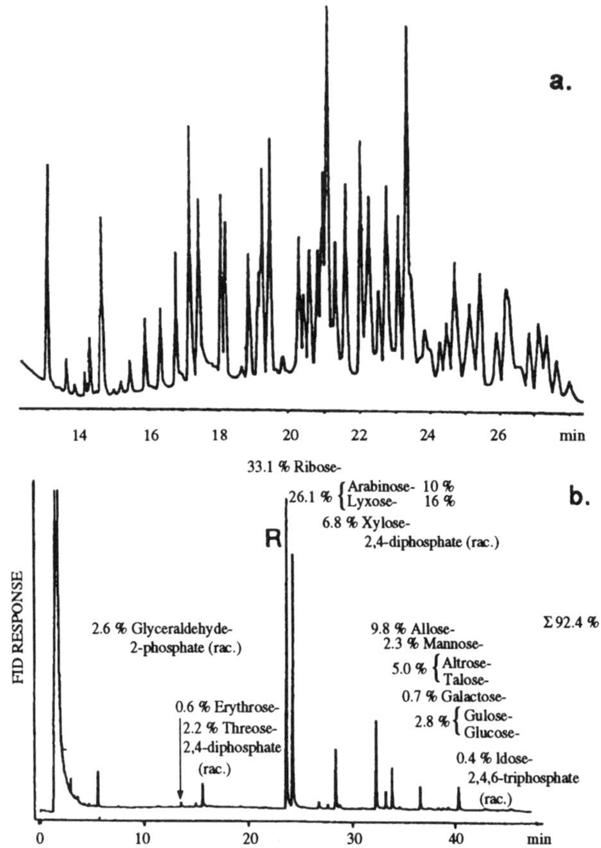

Figure 2.1. Gas chromatograms of (**a**) the large number of sugars and sugar derivatives with small amounts of ribose produced in the formose reaction compared to (**b**) the selective production of pentoses, primarily ribose (R) in the Eschenmoser glycolaldehyde phosphate–formaldehyde reaction (Müller *et al.*,1990).

The formose reaction (Butlerow, 1861; Fischer, 1888), representing the polymerization of formaldehyde, has long been resorted to as a primordial path toward the sugar phosphates that now form the backbone of RNA and DNA. This proposal has, however, been criticized for several reasons, summarized by Shapiro (1988). Among these are; the indiscriminate production of the large number of products, shown in **a**; lack of reaction in dilute solution; rapid hydrolysis of the potentially useful sugars, and lack of a geochemically plausible concentration mechanism. Eschenmoser's introduction of the phosphate ester of a simple aldehyde (glycolaldehyde phosphate) as the initial reactant leads (**b**) to highly selective production of a few sugar phosphate species, protected against rapid hydrolysis by the phosphate group, and, with formaldehyde as an additional reactant, dominated by ribose-2,4,-diphosphate. The reaction, represented by **b**, does, however, proceed only in extremely alkaline solution, and at high concentration of glycolaldehyde phosphate, and consequently lacks geochemical plausibility in this original form.

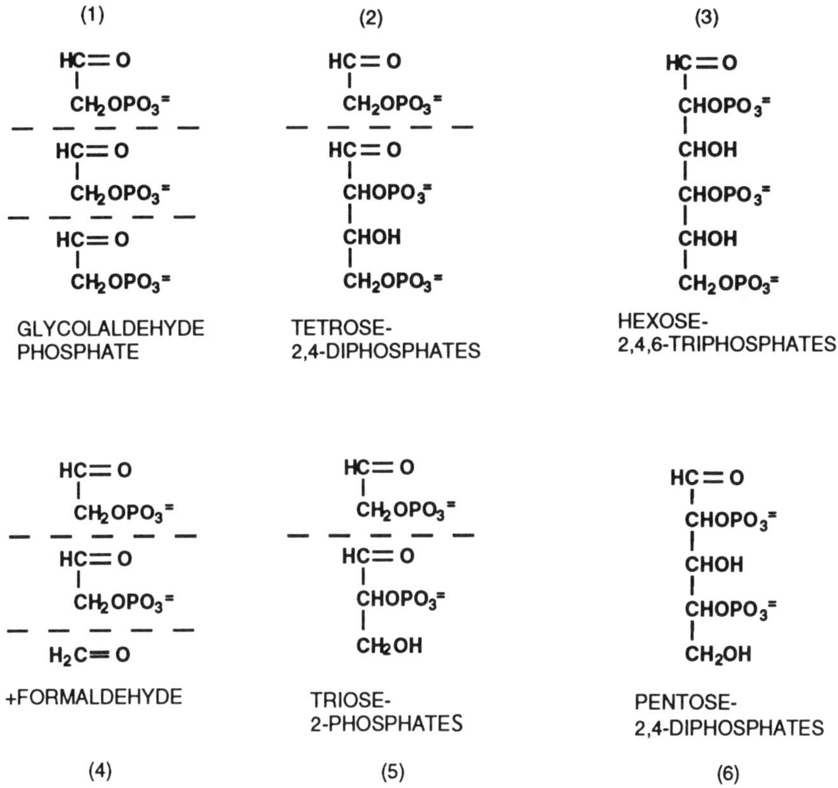

Figure 2.2. (1)–(3); mineral-induced selective sugar phosphate formation from glycolaldehyde phosphate in the interlayer of hydrotalcite-type minerals. Glycolaldehyde phosphate (three separate molecules shown in **1**) forms hexose- (primarily altrose) 2,4,6-triphosphate (**3**) via tetrose -2,4-diphosphate (**2**) (Pitsch *et al.* 1993). The altrose diastereoisomer is particularly effective in stabilizing duplexes between hexose based oligonucleotides (Eschenmoser and Loewenthal, 1992; Fischer *et al., op. cit.* footnote, page 3).

In absence of the sorbent double layer hydroxide mineral the aldolization reaction proceeds only at high reactant concentration, and under extremely alkaline conditions. In the presence of formaldehyde (**4**), it yields pentose– (mainly ribose–) 2,4-diphosphate (**6**) via triose-2-phosphate (**5**) in addition to hexose phosphate (**3**), in this case mainly allose-2,4,6-phosphate (cf. Fig. 2.1.b).

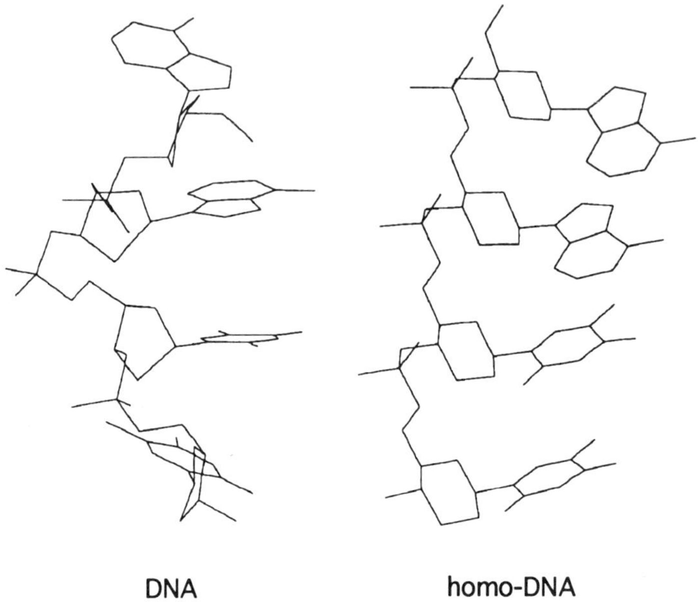

Figure 2.3. Structure of single stranded homo–DNA, compared to present–day natural DNA; the former with a hexapyranosyl moiety replacing the now naturally occurring pentose–furanose sugar ring. Homo–DNA assumes a near-planar, ladder–like ribbon structure in contrast to the DNA helix. The artificial, and comparatively easily synthesized homo-DNA serves as a conformational model for similar synthetic oligonucleotides derived from natural hexose sugar phosphates such as allose- or altrose phosphate. The synthesis of such oligomers has shown[a] that particularly favorable base pairing properties are conferred on the macromolecule when the hexose diastereoisomer altrose, rather than allose, is the sugar moiety.
The fact that the mineral induced aldolization reaction readily and preferentially produces the 2,4,6–phosphate of altrose has, together with the intriguing conformational properties, led to the suggestion that such hexose based polynucleotides may possibly have been mineral–induced precursors of modern DNA. (From Eschenmoser and Loewenthal, 1992.)

---

[a] *Op. cit.* footnote, page 3; R. Hammer (unpublished); G. Rehberg, K. Groebke and P. P. Adam (unpublished) and P. P. Adam (unpublished); ref. in Eschenmoser and Loewenthal, 1992.

work (Arrhenius, 1987; Arrhenius *et al.*, 1989; Kuma *et al.*, 1989; Gedulin and Arrhenius, 1993; Pitsch *et al.*, 1993). After the Mg(II) Fe(III) member, this mineral family is often referred to as the pyroaurite group. It includes the Fe(II) Fe(III) member, known as green rust (Table 3.1; page 11), which is thought to be the hydrous precursor of magnetite (Fig. 3.2) that constitutes the main component of the Archean banded iron formations. Except in minor dysoxic environmental enclaves, such as in waterlogged soils, these deposits lack modern counterparts due to the rise of oxygen in the Proterozoic atmosphere and hydrosphere. However, other species, insensitive to oxidation, such as the Mg (II) Al (III) mineral hydrotalcite (Fig. 3.3) are widespread also in modern ocean sediments and evaporites (Thomassin and Touray, 1982; Crovisier *et al.*, 1983; Drits *et al.*, 1985, 1987; Bookin *et al.*, 1993; Bookin and Drits, 1993).

The double layer hydroxide minerals (Fig. 3.4; page 12) owe their anion exchange properties and large, active internal surfaces to the excess positive charge introduced by partial substitution with trivalent cations in the simple $Fe(OH)_2$ or $Mg(OH)_2$ (brucite) structure. The general formula for the double layer hydroxide minerals and synthetic analogs, can be written:

$$[M^{2+}{}_{3-x}M^{3+}{}_{x}(OH)_6][\tfrac{x}{n}\ A^{n-}] \bullet mH_2O \quad (m\sim4)$$

where M^{2+} and M^{3+} are divalent and trivalent cations. A^{n-} are the exchangeable interlayer anions with n = 1 or 2, balancing the excess positive charge on the hydroxide sheets, introduced by substitution of M^{2+} with M^{3+}. Higher valence anions such as $PO_4{}^{3-}$, $P_2O_7{}^{4-}$, and $Fe(CN)_6{}^{4-}$ enter even more readily in the interlayer because of their high charge densities. These transformations of the structure are shown schematically in Fig. 3.4.

In the modern ocean, where phosphate is voraciously consumed by cellular organisms, and reduced organic and inorganic species are destroyed by oxidation, the most common interlayer anions are species such as carbonate, sulfate and chloride. The structure, exemplified by green rust (Fig. 3.1, Table 3.1; page 11), is stable in the range of M^{3+}/M^{2+} from 0.20 to 0.57 ($0.5 < x < 1.1$) (Bernal, et al., 1959). Large monovalent anions such as Cl^- are relatively weakly bound in the interlayer, as indicated by downward concave Langmuir isotherms (Miyata, 1983). At exchange with charged organic molecules, such ions thus provide good leaving groups. In contrast, the planar, divalent $CO_3{}^{2-}$ ion is strongly bound, and is the dominant anion in non-marine pyroaurite minerals, displaceable with high charge density species such as ferrocyanide and phosphate ion.

As a part of our past work, we have synthesized a range of double layer hydroxide mineral structures (Kuma *et al.*, 1989) and investigated their relevant physical and chemical properties, including anion exchange and sorption isotherms of cyanide and phosphate species. Perhaps the geochemically most interesting results of this work have been the formation of ferriferrocyanide (Prussian blue) (Fig. 3.5; page 13) at sorption of cyanide ion in green rust; furthermore, the spontaneous selective formation of DNA-RNA related sugar phosphates in the mineral interlayer (Fig. 2.2) (Pitsch *et al.*, 1993) and the display of cooperative interaction between nucleic acids, bound in the interlayer (Arrhenius, M. L., Frutos, A. P., Arrhenius, G., in preparation).

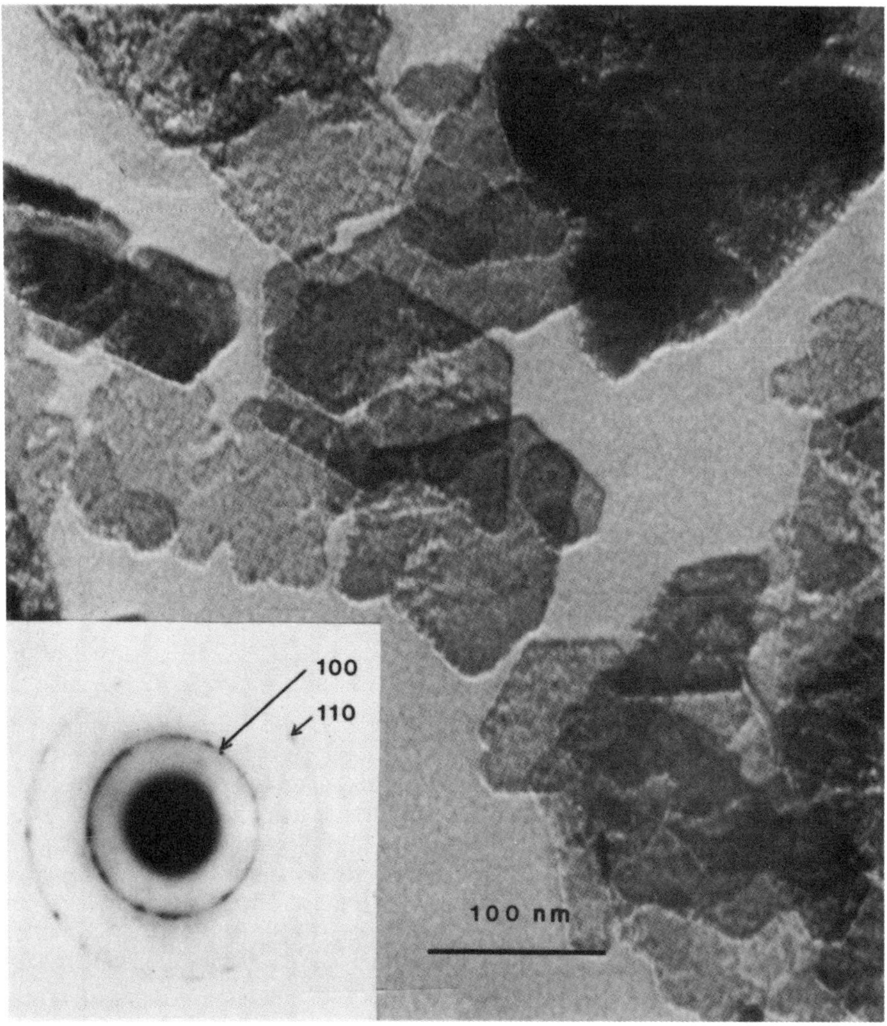

Figure 3.1: Synthetic green rust (low-temperature, rhombohedral modification). The thin, leaf-like crystals rest preferentially on basal faces and expose almost exclusively vertical planes (h00 and hk0) to diffraction of the electron beam (insert).

Green rust, which is an efficient concentrator of ferrocyanide and phosphate ions, is metastable against conversion to magnetite (Figure 3.2), which does not form directly from aqueous solution; green rust is thus probably the primary source of the large deposits of magnetite in the Archean banded iron-formations, although a bacterial origin has also been proposed (Vali and Kirschvink, 1988).

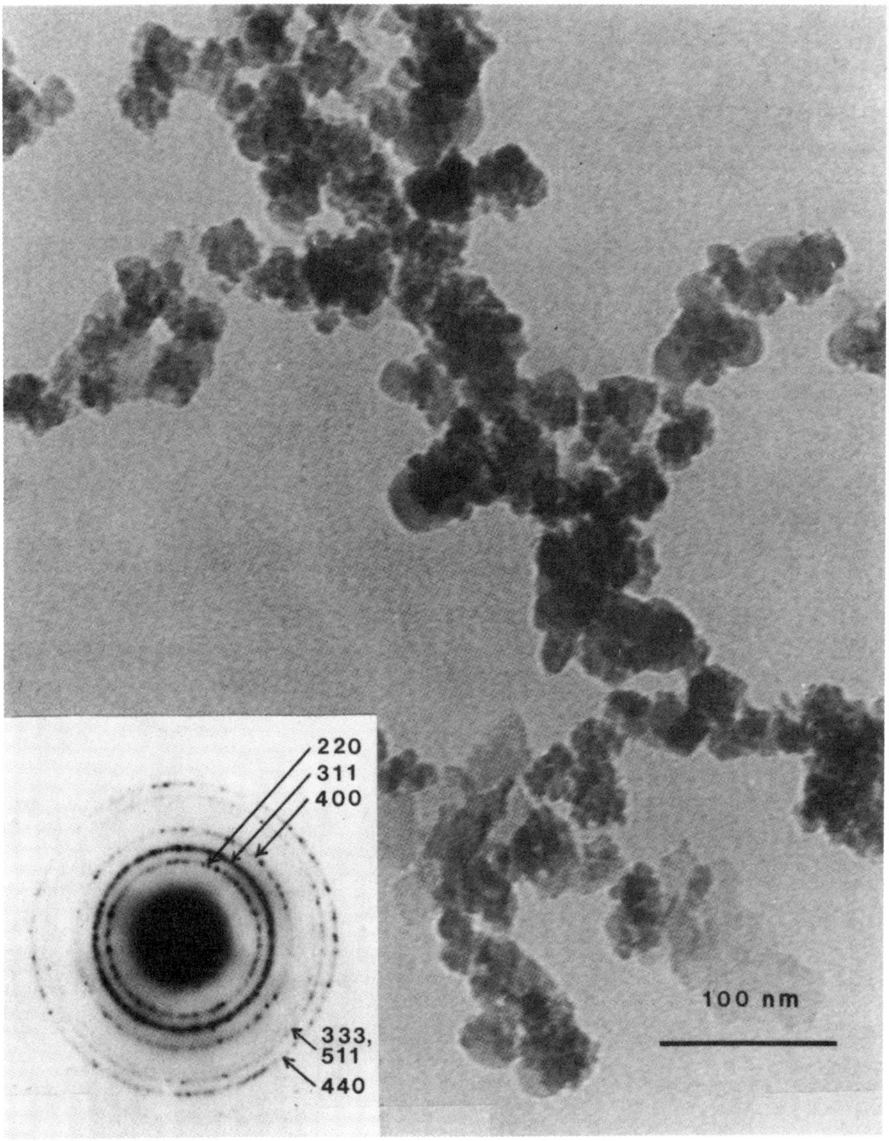

Figure 3.2: Microcrystalline magnetite formed by topotactic transformation of green rust. The magnetite crystals form as platy pseudomorphs, about 1/10 of the size of the parent green rust particles.

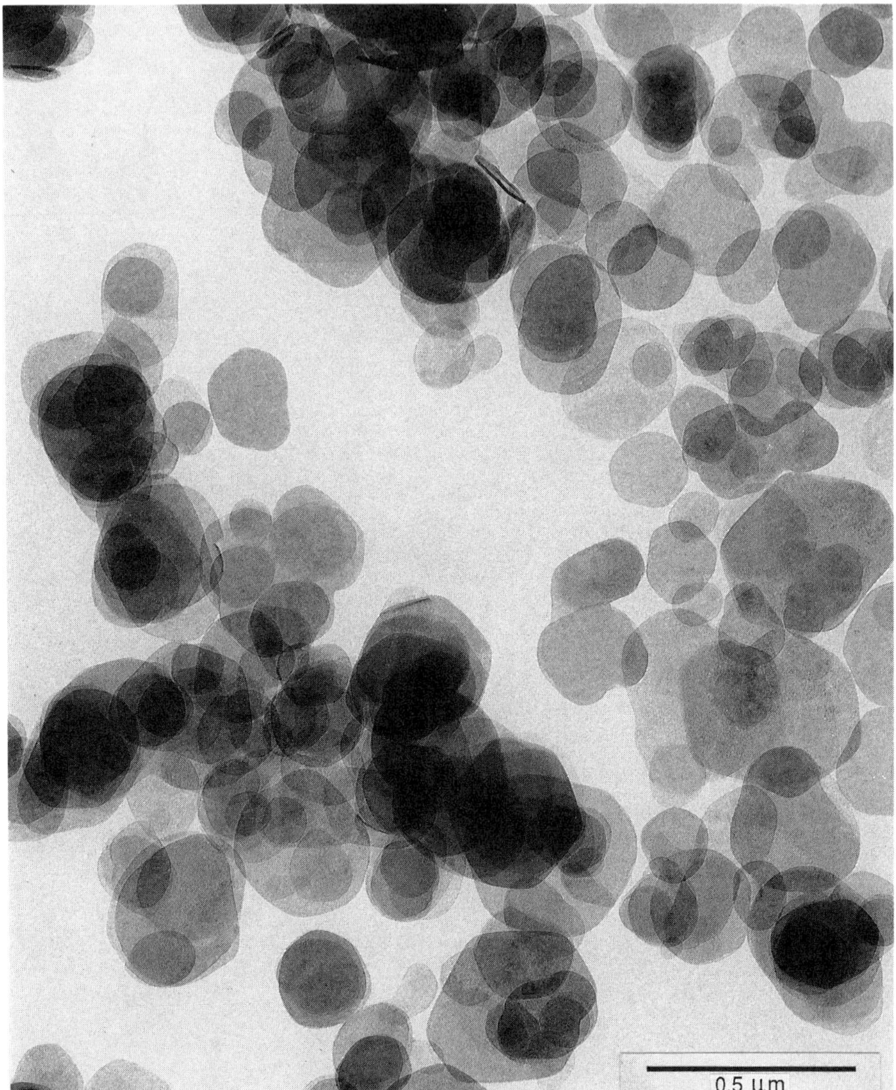

Figure 3.3. Thin platy crystals of hydrotalcite, magnesium aluminum hydroxide (chloride form) a common representative (Table 3.1) of the double layer hydroxide minerals.

TABLE 3.1. SOME NATURALLY OCCURRING M^{2+} M^{3+} HYDROXIDE SHEET STRUCTURES WITH EXCHANGEABLE INTERLAYER ANIONS.

Mineral name		M^{2+}	M^{3+}	Interlayer anion	Generic name
Rhombohedral[a]	Hexagonal[a]				
hydrotalcite[b]	manasseite	Mg	Al	CO_3^{2-}	MgAl hydroxide carbonate
stichtite	barbertonite	Mg	Cr	CO_3^{2-}	MgCr hydroxide carbonate
pyroaurite	sjögrenite	Mg	Fe	CO_3^{2-}	MgFe hydroxide carbonate
iowaite		Mg	Fe	Cl^-	MgFe hydroxide chloride
chlormagaluminite		Mg,Fe	Al	Cl^-,CO_3^{2-}	Mg-FeAl hydroxide chloride-carbonate
hydrocalumite[c]		Ca	Al	OH^-	CaAl hydroxide
green rust 1[d]	green rust 2[d]	Fe	Fe	CO_3^{2-}	Ferroferric hydroxide carbonate
berthierine[e]		Fe	Fe	SiO_3^{2-}	Ferroferric hydroxide silicate
	takovite	Ni	Al	CO_3^{2-}	NiAl hydroxide carbonate
reevesite		Ni	Fe	CO_3^{2-}	NiFe hydroxide carbonate
honessite		Ni	Fe	SO_4^{2-}	NiFe hydroxide sulfate
eardlyite		Zn,Ni	Al	CO_3^{2-}	Zn-NiAl hydroxide carbonate
8.85Å SO4 hydrotalcite 3R	8.85Å SO4 hydrotalcite 2H	Mg	Al	SO_4^{2-}	MgAl hydroxide sulfate

[a] Alternative stacking sequences of the hydroxide sheets give rise to either rhombohedral or hexagonal symmetry, the latter generally favored by high crystallization temperature

[b] Hydrotalcite with varying proportions in solid solution of other pyroaurite minerals, such as pyroaurite proper, iowaite, chlormagaluminite, 8.85Å SO_4 hydrotalcite 3R, and berthierine, occurs as a common product of submarine basalt weathering (Crovisier et al., 1983; Drits et al., 1987; Thomassin and Touray, 1979; Thomassin and Touray, 1982)

[c] Pyroaurite structure deformed (monoclinic), due to large size of Ca^{2+}.

[d] The various intercalation forms of green rust are, at the exposure to oxygen, rapidly transformed, initially to magnetite ($FeO{\bullet}Fe_2O_3$), and at a higher partial pressure of oxygen, mainly to γ–FeOOH (lepidocrocite). Since magnetite does not form inorganically from aqueous solution at low temperature, the large Archean deposits of this mineral are likely to be diagenetic products formed by disproportionation or partial oxidation of green rust. Alternatively, the Archean sedimentary magnetite may be of biogenic origin (Vali and Kirschvink, 1988), since bacteria are known to form this mineral intracellularly.

[e] A variety of ferroferric hydroxide silicates with 1:1 serpentine structures, e.g. greenalite (Bailey, 1980); minnesotaite (Ahn and Buseck, 1989), occur as important iron ore minerals in the Archean iron formations; they are probably metamorphic (dehydrated-dehydroxylated and partially ordered) green rust with silicates as interlayer anion.

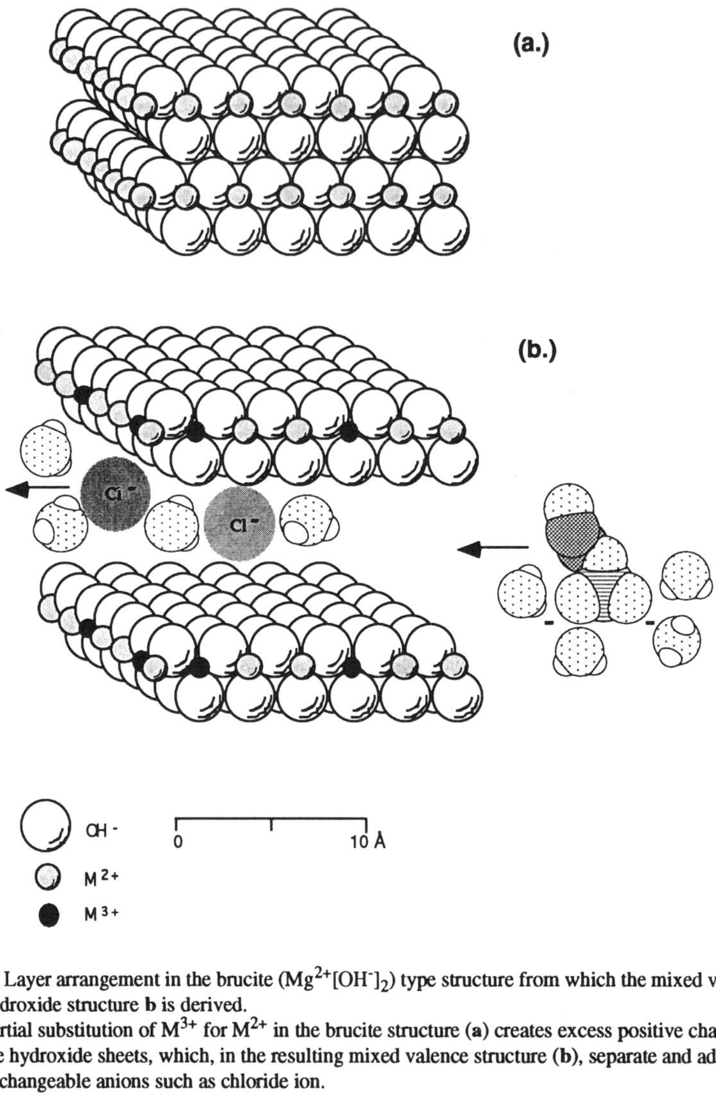

Figure 3.4. (**a**) Layer arrangement in the brucite ($Mg^{2+}[OH^-]_2$) type structure from which the mixed valence hydroxide structure **b** is derived.

Partial substitution of M^{3+} for M^{2+} in the brucite structure (**a**) creates excess positive charge on the hydroxide sheets, which, in the resulting mixed valence structure (**b**), separate and admit exchangeable anions such as chloride ion.

Low charge density ions such as Cl^- are effectively displaced by high charge density species such as ferrocyanide, orthophosphate, pyrophosphate and phosphate esters such as glycolaldehyde phosphate (GAP).

A GAP anion, coordinated with four water molecules is shown in **b** entering the interlayer, displacing chloride ions, and expanding the interlayer from 7.8 to 11.4Å. In the interlayer the anion with four water molecules forms a concentrated, two dimensional solution. In this highly concentrated medium, reactions such as condensation of simple aldehyde phosphates to sugar phosphates (Fig. 2.2) are induced by the adjacent metal hydroxide sheets.

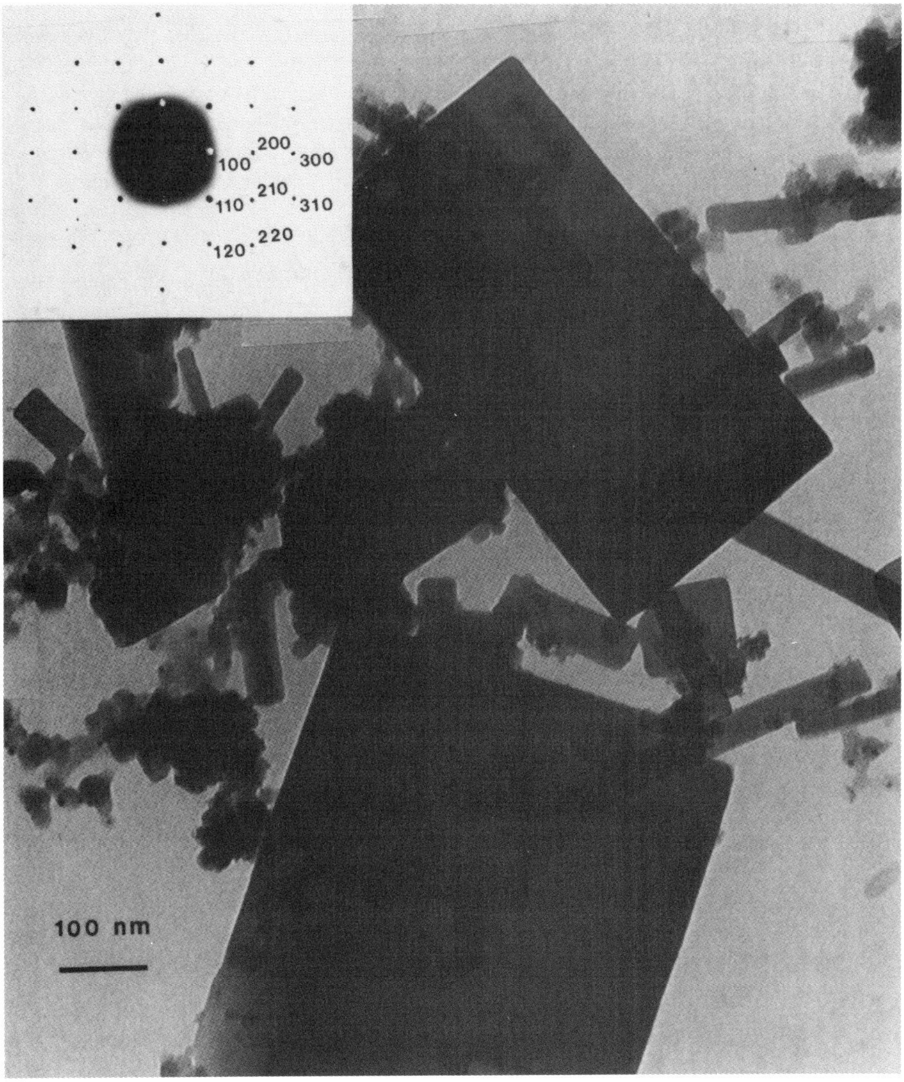

Figure 3.5. Large, perfect, platy and lamellar crystals of ferroferricyanide (Prussian blue) together with minor amounts of magnetite. The electron diffraction pattern, reflecting the cubic symmetry, was obtained from an isolated single crystal of ferroferricyanide not shown here, similar in size and shape to the large crystal in the upper part of the figure. The thin film texture of this cubic solid, normally with isometric crystal habit, is due to epitactic growth from the ferroferric hydrotalcite molecular sheet structure (green rust; Fig. 3.1) with interlayer ferrocyanide (hexacyanoferroate) ion sorbed from dilute solution.

4. PHOSPHATE MINERAL FORMATION IN A LIFELESS OCEAN

Another geochemically significant indication is that, in the absence of organisms, calcium hydroxyphosphate (apatite) with or without fluoride and /or carbonate substitution, does not form stably or metastably, from media of sea water or average ground water composition at pH below 8.5. The stability fields of the phosphate minerals that actually form from such media are determined by the high relative magnesium and hydrogen ion concentrations which dictate the formation of a series of magnesium- and calcium-hydrogen-magnesium phosphates instead of apatite, depending on pH (Table 4.1) (Gedulin and Arrhenius, 1993).

Conversely, other workers, particularly Lucas and his group in Strasbourg (Lucas and Prévôt, 1981; Prévôt *et al.*, 1989; Hirschler *et al.*, 1990) and also Gulbrandsen *et al.* (1983) and Rivadeneyra *et al.* (1992) have shown that organisms present in the system are uniquely capable of segregating the magnesium and hydrogen ions in their intercellular or circumcellular microenvironment, thus making possible the inorganically prohibited precipitation of apatite as exo- or endoskeletons or excretion products. Elimination of the unwanted magnesium ions is achieved by use of ion selective membrane channels, intracellular metal-organic complex formation, or, particularly among bacteria, formation of ammonia. In a sediment or tissue microenvironment or in a small-volume laboratory experiment ammonia can be used to locally exhaust magnesium ion from a limited magnesium ion reservoir by the precipitation of struvite, $NH_4MgPO_4{\bullet}6H_2O$.

These relationships are of geochemical importance in two respects. First, the presence of protons in the characteristic abiogenic orthophosphate crystal structures sets the stage for condensation of the organically unreactive orthophosphate groups in the minerals, to form reactive diphosphate (pyrophosphate) and higher oligomers, potentially with a prebiotic function analogous to those of ADP and ATP in life. This condensation reaction, incipient already below 100°C (Morey, 1953) has been expanded and studied quantitatively by Sales and collaborators (Sales *et al.*, 1992; Gedulin, 1993). Condensed phosphates were also obtained at significant yields at room temperature in presence of cyanate as a condensing agent (Orgel and Beck, 1965).

The two authigenic phosphate minerals that on the basis of these studies would be expected to be most prevalent in prebiotic Archean sediments, are brushite ($CaPO_4{\bullet}2H_2O$) and whitlockite ($Ca_{18}Mg_2H_2(PO_4)_{14}$; Table 4.1; Fig. 4.1; whitlockite structure in Gopal (1974)).

While brushite forms with long range order at room temperature, whitlockite is initially precipitated as a nanocrystalline gel (Fig. 4.2) with enhanced solubility due to the large surface area. Long range order, permitting diagnosis by diffraction in the X-ray wavelength region, is acquired by annealing for a long time, or at elevated temperature, typically a few days at 100°C (Fig. 4.1).

Both brushite and whitlockite, like newberyite, $MgHPO_4{\bullet}3H_2O$, the latter expected at higher Mg/Ca ratios but not produced in our current experiments, give rise to a series of condensed phosphates at heating to a few hundred degrees (Fig. 4.3 and Fig. 4.4).

TABLE 4.1. CRYSTALLINE PHASES FORMING IN THE RANGE OF COMPOSITIONS AND CONDITIONS SPECIFIED

Mineral	Composition	pH
brushite	$CaHPO_4 \cdot 2H_2O$	6–7.5
whitlockite	$Ca_{18}Mg_2H_2(PO_4)_{14}$	7-9
Mg-phosphate pentahydrate[a]	$Mg_3(PO_4)_2 \cdot 5H_2O$	7-9
apatite	$Ca_5[OH,F \mid PO_4,CO_3)_3]$	>8.5

[a] Mazghouni *et al.* (1981)

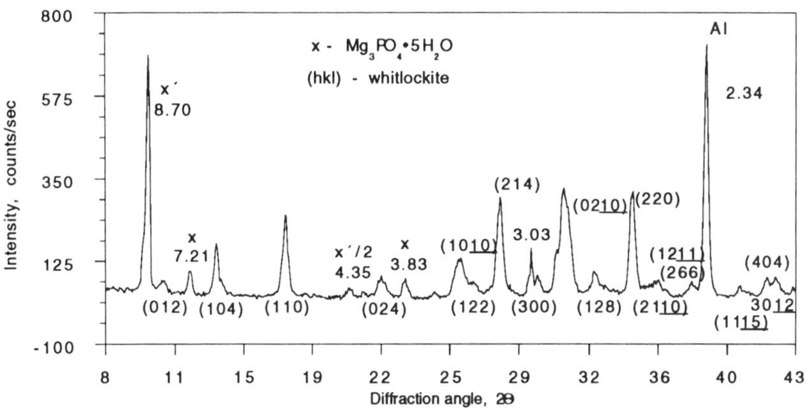

Figure 4.1. X-ray diffractogram of Ca-Mg-H phosphates, precipitated from aqueous solution of synthetic, sulfate-free sea water at pH 7.5 and annealed in saturated water vapor at 130°C.
Annealing of the initially X–ray amorphous gel (Fig. 4.2) enhances the growth of whitlockite (diffraction maxima marked with Miller indices (hkl)) together with magnesium phosphate pentahydrate (marked X). Interplanar spacings are indicated in Ångström units. X-ray penetration through the sample gives rise to diffraction marked (Al) from the (111) reflection of aluminum in the sample holder.
Whitlockite (Frondel, 1949) was earlier thought to be a calcium phosphate; the essential structural role of magnesium- and hydrogen ions was revealed in the detailed structural work by Gopal *et al.* (1974). The protons in the structure are essential for the partial condensation of the phosphate groups to mainly di- (pyro-) and triphosphate at heating as illustrated by Fig. 4.3b

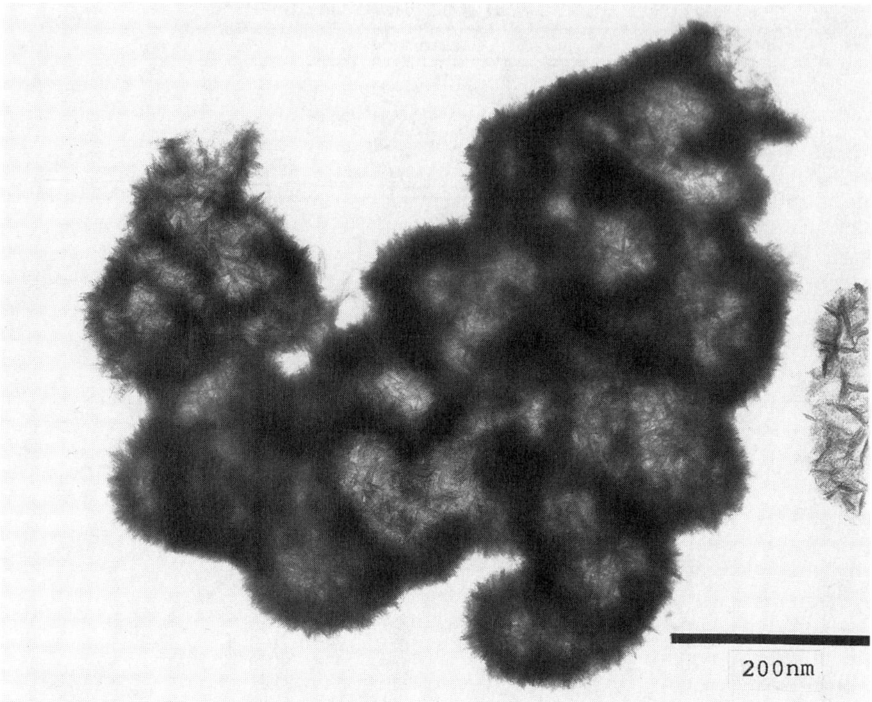

Figure 4.2. Calcium–magnesium–hydrogen phosphate formed by slow oxidation of phosphite ion to orthophosphate in artificial seawater at room temperature and pH 7.5. The crystals, aggregated in globules, are too small for identification by standard X-ray diffraction methods. Annealing in water vapor at 130°C reveals the presence of two phases (Fig. 4.1); $Ca_{18}Mg_2H_2(PO_4)_{14}$ (whitlockite) together with $Mg_3(PO_4)_2 \cdot 5H_2O$ (magnesium phosphate pentahydrate; so far not reported as a mineral).

In an abiotic Archean context, one would thus visualize a chain of common geological events leading to the production and release of condensed phosphate ions. This would begin with the ubiquitous deposition of protonated phosphates such as brushite and whitlockite in marine and lacustrine sediments, diluted to various degrees with other sedimentary components. Condensation of these phosphates would be expected to occur in regions of thermal metamorphism, such as in contact with volcanic intrusives or eruptive lava flows. Leaching of the soluble condensed phosphates could be followed by preferential sorption in the common double layer hydroxide minerals, where phosphorylation of organic anions or polar molecules could take place. Such a scenario is, as a complete sequence, totally speculative, but its individual steps are open to quantitative experimental and field evaluation.

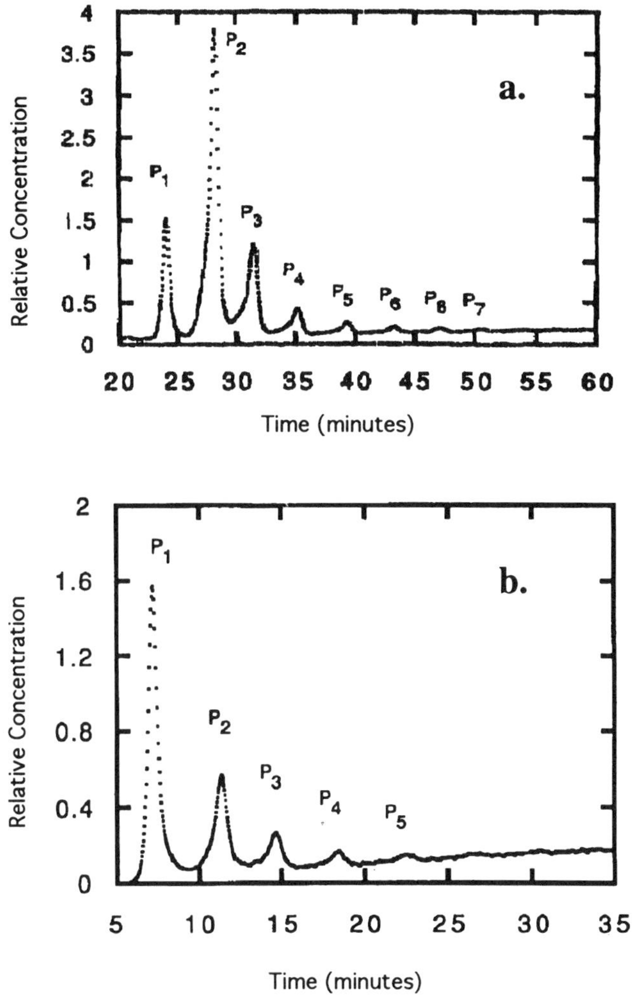

Figure 4.3. Chromatograms of oligophosphate species formed at heating in protonated phosphate, precipitated at room temperature from synthetic, sulfate-free artificial sea water. (**a**) brushite, crystallized at room temperature and pH 6.5; (**b**) whitlockite, together with $Mg_3(PO_4)_2 \cdot 5H_2O$, grown at 130°C from gel precipitated at room temperature at pH 7.5.

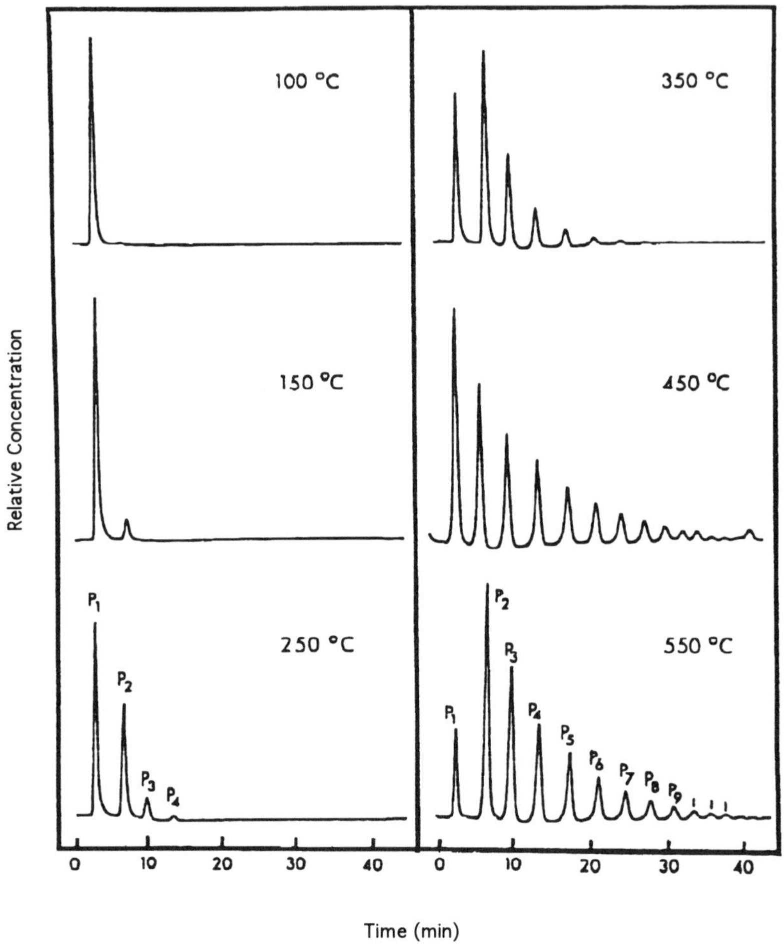

Figure 4.4. Ion chromatograms of oligophosphates formed from newberyite ($MgHPO_4 \cdot 3H_2O$) heated to equilibrium at temperatures between 100 to 550°C. Up to 12–membered linear oligophosphates can be distinguished after the highest temperature exposure.
Newberyite does not form in the range of conditions employed in our experiments, but could possibly be expected at higher Mg/Ca ratios suggested by Christophersen (1991, *op. cit.*, footnote, page 20). From Sales *et al.* (1992).

Even at a concentration estimated as low as 0.1 mM in the Archean ocean (Holland, 1984), ferrous ion may have a significant influence on the properties, including condensation of aqueous H-Mg-Ca orthophosphate minerals and gels. Exploration of the solid solubility of Fe^{2+} in the protonated phosphate minerals, and its effect on their solubilities and on phosphate condensation are important. Particularly ferric phosphates have low solubilities; in the present-day oxidized surface regions of the Earth, they form a substantial inorganic phosphate reservoir (Arrhenius, 1952; Berner, 1991) of often inconspicuous and complex minerals (Moore and Shen, 1983), in competition with the pervasive infestation by biogenic apatite.

Because of the low concentration of free oxygen implied for the Archean atmosphere-ocean system (Towe 1983, 1990), the sulfate ion concentration in the hydrosphere should have been low and mainly supported by photochemical oxidation of sulfur species in the atmosphere (Walker and Brimblecombe, 1985). In the studies of apatite stability referred to above, most experiments were therefore carried out in sulfate-free synthetic sea water. Controls using present-day sulfate ion concentration (0.028 M), however, confirm that this component does not markedly affect the (abiotic) stability field of apatite, e.g. by differential complexing of calcium and magnesium. Carbonate and fluoride ion were also found to lack any measurable effects outside of the stability field of apatite, which, even in their presence to saturation, is limited to the region above pH 8.5 (Gedulin, 1993).

The condensation studies, in collaboration with B. Sales, have only touched the surface of a complex phenomenon of importance in prebiotic geochemistry and condensed matter chemistry. They include exploration of the possibility of primary or metamorphically induced occurrences of hydrogen phosphate and pyrophosphate minerals in a suite of rocks representing the hydrothermal range outlined by Moore (1973). The results show pyrophosphate generation in mixed valence transition element orthophosphate minerals such as alluaudite and hureaulite, explainable by a redox mechanism proposed by Sales (1992). This points to the possibility of thermal condensation in common sedimentary transition element orthophosphates such as vivianite ($Fe_3(PO_4)_2 \cdot 8H_2O$) and other common sedimentary and hydrothermal ferrous and ferroferric orthophosphates, or their dehydration products.

5. APATITE AS A BIOMARKER

The exclusively biogenic origin of authigenic sedimentary apatite (under the specified geochemical conditions discussed above), has significant paleontological consequences - such apatite occurrence may serve as a chemical indicator of the presence of cellular life forms. This source of information is rather trivial when applied to late Archean and younger formations, where indisputable morphological fossil evidence and coherent (carbon) isotope fractionation patterns clearly show that life had taken over the surface of the planet.

In the oldest metasediments, now represented by the Isua formation, with an age of 3.8 b.y., the situation is, however, different; the isotopic record has been confused by metamorphism (Schidlowski, 1988; Appel, 1980; Monster *et al.*, 1979), and convincing morphological traces of biota are missing (Bridgwater *et al.*, 1981; Pflug, 1990). In this case, strong evidence comes from the phosphate phases that have been identified in these formations. Indeed the finding of rod-shaped aggregates of apatite, intergrown with graphite on the bedding planes in the Isua banded iron formation corroborates the opinion that life was present on Earth 3.8 b.y. ago, a notion also implied from the cellular complexity found in sedimentary rocks only 300 m.y. younger (Schopf and Walter, 1980; Schopf, 1983; 1993, this volume; Awramik *et al.*, 1988).

Authigenic, sedimentary apatite thus may provide an independent chemical indicator of the earliest recorded occurrence of life on Earth. This new tool is now awaiting further testing and applications to decipherable metasedimentary rocks on Earth older than 3.8 b.y., if and when they can be found, and to martian sedimentary rocks when they can be retrieved. In anticipation of a future martian sampling mission, current work aims at petrologic microanalysis of the sedimentary material in certain SNC meteorites (shergottite, nakhlite and chassignite achondrites) (Gooding *et al.*, 1988; 1992). These are believed to be derived from Mars because of the characteristic heavy nitrogen isotope enrichment in the occluded gases (Becker and Pepin, 1984; Bogard *et al.*, 1984; Swindle *et al.*, 1984).

In all these applications, the limiting conditions for the method have to be born in mind; for sedimentary apatite to be considered biogenic, the Mg/Ca ratio of the source medium must be assumed to have been larger than ~ 0.2; (in comparison, present day sea water has a ratio of 5.0, and fresh water 0.4; Broecker and Peng, 1982). Furthermore, pH would have to be less than 8.5 for exclusion of inorganically formed apatite, or the concentration of ammonia to have been less than isomolar with magnesium. It is unlikely that these extreme geochemical limits were ever approached. Indeed, the magnesium concentration in the hydrosphere is believed initially to have been higher than today (Christophersen, 1991)[3] and pH to have been depressed below neutrality by HCl, HF and H_2CO_3, initially uncompensated by weathering (Holland, 1984; Chang, 1993). Also, extreme deviations from the range of conditions that are currently considered possible, would be reflected by characteristic sedimentary mineral associations. There is reason to believe that some of these primary petrogenetic features are sufficiently resistant to low grade metamorphism to remain decipherable. In contrast, we know very little about the ancient hydrosphere on Mars (Carr and Wänke, 1992), and a thorough phase analysis of the presumed martian sediments in SNC meteorites is therefore required, to reveal indications of pH, Mg/Ca ratio, and ammonia content of the source solutions.

[3] Christophersen, O. A., 1991: A comparison of the calcium and magnesium budgets of sea water today, as seen in relation to the historical development of the sea water Ca/Mg ratio and the problem of apatite solubility when life originated. (Unpublished notes – personal communication).

REFERENCES

Ahn, H. J. and R. P. Buseck, 1989: Microstructures and tetrahedral strip-width order and disorder in Fe-rich minnesotaites, American Mineralogist, 74, 384-393.

Allmann, R., 1970: Doppelschichtstrukturen mit brucitähnlichen Schichtionen $[(Me(II)_{1-x}Me(III)_x(OH)_2]^{x+}$, Chimia, 24, 99-108.

Appel, P. W. U., 1980: On the early Archaean Isua iron-formation, West Greenland, Precambrian Res., 11, 73-87.

Arrhenius, G. 1952: Sediment cores from the East Pacific, 1.5.3. In H. Pettersson (Ed.), Reports of Swedish Deep Sea Expedition Göteborg: Elanders.

Arrhenius, G., 1987: Interaction of hydrous minerals with bioorganic precursor molecules, NASA Space Life Sciences Symposium, June 21-26, 260-261.

Arrhenius, G., J. Bachman, B. Gedulin, S. Hui and W. Paplawsky, 1989: Anion selective minerals as concentrators and catalysts for RNA precursor components, Orig. Life, 19, 235-236.

Awramik, S. M., J. W. Schopf and M. R. Walter, 1988: Carbonaceous filaments from North Pole, Western Australia: are they fossil bacteria in ancient stromatolites? Precambrian Res., 39, 303-309.

Bailey, S. W. 1980: Structures of layer silicates. In G. W. Brindley and G. Brown (Eds.), Crystal Structures of Clay Minerals and their X-ray Identifcation (pp. 1-123). London: Mineralogical Society.

Beaudry, A. A. and G. F. Joyce, 1992: Directed evolution of an RNA enzyme, Science, 257, 635-641.

Beck, A. and L. E. Orgel 1965: The formation of condensed phosphate in aqueous solution, Proc. Natl. Acad. Sci., 54, 664-667.

Becker, R. H. and R. O. Pepin, 1984: The case for a martian origin of the shergottites: nitrogen and noble gases in EETA 79001, Earth and Planetary Sci. Ltrs., 69, 225-242.

Bernal, J. D., 1949: The physical basis of life, Proc. Phys. Soc. London, 62(A), 537-558.

Bernal, J. D., D. R. Dasgupta and A. L. Mackay, 1959: The oxides and hydroxides of iron and their structural inter-relationships, Clay Miner. Bull., 4, 15-30.

Berner, R. A. 1991: Diagenesis of phosphorus in sediments from non-upwelling areas. In W. C. Burnett and S. R. Riggs (Eds.), Phosphate deposits of the world (pp. 27-32). New York: Cambridge University Press.

Bogard, D. D., L. E. Nyquist and P. Johnson, 1984: Noble gas contents of shergottites and implications for the Martian origin of SNC meteorites, Geochim. Cosmochim. Acta, 48, 1723-1739.

Bookin, A. S. and V. A. Drits, 1993: Polytype diversity of the hydrotalcite-like minerals I. Possible polytypes and their diffraction features, Clays and Clay Minerals, (in press).

Bookin, A. S., V. I. Cherkashin and V. A. Drits, 1993: Polytype diversity of the hydrotalcite-like minerals II. Determination of the polytypes of experimentally studied varieties, Clays and Clay Minerals, (in press).

Bridgwater, D., J. H. Allaart, J. W. Schopf, C. Klein, M. R. Walter, E. S. Barghoorn, P. K. Strother, A. H. Knoll and B. E. Gorman, 1981: Microfossil-like objects from the Archaean of Greenland: a cautionary note, Nature, 289, 51-53.

Brindley, G. W. and D. L. Bish, 1976: Green rust: A pyroaurite type structure, Nature, 263, 353.

Broecker, S. W. and T. H. Peng, 1982: Tracers in the sea, Lamont-Doherty Geological Observatory, Palisades, New York.

Butlerow, A., 1861: Bildung einer zuckerartigen Substanz durch Synthese, Annalen d. Chem. und Pharm., 120, 295-298.

Carr, M. and H. Wänke, 1992: Earth and Mars: water inventories as clues to accretional histories, Icarus, 98, 61-71.

Chang, S., 1993: The planetary setting of prebiotic evolution, in Early Life on Earth, Nobel Symposium No. 84, S. Bengtson(Eds.), Columbia University Press (in press).

Crovisier, J. L., J. H. Thomassin, T. Juteau, J. P. Eberhart, J. C. Touray and P. Baillif, 1983: Experimental seawater-basaltic glass interaction at 50°C: Study of early developed phases by electron microscopy and X-ray photoelectron spectrometry, Geochim. Cosmochim. Acta, 47, 377-387.

Cairns-Smith, A. G., 1982: Genetic Takeover and the Mineral Origins of Life, Cambridge University Press, Cambridge.

Drits, A. V., A. N. Lisitsyna and I. V. Cherkashin, 1985: New mineral varieties of the hydrotalcite-manasseite group, low temperature alteration products of basalt and volcanogenic sediments on the ocean floor, Doklady Akademii Nauk SSSR, 284, 443-447.

Drits, A. V., T. N. Sokolova, G. V. Sokolova and I. V. Cherkashin, 1987: New members of the hydrotalcite-manasseite group, Clays Clay Miner., 35, 401-417.

Eschenmoser, A. and M. Dobler, 1992: Warum Pentose- und nicht Hexose-Nukleinsäuren? Teil I. Einleitung und Problemstellung, Konformationsanalyse für Oligo-2',3'-dideoxy-glucopyranosyl-nukleotide ('Homo-DNS') sowie Betrachtungen zur Konformation von A- und B-DNS', Helv. Chim. Acta, 75, 218-259.

Eschenmoser, A. and E. Loewenthal, 1992: Chemistry of potentially prebiological natural products, Chemical Society Reviews, 21, 1-16.

Fischer, E., 1888: Ber. D. Chem. Ges., 21, 988.

Frondel, C., 1949: Wolfeite, xanthoxenite and whitlockite from the Palermo mine, New Hampshire, Am. Miner., 34, 692-705.

Gedulin, B., 1993: Ph.D. Thesis, University of California, San Diego, (in preparation).

Gedulin, B. and G. Arrhenius, 1993: Sources and geochemical evolution of RNA precursor molecules; the role of phosphate, Early Life on Earth, Nobel Symposium No. 84. S. Bengtson, Ed. Columbia Univ. Press.

Gilbert, W., 1986: The RNA world, Nature, 319, 618.

Goldschmidt, V. M., 1952: Geochemical aspects of the origin of complex organic molecules on the Earth, as precursors to organic life, New Biology, 12, 97-105.

Gooding, J. L., 1992: Soil mineralogy and chemistry on Mars: possible clues from salts and clays in SNC meteorites, Icarus, 99, 28-41.

Gooding, J. L., S. J. Wentworth and M. E. Zolensky, 1988: Calcium carbonate and sulfate of possible extraterrestrial origin in the EETA 79001 meteorite, Geochimica et Cosmochimica Acta, 52, 909-915.

Gopal, R., C. Calvo, J. Ito and K. W. Sabine, 1974: Crystal structure of synthetic Mg-whitlockite $Ca_{18}Mg_2H_2(PO_4)_{14}$, Can. J. Chem., 52, 1155-1164.

Guerrier-Takada, C., K. Gardner, T. Marsh, N. Pace and S. Altman, 1983: The RNA moiety of ribonuclease P is the catalytic subunit of the enzyme, Cell, 35, 849-857.

Gulbrandsen, R. A., C. E. Roberson and S. T. Neil, 1983: Time and the crystallization of apatite in seawater, Geochim. Cosmochim. Acta, 48, 213-218.

Hirschler, A., J. Lucas and J.-C. Hubert, 1990: Bacterial involvement in apatite genesis, FEMS. Microbiol. Ecol., 73, 211-220.

Holland, D. H. 1984: The Chemical Evolution of Atmosphere and Oceans, Princeton University Press, Princeton.

Joyce, G. F., 1989: RNA evolution and the origins of life, Nature, 338, 217-224.

Joyce, G. F., 1992: Directed molecular evolution, Scientific American, 267(6), 90-97.

Kruger, K., P. J. Grabowsky, A. J. Zang, D. E. Gottschling and T. R. Cech, 1982: Self splicing RNA: autoexcision and autocyclization of the ribosomal RNA intervening sequence, Cell, 31, 143-157.

Kuma, K., B. Gedulin, B. Paplawsky and G. Arrhenius, 1989: Mixed-valence hydroxides as bio-organic host minerals, Orig. Life, 19, 573-602.

Lucas, J. and L. Prévôt, 1981: Synthèse d'apatite à partir de matière organique phosphorée (ARN) et de calcite par voie bactérienne, C. R. Acad. Scí. II, 292, 1203-1208.

Mazghouni, M., N. Kbir-Ariguib, J. J. Counioux, and A. Sebaoun, 1981: Étude des equilibres solide–liquide–vapeur des systèmes binaires K_3PO_4–H_2O et $Mg_3(PO_4)_2$–H_2O, Thermochimica Acta, 47, 125-139.

Miyata, S., 1983: Anion-exchange properties of hydrotalcite-like compounds, Clays and Clay Minerals, 31(4), 305-311.

Monster, J., Appel, P. W. U., Thode, H. G., Schidlowski, M., Carmichael, C. M. and Bridgwater, D., 1979: Sulfur isotope studies in early Archean sediments from Isua, west Greenland: implictions for the antiquity of bacterial sulfate reduction, Geochimica et Cosmochimica Acta, 43, 405-413.

Moore, P. B., 1973: Pegmatite phosphates: descriptive mineralogy and crystal chemistry, Mineralogical Record, 105-130.

Moore, P. B., and J. Shen, 1983: An X-ray structural study of cacoxenite, a mineral phosphate, Nature, 306, 356-358.

Morey, G. W., 1953: The System H_2O–$NaPO_3$, J. Am. Chem. Soc., 75, 5794-5797.

Müller, D., S. Pitsch, A. Kittaka, E. Wagner, C. E. Wintner and A. Eschenmoser, 1990: Chemie von Alpha-Aminonitrilen. Aldomerisierung von Glycolaldehyd-phosphat zu racemischen Hexose-2,4,6-triphosphaten und (in Gegenwart von Formaldehyd) racemischen Pentose-2,4-diphosphaten: rac-Allose-2,4,6-triphosphat und rac-Ribose-2,4-diphosphat sind die Reaktionshauptprodukte, Helv. Chim. Acta, 73, 1410-1468.

Pflug, H. D., 1990: History of life in the Precambrian - a review, Geological Institute, Special Publication 70, Cologne University Press.

Pitsch, S., A. E. Eschenmoser, B. Gedulin, S. Y. Hui and G. Arrhenius, 1993: Mineral induced selective formation of sugar phosphates, Helv. Chim. Acta (in preparation).

Pitsch, S. 1992: Ph.D. thesis, ETH, Zürich.

Prévôt, L., E. M. El Faleh and J. Lucas, 1989: Details on synthetic apatites formed through bacterial mediation. Mineralogy and chemistry of the products, Sci. Géol. Bull., 42(3), 237-254.

Rabinowitz, J., S. Chang and C. Ponnamperuma, 1968: Phosphorylation by way of inorganic phosphate as a potential prebiotic process. Nature, 218, 442-443.

Rivadeneyra, M. A., I. Pérez–García and A. Ramos-Cormenzana, 1992: Struvite precipitation by soil and fresh water bacteria. Curr. Microbiology, 24, 343-347.

Sales, B. C., B. C. Chakoumakos, L. A. Boatner and J. O. Ramey, 1992: Structural evolution of the amorphous solids produced by heating crystalline $MgHPO_4 \cdot 3H_2O$, J. Mater. Res., 7(10), 1-5.

Schidlowski, M., 1988: A 3,800-million-year isotopic record of life from carbon in sedimentary rocks, Nature, 333, 313-318.

Schopf, J. W. and M. R. Walter, 1980: Archean microfossils and "microfossil-like" objects - a critical appraisal, Glover, J.E. and Groves, D.I. (Eds.), Interml. Archaean Symp., Australian Academy of Sciences.

Schopf, J. W. and M. R. Walter, 1983: Archaean microfossils: new evidence for ancient microbes, J. W. Schopf (Ed.), Earth's earliest biosphere, its origin and evolution, Princeton University Press.

Shapiro, R., 1988: Prebiotic ribose synthesis: A critical analysis, Orig. Life, 17, 1-14.

Szostak, J. W., 1992: In vitro genetics, Trends in Biochemical Sciences, 17(3) 89-93.

Swindle, T.D., M.W. Coffee, C.M. Hohenburg, G.B. Hudson and R.J. Rajan, 1984: Noble gases in SNC meteorites (abstract), Meteoritics, 19, 318-319.

Taylor, R. M., 1982: Stabilization of colour and structure in the pyroaurite-type compounds Fe(II)Fe(III)Al(III)-hydroxycarbonates, Clay Miner., 17, 369-372.

Thomassin, J.-H. and J.-C. Touray, 1979: Étude des premiers stades de l'interaction eau-verre basaltique: données de la spectrométrie de photoélectrons (XPS) et de la microscopie électronique à balayage, Bull. Mineral., 102, 594-599.

Thomassin, J.-H. and J.-C. Touray, 1982: L'hydrotalcite, un hydroxycarbonate transitoire précocement formé lors de l'interaction verre basaltique-eau de mer, Bull. Mineral., 105, 312-319.

Towe, K., 1983: Precambrian atmospheric oxygen and banded iron formations: a delayed ocean model, Precambrian Res., 20, 161-170.

Towe, K. M., 1990: Aerobic respiration in the Archean? Nature, 348, 54-56.

Walker, J. C. A. and P. Brimblecombe, 1985: Iron and sulphur in the pre-biological ocean, Precambrian Res., 28, 205-222.

Westheimer, F. H., 1987: Why nature chose phosphates, Science, 235, 1173-1178.

Vali, H. and J. Kirschvink, 1989: Magnetofossil dissolution in a palaeomagnetically unstable deep-sea sediment, Nature, 339, 203-206.

EVOLUTION IN AN RNA WORLD

Peter Schuster
Institut für Molekulare Biotechnologie
D O-6900 Jena, Germany
and
Institut für Theoretische Chemie der Universität Wien
A 1090 Wien, Austria

ABSTRACT

The concept of an RNA world has been introduced for a prebiotic scenario that is dominated by RNA molecules and their properties, in particular their capabilities to act as templates for reproduction and as catalysts for several cleavage and ligation reactions of polynucleotides and polypeptides. This notion is used here also for simple experimental assays which are well suited to study evolutionary phenomena in the test tube. The molecular structures and other properties of RNA molecules that are relevant in molecular evolution are discussed. Evidence for adaptation to the environment in cell-free media is presented. In the case of RNA molecules adaptation can be studied and interpreted by considering secondary structures. RNA based molecular evolution experiments have led to interesting spin-offs in biotechnology, commonly called "applied molecular evolution" or "evolutionary biotechnology", which makes use of the Darwinian trial-and-error strategies in order to synthesize new pharmacological compounds and other biologically based advanced materials. Error-propagation in RNA replication leads to formation of mutant distributions calles "quasispecies". An increase in the error rate broadens the mutant spectrum. There exists a sharply defined threshold beyond which heridity breaks down and evolutionary adaptation becomes impossible. Almost all RNA viruses studied so far operate at conditions very close to the error threshold. Quasispecies and error thresholds are important for an understanding of RNA virus evolution and for the development of novel antiviral strategies.

The notion of sequence space allows to introduce a distance between pairs RNA sequences which is tantamount to counting the minimal number of point mutations required to convert the sequences into each other. The mean sensitivity of RNA secondary structures to mutation depends strongly on the base pairing alphabet: structures from sequences which contain only one base pair (**GC** or **AU**) are much less stable against mutation than those derived from the natural (**AUGC**) sequences. Evolutionary optimization of two-letter sequences is thus much more difficult than optimization in a world of natural RNA sequences with two base pairs. This fact might provide a rationale for the usage of four bases in the genetic language of nature. Finally we analyze the topological properties of the RNA shape space and demonstrate that "neutral paths" connecting neighbouring sequences with identical structures which go through the entire sequence space are very frequent. In addition, we show that common structures are found everywhere in sequence space. Hence, evolution can migrate to almost every part of sequence space without "hill climbing" and only small fractions of the entire numbers of sequences have to be searched in order to find suitable structures.

ISBN 0–937194–31–X

1. THE CONCEPT OF AN RNA WORLD

RNA molecules were found to be unique in chemistry and biology since they can not only act as templates for their reproduction but are also catalysts for several reactions involving other RNA molecules or oligopeptides. Template induced reproduction of RNA molecules occurs in cells that were infected by RNA viruses or viroids (Riesner and Gross, 1985). Enzyme free template induced RNA synthesis has been studied extensively by Leslie Orgel (1992). Replication of RNA molecules under non-equilibrium conditions provides the basis for evolutionary adaptation in the sense of Darwin's mechanism as we shall see in the next two sections. The discoveries of Thomas Cech and Sidney Altman (Cech, 1986; Symons, 1992) have shown that, in contrast to previous belief, RNA molecules can catalyze several classes of reactions with similar high efficiency and specificity as protein enzymes. The notion of "ribozymes" was created for such RNA based catalysts. The catalytic activities of RNA molecules are all related to cleavage or formation of nucleotide or peptide bonds. For RNA substrates ligation and cleavage are highly sequence specific. The capability of RNA to catalyze peptide bond cleavage and formation (Noller, 1991; Noller *et al.*, 1992; Piccirilli *et al.*, 1992) is of particular interest since it can be interpreted as a hint that primordial ribosomes (the present day cellular particles that synthesize proteins) might have worked without proteins. Although the catalytic repertoire of ribozymes is rather poor compared to that of protein enzymes, it comprises the key reactions that are necessary to produce polynucleotides from oligomers and monomers. A world of RNA molecules driven by a steady supply of suitable energy rich compounds thus represents a kind of biological toy universe which allows to study the evolutionary process in its simplest form. Such an **RNA world** is an adaptive system of minimal complexity.

Is there room for an RNA world in the current view of chemical evolution on the primordial Earth? In figure 1 we show a sketch of a plausible sequence of scenarios leading to the onset of RNA based Darwinian evolution (for more details see e.g. Schuster, 1981; Mason, 1991). The major problem with RNA in chemical evolution is to be seen in its highly elaborate molecular structure: for RNA template induced reactions molecules of high stereochemical purity are required, and chiral compounds such as ribose have to be present as pure single antipods (Racemic mixtures of D- and L-ribose incorporated into polynucleotides are prohibitive for template induced replication). It was suggested therefore that less complicated molecules which could nevertheless act as templates preceded the RNA world (Orgel, 1986; Joyce, 1989). Recent progress in **template chemistry** (Orgel, 1992) makes this suggesetion more and more plausible. RNA molecules, on the other hand, are the first template molecules which share most of their chemical properties with DNA the predominant genetic information carrier at present. Information transfer from DNA to RNA and vice versa, known as transcription and reverse transcription, is routine in present day biochemistry. We can speculate therefore that a phylogenetic extrapolation of present day genetic information may go back as far as to the first efficiently replicating species of a primordial RNA world. In short RNA molecules seem to appear late in chemical evolution and very early in biological evolution.

According to the analysis of microfossils carried out by William Schopf (Schopf and Packer, 1992) the first sure remnants of microorganisms are dated 3.4×10^9 years ago. These microorganisms were most likely photosynthetic and related to present day cyanobacteria. Needless to say, there was a long way with many intermediate steps to go

2. MOLECULAR GENOTYPES AND PHENOTYPES

The dichotomy between genotypes and phenotypes is basic to biological evolution. Molecular biology has shown that the genetic information for the unfolding of organisms (which represent the phenotypes) is contained in the DNA (which is the genotype). The Darwinian priciple of evolution is based on heridity, variation and selection. Acccording to our present day knowledge all inheritable variations are caused by changes of the genotype (i.e. by changes in the sequence of nucleotide bases of the DNA), and selection acts through differential reproductive success of phenotypes. RNA molecules in cell-free evolution experiments are both genotypes and phenotypes. As we see in figure 2 the genotype is again a sequence of nucleotide bases, this time in the form of an RNA molecule. Following Sol Spiegelman we consider the spatial structure of the RNA molecule as its phenotype. Indeed, this structure is evaluated by the selection process. Systematic kinetic studies on RNA replication and test tube evolution carried out by Christof Biebricher in Manfred Eigen's laboratoty (Biebricher *et al.* 1983, 1984, 1985) revealed the mechanism of RNA replication. The rate and equilibrium constants which determine the outcome of selection under different conditions are known now. They can be determined independently from selection experiments.

Selection experiments may be carried out by discontinuous renewal of the consumed material as in the serial transfer technique (Spiegelman, 1971) or under the continuous constraint maintained in elaborate evolution reactors (Husimi *et al.*, 1982; Husimi and Keweloh, 1987). An new selection technique was used by John McCaskill in his molecular evolution experiments in capillaries (Bauer *et al.*, 1989): the capillaries contain a medium suitable for replication, RNA is injected and a wave front spreads through the medium. The front velocity of the traveling wave increases with the replication rate and hence, faster replicating mutants are selected by the wave propagation mechanism.

The capability of RNA molecules to evolve by natural selection is used now in applied molecular evolution to produce biomolecules with new properties. This novel evolutionary technology was originally suggested by Eigen and Gardiner (Eigen and Gardiner, 1983; for a recent review see Kauffman, 1992). The essential trick of this technique is to encode the desired functions into the selection constraint. As indicated in figure 3 sufficient variation is introduced into populations either by artificially increased mutation rates or by partial randomization of sequences. The synthesis of oligonucleotides with random sequences is routine nowadays. Several examples of successful applications of molecular selection techniques to biochemical problems are found in the current literature (Horwitz *et al.*, 1989; Tuerk and Gold, 1990; Ellington and Szostak, 1990; Beaudry and Joyce, 1992).

3. ADAPTATION TO ENVIRONMENTAL CHANGES

The most frequently occurring genotypes (commonly called "master sequences") can be isolated from populations and analyzed during the course of molecular evolution experiments. These investigations show how the structure of RNA molecules is adjusted in order to cope with changes in the environment. Such environmental changes are caused by adding suitable substances to the replication medium which deteriorate the conditions for reproduction. For example, heterocyclic dies (for example ethidium bromide or acridinium

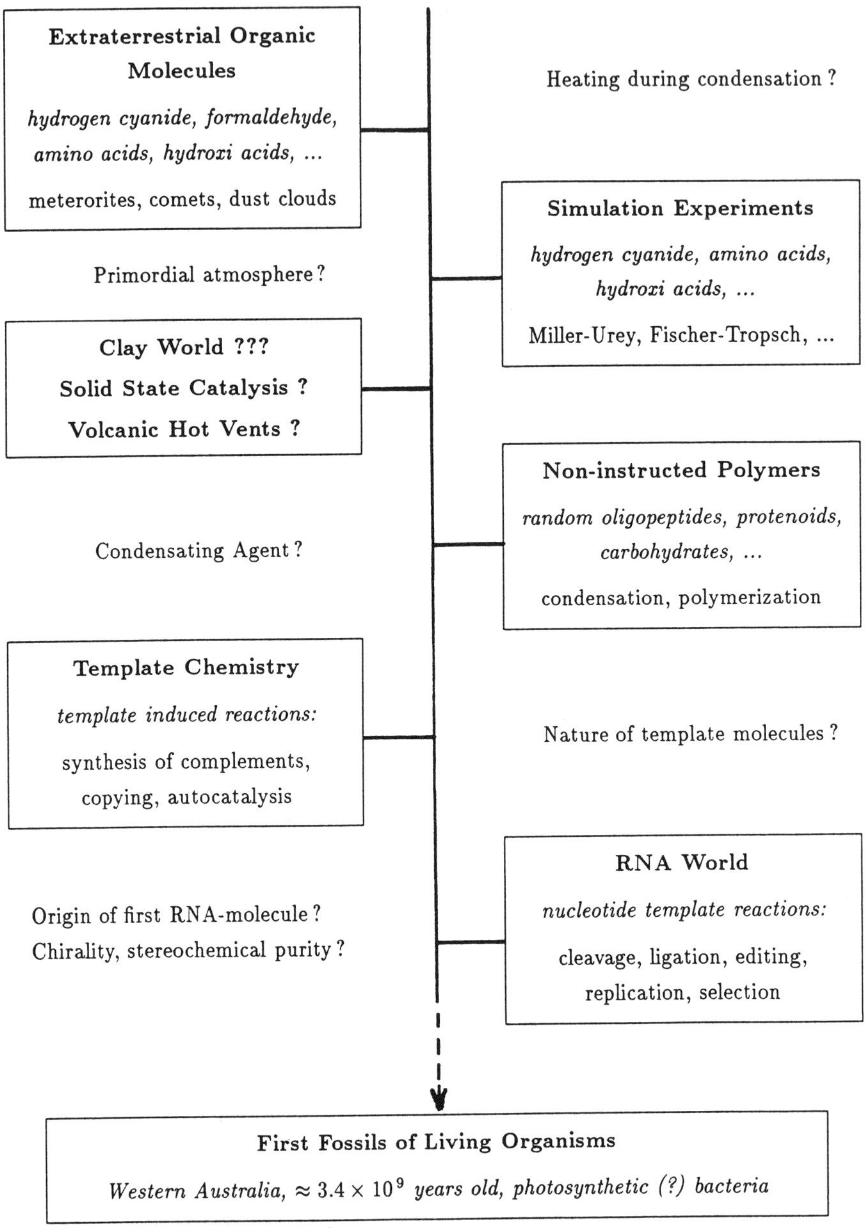

Fig.1: Some facts and open questions about the origin of life.

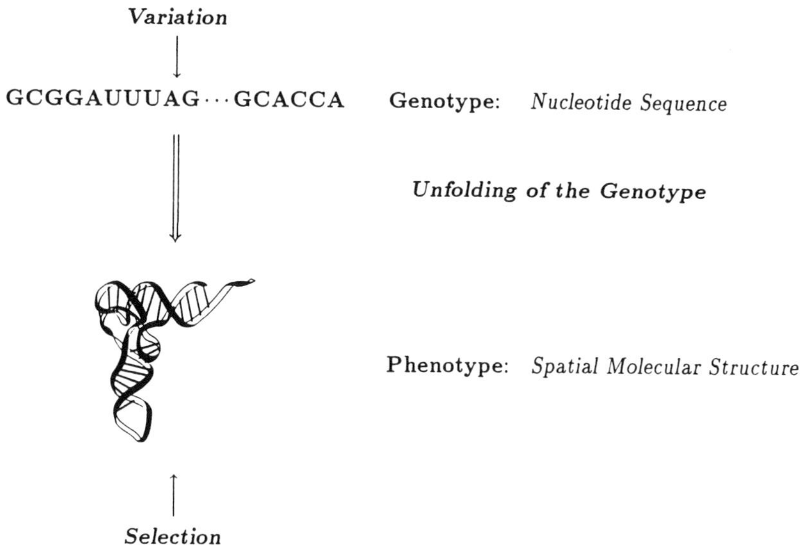

Fig.2: Molecular genotypes and phenotypes in RNA evolution experiments.

from a primitive Darwinian system like the RNA world to a fully developed photosynthetic organism (Eigen and Schuster, 1985). The flow chart sketched in figure 1 does not refer explicitly to these other "worlds" which inevitably lay in between.

Evolution experiments with pure RNA world have to face severe problems: template induced RNA synthesis is slow, does not work for many templates under the conditions applied so far, and the accuracy of base incorporation is low (of the order of one error in hundred nucleotides). In order to explore the capabilities of a minimal complexity adaptive system we are, however, not restricted to a pure RNA world. Why not use protein catalysts a highly specific and efficient environmental factors? Addition of enzymes leads to a new toy universe for studies on evolution that might be characterized as a **protein assisted RNA world**. Sol Spiegelman (1971) has indeed shown that such a universe can be created in the test tube, and indeed, it shows all features of the Darwinian scenario. RNA dependent RNA polymerases, commonly called RNA replicases, occur in nature. Spiegelman used an enzyme isolated from bacterial cells of *E. coli* which were infected by the RNA bacteriophage $Q\beta$. This enzyme replicates RNA fast and with much higher accuracy than achieved without a protein catalyst: many thousands of generations can be studied in experiments lasting less than a day, and the accuracy of replication is about one error in 3000 nucleotides.

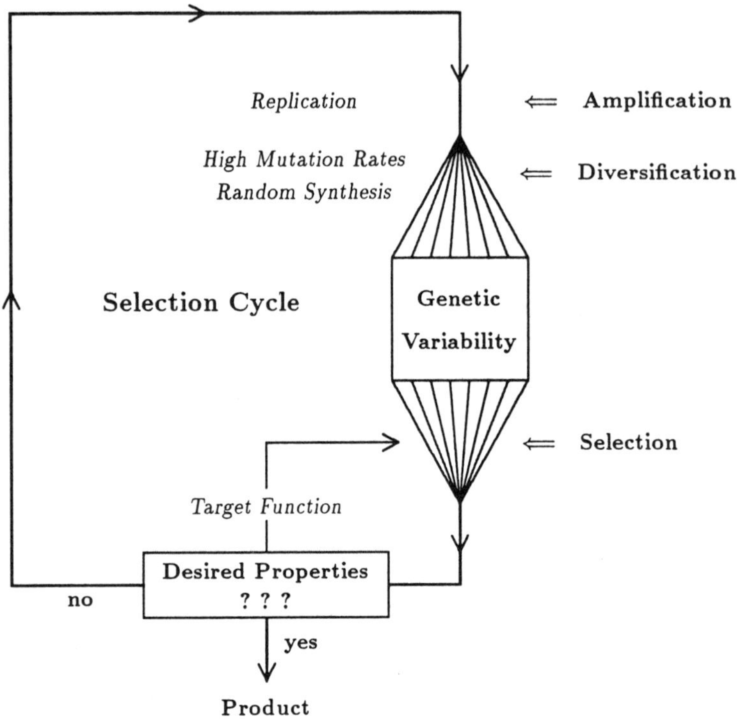

Fig.3: Selection techniques of applied molecular evolution.

orange), which intercalate between Watson-Crick base pairs, and interfere with replication therefore, can be used (Kramer *et al.*, 1974), or ribonucleases (RNA cleaving enzymes) can be applied (Strunk, 1992). Mutants appear and are enriched in the populations which compensate for the change in the environment by having fewer binding or cleavage sites than the wild type.

Replication errors lead to new molecular species whose replication efficiency is evaluated by the selection mechanism. The higher the error rate, the more mutations occur and the more viable mutants appear in the population. The stationary mutant distribution is characterized as "quasispecies" since it represents the genetic reservoir of asexually replicating populations. An increase in the error rate thus leads to a broader spectrum of mutants and makes evolutionary optimization faster and more efficient in the sense that

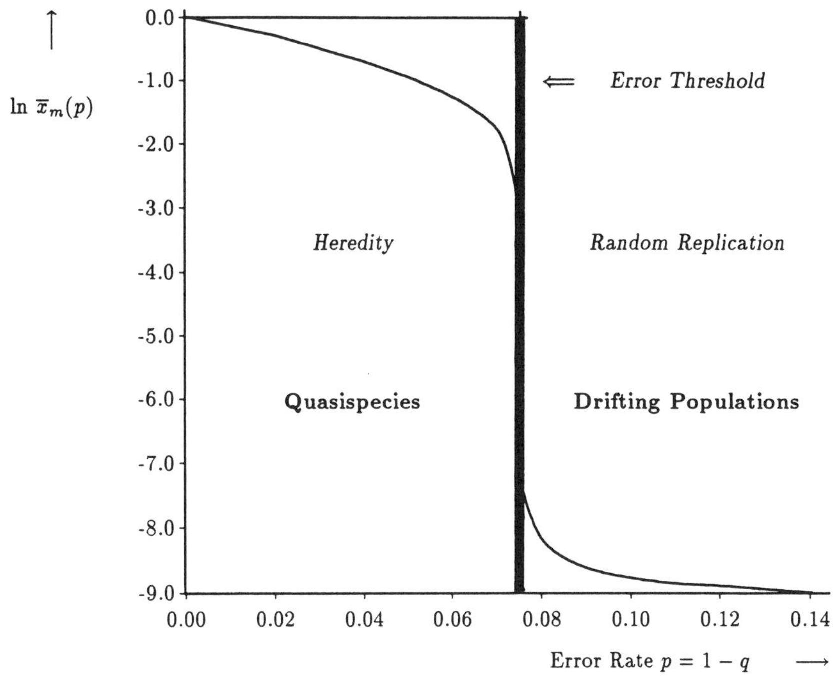

Fig. 4: Evolution at the error threshold of replication. The fraction of the most frequent species in the population, called the **master sequence**, is denoted by $x_m(p)$. It becomes very small at the error threshold. Accordingly, the total fraction of all mutants, $1 - x_m(p)$, approaches one at the critical mutation rate.

populations are less likely caught in local fitness optima. There is, however, a critical error threshold (Eigen, 1971; Eigen and Schuster, 1979; Swetina and Schuster, 1982): if the error rate exceeds the critical limit heridity breaks down, populations are drifting in the sense that new RNA sequences are formed steadily, old ones dissapear, and no evolutionary optimization according to Darwin's principle is possible (Fig.4). Variable environments require sufficiently fast adaptation and species with tunable error rates will adjust their quasispecies to meet the environmental challenge. In constant environments, on the other hand, such species will tune their error rates to the smallest possible values in order to maximize fitness.

Viruses are confronted with extremely fast changing environments since their hosts developed a variety of defense mechanisms ranging from the restriction enzymes of bacteria to the immune system of mammals and man. RNA viruses have been studied extensively. Their multiplication is determined by enzymes that do not allow large scale variations of the replication accuracies. They vary the error rate by changing the length of their genomes, and adjust the RNA chain lenghts to optimal values, which correspond to maximal chain lengths (Eigen, 1971; Eigen and Schuster, 1977):

$$\nu_{max} = \frac{\ln \sigma}{p} . \tag{1}$$

Herein $\sigma \leq 1$ is the superiority of the master sequence in the stationary population and p is the error rate per (newly incoporated) base and replication event. The superiority expresses differential fitness between the master sequence and the average of the remaining population. In the limit lim $\sigma \rightarrow 1$ we are dealing with "neutral evolution" (Kimura, 1983). Experimental analysis of several RNA virus populations has shown, that almost all chain lengths are adjusted to yield error rates close to the threshold value. Thus, RNA viruses adapt to their environments by driving optimization efficiency towards the maximum.

4. EVOLUTIONARY STABILITY OF RNA STRUCTURES

No physical process can occur with ultimate accuracy and hence, mutations are unavoidable. Given a certain mutation rate, we may ask what are the differences in stability of RNA structures against mutation. How likely does a change in the sequence result in an actual change in the structure? In other words we try to estimate the fraction of neutral mutants in the neighbourhood of a typical sequence ("neutral" is commonly used for sequences which have properties that are indistinguishable for the selection process; we shall use the term here in the narrower sense of identical structures). This question is of statistical nature and can be answered only by proper application of statistical techniques.

Firstly, the RNA sequences have to be ordered in a natural way. In case point mutations (single base exchanges) are predominant, the Hamming distance (d_h, counting the number of positions in which two aligned sequences differ) is an appropriate measure of the relationship between two sequences, since it counts the minimum number of point mutations required to convert one sequence into another. Moreover, the Hamming distance d_h induces a metric on the sequence space. The space of binary sequences of chain length $\nu = 4$ is shown in figure 5 as an example.

RNA secondary structures are first approximations to the spatial structures of RNA molecules. They are understood as a listing of the Watson-Crick-type base pairs in the actual structure and may be represented as planar graphs (Fig.6). We consider RNA secondary structures as elements of an abstract "shape space". Again a measure of relationship of RNA structures which induces an metric on the shape space can be found (Fontana *et al.*, 1991, 1992). We derived this distance measure from trees which are equivalent to the structure graphs, and accordingly it is called the "tree distance", d_t. RNA folding thus can be understood as a mapping from one metric space into another, in particular from sequence space into shape space. A path in sequence space corresponds uniquely to a path in shape space (The inversion of this statement is not true as we shall see in the next section 5).

The whole machinery of mathematical statistics and time series analysis can now be applied to RNA folding. In particular, an autocorrelation function of structures based on tree distances (d_t) is computed from the equation

$$\varrho_t(h) = 1 - \frac{< d_t^2(h) >}{< d_t^2 >} . \tag{2}$$

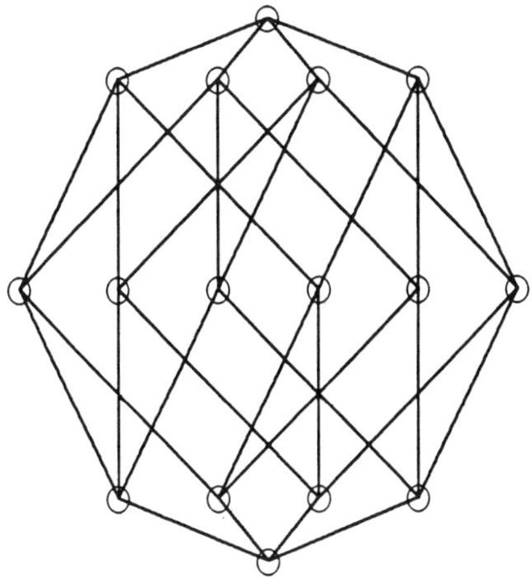

Fig. 5: The sequence space of binary (**AU** or **GC**) sequences of chain length $\nu = 4$. Every circle represents a single sequence of four letters. All pairs of sequences with Hamming distance $d_h = 1$ (these are pairs of sequences that differ in one position) are connected by a straight line. The geometric object obtained is a hypercube in four-dimensional space and hence, all positions (and all sequences) are topologically equivalent.

Mean square averages are taken here over all sequences in squence space ($< d_t^2 >$), or over all sequences in the mutant class h of the reference sequence ($< d_t^2(h) >$, i.e. over all sequences at Hamming distance h from the reference). The autocorrelation functions can be approximated by exponential functions, and correlation lengths (ℓ_t) are estimated from the relation: $\ln\big(\varrho_t(\ell_t)\big) = -1$.

The correlation length is a statistical measure of the hardness of optimization problems (see e.g. Eigen *et al.*, 1989). The shorter the correlation length, the more likely is a structural change occurring as a consequence of mutation. The correlation length thus measures stability against mutation. In figure 7 correlation lengths of RNA structures are plotted against the chain length. An almost linear increase is observed. Substantial differences are found in the correlation lengths derived from different base pairing alphabets. In particular, the structures of natural (**AUGC**) sequences are much more stable against mutation than pure **GC**-sequences or pure **AU**-sequences. This observation is in agreement with structural data obtained for ribosomal RNA (Wakeman and Maden, 1989). It provides also a plausible explanation for the use of two base pairs in nature: optimization in an RNA world with only one base pair would be very hard, and the base pairing probabilty in sequences with three base pairs is too low to form thermodynamically stable structures for short chain lengths. The choice of two base pairs thus appears to be a comprise between stability against mutation and thermodynamic stability.

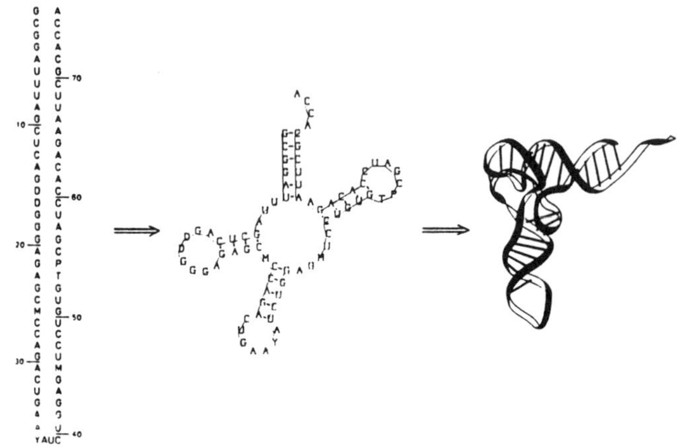

Fig. 6: Folding of an RNA sequence into its spatial structure. The process is partitioned into two phases: in the first phase only the Watson-Crick-type base pairs are formed (which constitute the major fraction of the free energy), and in the second phase the actual spatial structure is built by folding the planar graph into a three-dimensional object. The example shown here is phenylalanyl-transfer-RNA (t-RNA^{phe}) whose spatial structure is known from X-ray crystallography.

5. HOW TO SEARCH FOR RNA STRUCTURES

The sequence space is a bizarre object: it is of very high dimension (its dimension coincides with the chain length of RNA: $25 < \nu < 500$ for RNA molecules in test tube experiments, $250 < \nu < 400$ for viroids, and $3500 < \nu < 20\,000$ for (most) RNA viruses), but there are only few points on each coordinate axis (κ points; κ is the number of digits in the alphabet: $\kappa = 2$ for **AU** and **GC**, $\kappa = 4$ for **AUGC**). The number of secondary structures which are acceptable as minimum free energy structures of RNA molecules is much smaller than the number different sequences: in case of natural (**AUGC**) molecules we have $1.485 \times \nu^{-3/2}(1.849)^{\nu}$ structures for 4^{ν} sequences (Schuster *et al.*, 1993). The mapping from sequence space into shape space thus cannot be invertible. We cannot expect therefore that our intuition which is well trained with invertible maps in three-dimensional space will guide us well through sequence and shape spaces. What we need is an appropriate statistics to deal with these abstract objects.

The information contained in the mapping from sequence space into shape space is condensed into a two-dimensional, conditional probability density surface

$$S(t,h) \;=\; \mathrm{Prob}\left(\, d_t = t \;\middle|\; d_h = h \,\right) . \tag{3}$$

This structure density surface (SDS) expresses the probability that the secondary structures of two randomly chosen sequences have a structure distance t provided their Ham-

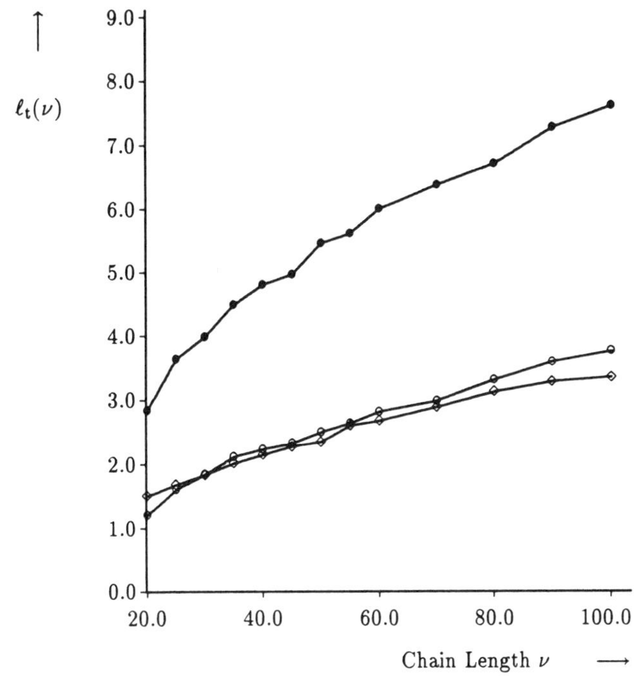

Fig. 7: Correlation lengths of structures (ℓ_t) of RNA moleucles in their most stable secondary structures as functions of the chain length ν. Values are shown for binary pure **GC**-sequences ($\diamond$), for binary pure **AU**-sequences ($\circ$), and for natural **AUGC**-sequences ($\bullet$). The correlation lengths are computed from $(\ln \varrho_t(h), h)$-plots by means of a least root mean square deviation fit.

ming distance is h. An example of a structure density surface for natural sequences of chain length $\nu = 100$ is shown in figure 8. We recognize an overall shape that corresponds to one half of a horseshoe and superimposed on it rugged details. The contour plot illustrates an important property of the structure density surface: at short Hamming distances $(1 \leq \nu < 16)$ the probability density changes strongly with increasing Hamming distance, but further away from the reference sequence $(16 < \nu < 100)$ this probality density is essentially independent of the Hamming distance h. The first part reflects the local features of sequence-structure relations. Up to a Hamming distance of $h = 16$ there is still some memory of the reference sequence. Then, at larger Hamming distances the structure density surface contains exclusively global information which is independent of the reference.

In order to gain more information on the relation between sequences and structures an inverse folding algorithm which determines the sequences that share the same minimum free energy secondary structure was conceived and applied to a variety of different structures (Schuster *et al.*, 1993). The frequency distribution of structures has a very sharp

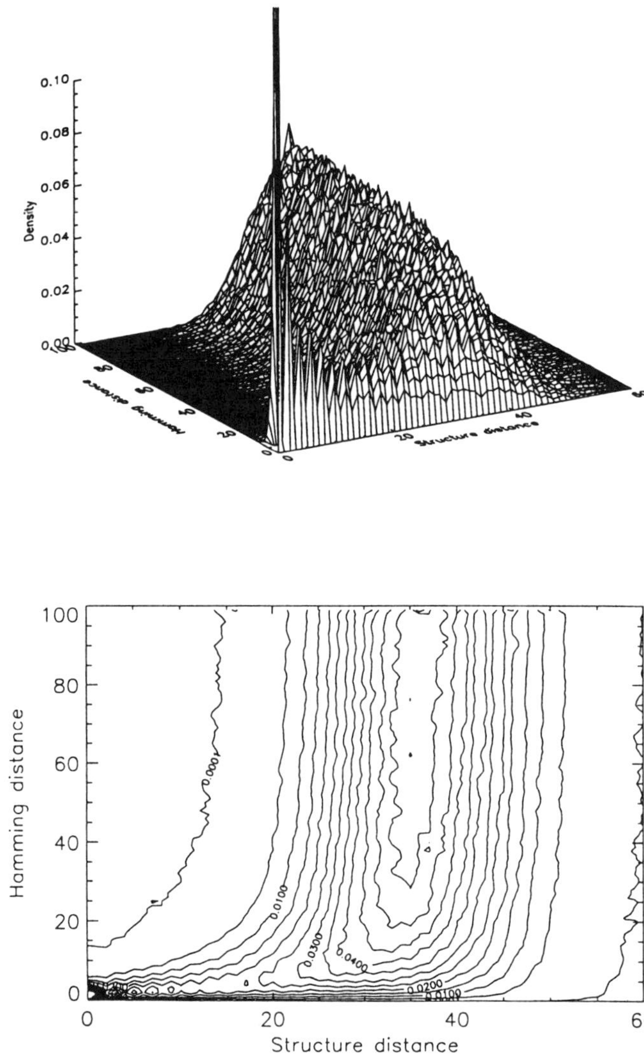

Fig. 8: The structure density surface $S(t,h)$ of natural **AUGC**-sequences of chain length $\nu = 100$. The density surface (upper part) is shown together with a contour plot (lower part). In order to dispense from confusing details the contour lines were smoothened. In this computation a sample of 1000 reference sequences was used which amounts to a total sample size of 10^6 individual RNA foldings.

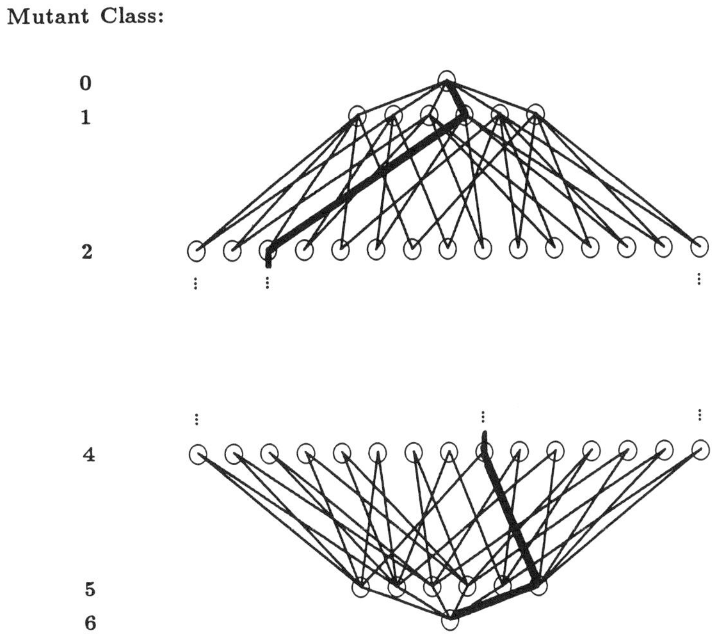

Fig. 9: Percolation of the sequence space by neutral networks. The example shows the space (or subspace) of binary (**AU** or **GC**) sequences with chain length $\nu = 6$. The emphasized path corresponds to one complete neutral trajectory which connects sequences folding into identical structures through the entire seqeunce space.

peak: relatively few structures are very common, many structures are rare and play no statistically significant role. The results obtained show in addition that sequences folding into the same secondary structure are, in essence, randomly distributed. Consequently all common structrues are found in relatively small patches of sequence space. For natural sequences of chain length $\nu = 100$ a sphere of radius $h = 16$ (in Hamming distance) is sufficient to yield a global distribution of structure distances.

In order to complement this illustration of the RNA shape space, a computer expriment was carried out which allows an estimate of the degree of selective neutrality (two sequences are considered neutral here if the fold into the same secondary structure). As indicated in figure 9 we search for "neutral paths" through sequence space. The Hamming distance from the reference increases monotonously along such a neutral path but the structure remains unchanged. A neutral path ends when no further neutral sequence is found in the neighbourhood of the last sequence. The length ℓ of a path is the Hamming distance between the reference sequence and the last sequence. Clearly, a neutral path cannot be longer than the chain length ν ($\ell \leq \nu$). The length distribution of neutral path in

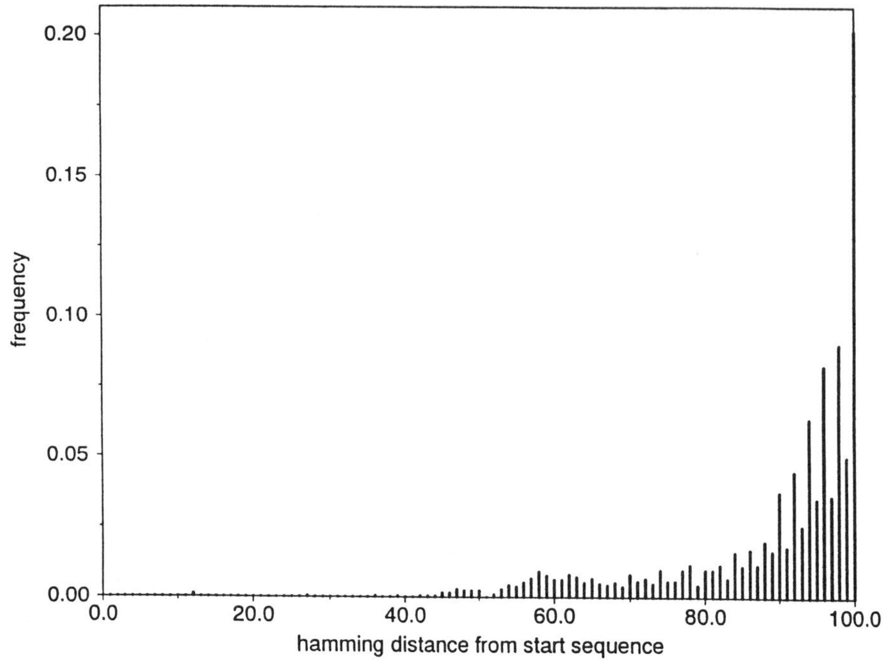

Fig. 10: Length distribution of neutral paths starting from random **AUGC**-sequences of chain length $\nu = 100$. A neutral path connects pairs of sequences with identical structures and Hamming distance $d_h = 1$ (single base exchange) or $d_h = 2$ (base pair exchange). The Hamming distance to the reference sequence is monotonously increasing along the path.

the sequence space of natural RNA molecules of chain length $\nu = 100$ is shown in figure 10. It is remarkable that about 20 % of the neutral paths have the maximum length, and thus lead through the whole sequence space to a sequence which differs in all positions from the reference, but shares its structure.

Combination of information derived from figures 8 and 10 provides insight into the structure of the shape space of RNA secondary structures which may be casted into four statements:

(1) sequences folding into one and the same structure are distributed randomly in sequence space,

(2) the frequency distribution of structures is sharply peaked (there are comparatively few common structures and many rare ones),

(3) sequences folding into all common structures are found within (relatively) small neighbourhoods of any random sequence, and

(4) the shape space contains extended "neutral networks" joining sequences with identical structures (a large fraction of neutral path leads from the initial sequence through the entire sequence space to a final sequence on opposite side – there are $(\kappa - 1)^{\nu}$ sequences which differ in all positions from an initial sequence).

These results suggest straightforward strategies in the search for new RNA structures. It provides little advantage to start from natural or other preselected sequences since any random sequence would do equally well as the starting molecule for the selection cycles shown in figure 3. Any common secondary structure with optimal functions is accessible in a few selection cycles. The secondary structure of RNA is understood as a crude first-order approximation to the actual spatial structure. Fine tuning of properties by choosing from a variety of molecules sharing the same secondary structure will often be necessary. In order to achive this goal it is of advantage to adopt alternations of selection cycles with low and high error rates. At low error rates the population performs a search in the vicinity of the current master sequence (the most common sequence which is usually also the fittest sequence). If no RNA molecule with satisfactory properties is found, a change to high error rate is adequate. Then the population spreads along the neutral network to other regions in sequence space which can be explored in detail after tuning the error rate low again.

The structure of shape space is highly relevant for evolutionary optimization in nature too. It provides a firm answer to the old probability argument against the possibility of successful adaptive evolution (Wigner, 1961). How should nature find a given biopolymer by trial and error when the chance to guess it is as low as $1:\kappa^{\nu}$? Previously given answers (Eigen, 1971; Eigen and Schuster, 1979) can be supported and extended by precise data on the RNA shape space. The numbers of sequences that have to be searched in order to find adequate solutions are many orders of magnitude smaller than those guessed on naive statistical grounds. Since long neutral paths are common, populations drift readily through sequence space whenever selection constraints are absent. This is precisely what is predicted by the "neutral theory of evolution" (Kimura, 1983), and what is observed in molecular phylogeny by sequence comparisons of different organisms.

ACKNOWLEDGMENTS

The statistical analysis of the RNA shape space presented here is joint work with Drs. Walter Fontana and Peter F. Stadler and will be described elsewhere in detail. Financial support by the Austrian *Fonds zur Förderung der wissenschaftlichen Forschung* (Projects S 5305-PHY and P 8526-MOB) is gratefully acknowledged.

REFERENCES

Bauer, G.J., McCaskill, J.S. and Otten, H., 1989: Traveling waves of *in vitro* evolving RNA, **Proc.Natl.Acad.Sci.USA**, **86**, 7937-7941.

Beaudry, A.A. and Joyce, G.F., 1992: Directed evolution of an RNA enzyme, **Science**, **257**, 635-641.

Biebricher, C.K., Eigen, M. and Gardiner, W.C., jr., 1983: Kinetics of RNA replication, **Biochemistry**, **22**, 2544-2559.

Biebricher, C.K., Eigen, M. and Gardiner, W.C., jr., 1984: Kinetics of RNA replication: Plus-minus asymmetry and double-strand formation, **Biochemistry**, **23**, 3186-3194.

Biebricher, C.K., Eigen, M. and Gardiner, W.C., jr., 1985: Kinetics of RNA replication: Competition and selection among self-replicating RNA species, **Biochemistry**, **24**, 6550-6560.

Cech, T.R., 1986: RNA as an enzyme, **Sci.Am.**, **255/5**, 76-84.

Domingo, E., 1990: Virus quasispecies: inpact for disease control, **Futura**, **3/90**, 6-8.

Eigen, M., 1971: Selforganization of matter and the evolution of biological macromolecules, **Naturwissenschaften**, **58**, 465-523.

Eigen, M., and Gardiner, W., 1984: Evolutionary molecular engineering based on RNA replication, **Pure & Appl.Chem.**, **56**, 967-978.

Eigen, M., McCaskill, J. and Schuster, P., 1988: The molecular quasispecies – An abridged account, **J.Phys.Chem.**, **92**, 6881-6891.

Eigen, M., McCaskill, J. and Schuster, P., 1989: The molecular quasispecies, **Adv.Chem.Phys.**, **75**, 149-263.

Eigen, M. and Schuster, P., 1977: The hypercycle. A principle of natural selforganization. Part A: Emergence of the hypercycle, **Naturwissenschaften**, **64**, 541-565.

Eigen, M. and Schuster, P., 1979: **The hypercycle – A principle of natural selforganization**, Springer-Verlag: Berlin.

Eigen, M. and Schuster, P., 1985: Stages of emerging life - Five principles of early organization., **J.Mol.Evol.**, **19**, 47-61.

Ellington, A.D. and Szostak, J.W., 1990: *In vitro* selection of RNA molecules that bind to specific ligands, **Nature**, **346**, 818-822.

Fontana, W., Griesmacher, T., Schnabl, W., Stadler, P.F. and Schuster, P., 1991: Statistics of landscapes based on free energies, replication and degradation rate constants of RNA secondary structures, **Mh.Chem.**, **122**, 795-819.

Fontana, W., Konings, D.A.M, Stadler, P.F. and Schuster, P., 1992: **Statistics of RNA secondary structures**, Preprint No.90-02-008, Santa Fe Institute: Santa Fe (N.M.). Submitted to Biopolymers.

Horwitz, M.S.Z. and Dube, D.K. and Loeb, L.A., 1989: Selection of new biological activities from random nucleotide sequences: evolutionary and practical considerations, **Genome**, **31**, 112-117.

Husimi, Y., Nishgaki, K., Kinoshita, Y. and Tanaka, T., 1982: Cellstat – A continuous culture system of a bacteriophage for the study of the mutation rate and the selection process at the DNA level, **Rev.Sci.Instrum.**, **53**, 517-522.

Husimi, Y. and Keweloh, H.-C., 1987: Continuous culture of bacteriophage Qβ using a cellstat with a bubble wall-growth scraper, **Rev.Sci.Instrum.**, **58**, 1109-1111.

Joyce, G.F., 1989: RNA evolution and the origins of life, **Nature**, **338**, 217-224.

Kauffman, S.A., 1992: Applied molecular evolution, **J.theor.Biol.**, **157**, 1-7.

Kimura, M., 1983: **The neutral theory of molecular evolution**, Cambridge University Press: Cambridge (U.K.).

Mason, S.F., 1991: **Chemical evolution. Origin of the elements, molecules and living systems**, Clarendon Press: Oxford (U.K.).

Noller, H.F., 1991: Ribosomal RNA and translation, **Ann.Rev.Biochem.**, **60**, 191-227.

Noller, H.F., Hoffarth, V. and Zimniak, L., 1992: Unusual resistence of peptidyl transferase to protein extraction procedures, **Science, 256**, 1146-1419.

Orgel, L.E., 1986: RNA catalysis and the origins of life, **J.theor.Biol.**, **123**, 127-149.

Orgel, L.E., 1992: Molecular replication, **Nature, 358**, 203-209.

Piccirilli, J.A., McConnell, T.S., Zaug, A.J., Noller, H.F. and Cech, T.R., 1992: Aminoacyl esterase activity of the *tetrahymena* ribozyme, **Science, 256**, 1420-1424.

Riesner,D., and Gross, H.J., 1985: Viroids, **Ann.Rev.Biochem.**, **54**, 531-564.

Schopf, J.W., and Packer, B.M., 1992: Early archean (3.3-billion to 3.5-billion-year-old) microfossils from Warrawoona group, Australia, **Science, 237**, 70-73.

Schuster, P., 1981: **Prebiotic Evolution**, in: Gutfreund, H., ed. Biochemical Evolution, pp.15-87. Cambridge Univ.Press: Cambridge (U.K.).

Schuster, P., Fontana, W., Stadler, P.F., and Hofacker, I.L., 1993: **How to search for RNA structures**, Preprint.

Spiegelman, S., 1971: An approach to the experimental anlysis of precellular evolution, **Quart.Rev.Biophys.**, **4**, 213-253.

Symons, R.H., 1992: Small catalytic RNAs, **Ann.Rev.Biochem.**, **61**, 641-671.

Strunk, G., 1992: Doctoral Thesis, Universität Braunschweig.

Swetina, J., and Schuster, P., 1982: Self-replication with errors - A model for polynucleotide replication, **Biophys.Chem.**, **16**, 329-345.

Tuerk, C. and Gold, L., 1990: Systematic evolution of ligands by exponential enrichment: RNA ligands to bacteriophage T4 DNA polymerase, **Science, 249**, 505-510.

Wakeman, J.A., and Maden, B.E.H., 1989: 28 S Ribosomal RNA in vertebrates. Location of large-scale features revealed by electron microscopy in relation to other features of the sequences, **Biochem.J.**, **258**, 49-56.

Wigner, E., 1961: **The logic of personal knowledge**, in: Shils, E., ed. Essays presented to Michael Polanyi on his seventieth birthday 11th March 1961. Free Press: Glencoe (Ill.).

SMALL PATHOGENIC RNAs OF PLANTS: LIVING FOSSILS OF THE RNA WORLD?

T. O. Diener
Center for Agricultural Biotechnology and
Department of Botany, University of Maryland
College Park, Maryland 20742, and Agricultural
Research Service, U. S. Department of Agriculture
Beltsville, Maryland 20705, U. S. A.

ABSTRACT

The hypothesis is advanced that viroids, the smallest known autonomously replicating agents of infectious disease, and certain similar RNAs occurring in plants represent relics of a precellular RNA world that has been posited to exist before the evolution of DNA or protein. With the discovery that certain RNAs possess catalytic capabilities, it has become evident that RNA is the only biological macromolecule that can function both as genotype and phenotype, thus raising the possibility of Darwinian selection and precellular evolution at the RNA level in the absence of DNA or protein. Recent models of precellular RNAs are patterned after the properties of intron-derived ribozymes. On the basis of a phylogenetic analysis and known properties of viroids and viroidlike satellite RNAs (RNAs similar to viroids but requiring a specific helper virus for their replication), I suggest that these plant RNAs are as plausible as introns to represent "living fossils" of a precellular RNA world. Comparison of the molecular properties of the two types of RNA suggests that viroids and viroidlike RNAs are relics of an earlier stage of precellular evolution than are introns and that, contrary to earlier assumptions, introns may have evolved from "free-living" viroidlike RNAs after being captured by early cells.

1. INTRODUCTION

With the demonstration that certain RNAs possess catalytic capabilities (Kruger et al., 1982; Guerrier-Takada et al., 1983), earlier suggestions that RNA preceded DNA as the carrier of genetic information during evolution have gained much plausibility. Most recent models for self-replicating, precellular RNAs assume the existence of primitive RNA enzymes (ribozymes) with properties similar to known splicing mechanisms of certain introns (Sharp, 1985; Cech, 1986), specifically the intron of the *Tetrahymena thermophila* rRNA gene (Cech, 1986).

Introns, however, are not the only catalytic RNAs

ISBN 0-937194-31-X

known. Additional examples are the RNA component of RNase P (Guerrier-Takada et al., 1983) and certain small plant pathogenic RNAs, namely the viroids and viroidlike satellite RNAs (Diener, 1991). Here I summarize pertinent properties of these RNAs and propose that they are at least as plausible representing "living fossils" of the precellular RNA world as are introns and that viroids and viroidlike RNAs may be remnants of a more archaic stage of precellular evolution than that represented by introns.[a]

2. VIROIDS AND VIROIDLIKE RNAs

Although the existence of viroids and other small plant-pathogenic RNAs is well established, they are often regarded as oddities of nature that bear little, if any, significance for the central themes of biology, including the evolution of life. Hence they occupy a marginal position at the fringes of molecular biology and are little known outside of plant biology. A summary of their properties therefore is in order.

Viroids are small (246-375 nucleotides), single-stranded, covalently closed circular RNAs that, in contrast to viruses, are not encapsidated (reviewed by Diener, 1987). Like viruses, viroids replicate autonomously in susceptible plant cells despite their small size and severely limited amount of genetic information (Diener, 1971). Viroid replication occurs via RNA intermediates of opposite polarity and does not involve DNA. In sharp contrast to viral RNAs, viroids do not function as mRNAs; hence viroids must be replicated by preexisting host enzymes. The presence of oligomeric viroid forms (usually of opposite polarity) in infected plants suggests that replication occurs by rolling-circle transcription (Branch and Robertson, 1984).

Plant virus satellite RNAs, and particularly representatives of one group, termed 'virusoids' or viroidlike satellite RNAs, are similar in many respects to viroids, but are encapsidated and can replicate only in cells that are infected with specific helper viruses (Francki, 1987). Viroidlike satellite RNAs are circular single-stranded RNAs of about 350 nucleotides that are found, together with linear single-stranded viral RNAs of about 4500 nucleotides, in the isometric particles of several plant viruses of the sobemovirus group (their helper viruses). Little, if any, sequence similarity exists between viroidlike satellite RNAs and the RNA(s) of their helper viruses. These satellite RNAs, like viroids, lack mRNA activity and possess highly base-paired rod-like structures.

The encapsidated forms of another group of plant virus satellite RNAs that are associated with nepoviruses [among others, the satellite RNA of tobacco ringspot virus

---

[a]Some of the ideas expressed here have previously been discussed (Diener, 1989).

(sTobRV RNA)], are mostly linear (Schneider, 1969), but circular molecules can be isolated from infected tissue.

Because linear repetitive sequence dimers, trimers, and higher-order multimers occur in infected tissue (or within helper virus capsids), satellite RNAs, like viroids, appear to be replicated by a rolling-circle type process. Implicit in this type of replication is a precise mechanism by which viroid or satellite RNA monomers are excised from oligomeric replication intermediates and ligated to form the circular, monomeric progeny RNAs.

2.1 OLIGOMER CLEAVAGE

Experiments to elucidate the mechanisms of oligomeric viroid and satellite RNA processing have demonstrated that several disparate cleavage processes operate. Both plus and minus polarity oligomeric molecules of most satellite RNAs (Prody et al., 1986; Forster et al., 1987) and of two viroids [the avocado sunblotch viroid (ASBVd) (Hutchins et al., 1986) and the peach latent mosaic viroid (PLMVd) (Hernandez and Flores, 1992) are self-cleaving, but other viroids appear to require a factor present in cell nuclei (Tsagris et al., 1987). Structural requirements also are disparate. Whereas most self-cleaving plant pathogenic RNAs contain a highly conserved series of short nucleotide sequences that can assume, by base pairing, a characteristic secondary structure called a "hammerhead", viroids (except ASBVd and PLMVd) cannot assume such a structure (Forster and Symons, 1987).

Although their ability to assume the hammerhead structure appears to be a prerequisite for the self-cleavage of most plant satellite RNAs, multimers of the minus sequence of the sTobRV RNA self-cleave, but cannot form a hammerhead structure (Forster and Symons, 1987).

With viroids (except ASBVd and PLMVd), the mechanism of cleavage is less well known, except that certain host RNases have been implicated recently in the process (Tsagris et al., 1991). Infectivity studies of viroid-specific cDNAs and RNA transcripts therefrom have given clues as to a possible cleavage-ligation site. Such studies have shown that whereas cDNAs containing a monomeric viroid sequence are not infectious, constructs containing tandem repeats of the viroid sequence are highly infectious. Further analysis has shown that (as judged by infectivity) far less than a complete dimer of the viroid sequence is sufficient for cleavage to take place, provided that the sequence duplication occurs in the central portion of the upper strand of the viroid molecule (in its conventional representation). Because in representatives of the potato spindle tuber viroid (PSTVd) group this region is highly conserved, it has become known as the "central conserved region" (CCR). Recently, viroids have come to light [the apple scar skin viroid (ASSVd) group (Koltunow and Rezaian, 1989)] whose central region differs from that of the PSTVd group, but is nevertheless highly conserved within the group. With all viroids (except ASBVd and PLMVd), the CCR is flanked by inverted repeats of about nine nucleotides (hairpin I). ASBVd and PLMVd differ from all other viroids also by containing only a small portion of the PSTVd viroid CCR.

On the basis of these findings a model for a putative cleavage-ligation site has been proposed (Diener, 1986). The model has identified a thermodynamically highly stable palindrome that viroid constructs with a duplication of the upper CCR and the flanking hairpin I sequences (as well as oligomeric viroids) can assume and posits that formation of this structure is required for cleavage-ligation to occur. Recent results generally are compatible with this model (Candresse et al., 1990), but also point to the existence of less favored secondary processing sites in different regions of viroids (Hammond et al., 1989).

As discussed previously (Diener, 1989), sequence similarities exist between the upper portion of the viroid CCR and sequences involved in forming the hammerhead structure of the self-cleaving ASBVd. It is possible, therefore, that those sequences that are involved in forming the putative cleavage structure of most present-day viroids have evolved from sequences originally involved in forming an autocatalytic cleavage structure, such as is still present in ASBVd, PLMVd, and most satellite RNAs.

2.2 CLEAVAGE CASSETTES

Only relatively small portions of satellite RNA sequences are required as signals for self-cleaving. Thus, deletion of RNA sequences from both the 3' and 5' termini of one satellite RNA (that of the lucerne transient streak virus) has shown that a 51-nucleotide remnant of the original RNA, containing little more than the sequences comprising the hammerhead, is capable of rapid and complete self-cleavage (Forster et al., 1987); and a synthetic 19-nucleotide RNA fragment can cause rapid, highly specific cleavage of a 24-nucleotide RNA fragment under physiological conditions (Uhlenbeck, 1987). Because the 19-mer can participate in many cleavage reactions, it fulfills all criteria for a genuine ribozyme. The nucleotide sequences of the two fragments were constructed such that together they can form the hammerhead structure and hence conform to the model proposed for the self-cleaving domain of satellite RNAs. Similarly, the two mutationally sensitive regions of the minus strand of sTobRV have been produced as two separate oligonucleotides whose minimal sizes were determined to be 10 and 46 nucleotides for the substrate and ribozyme, respectively (Feldstein et al., 1989).

2.3 VIROID AND SATELLITE RNA EVOLUTION

A phylogenetic analysis of viroids and satellite RNAs (based on their computer-aligned nucleotide sequences) has shown that all viroids examined (except ASBVd) constitute one evolutionarily related cluster of RNAs and that satellite RNAs constitute another. ASBVd does not belong to either cluster and is further removed from both than are the RNAs in either cluster (Elena et al., 1991). The results are consistent with the concept that these RNAs have a common origin and that ASBVd represents a connecting link between satellite RNAs and viroids proper (Diener, 1989).

3. PUTATIVE VIROID RELATIVES

3.1 TRANSPOSONS AND PLANT-PATHOGENIC RNAs

Comparative sequence analysis of members of the PSTVd viroid group has revealed striking similarities with the ends of transposable genetic elements, such as the presence of inverted repeats often ending with the dinucleotides U.G and C.A, and flanking imperfect direct repeats; and it has been suggested that viroids may have originated from transposable elements or retroviral proviruses by deletion of interior portions of the viral or element DNA (Kiefer et al. 1983).

3.2 INTRONS AND PLANT-PATHOGENIC RNAs

Comparison of the small plant pathogenic RNAs with introns reveals that the two types of RNA possess a number of structural and functional properties in common. Structurally, both viroids and satellite RNAs contain cognates of the consensus sequences of group I introns that are required for self-splicing (Dinter-Gottlieb, 1986).

Functionally, the overall processes by which viroid and satellite RNA monomers are cleaved out of oligomeric replication intermediates and ligated to form circular progeny molecules are conceptually analogous to the cleavage-ligation reactions by which introns are spliced out of precursor RNAs and exons joined to form functional RNAs [although the chemistry of these processes differs in many cases (see below)]. Most significantly, the processing of self-splicing introns, like that of self-cleaving satellite RNAs (and of ASBVd and PLMVd), involves RNA-catalyzed processes; viroidlike satellite RNAs, ASBVd, and PLMVd, like self-splicing introns, are catalytic RNAs.

Also, the existence, among these small plant pathogenic RNAs, of two disparate cleavage processes, one requiring a host factor,the other being self sufficient, is reminiscent of splicing mechanisms operative among different groups of introns, in which the splicing of mRNA precursors requires external factors, whereas that of many group

I and group II precursor RNAs is RNA catalyzed.

Although viroids and satellite RNAs contain sequences resembling conserved elements of group I introns, these sequence elements cannot be of functional significance in the self-cleavage process of satellite RNAs: certain satellite RNA fragments from which all intron-like elements have been removed self-cleave efficiently (Buzayan et al., 1986a).

In view of the structural and functional similarities between introns and the small plant-pathogenic RNAs, as well as their possible evolutionary connection, it appears reasonable to include the latter in speculations regarding the evolutionary origin of introns; and, as I will show, if one accepts the idea that introns represent "living fossils" of an hypothetical RNA world, viroids and viroidlike satellite RNAs may be regarded as such "fossils" with equal, if not better, justification.

4. VIROIDS AND VIROIDLIKE RNAs AS EVOLUTIONARY RELICS

4.1 EVOLUTIONARY AGE OF INTRONS AND VIROIDS

If there exists one principle in evolutionary thought that enjoys nearly universal acceptance it is the concept that evolution generally proceeds from more simple to more complex entities and that therefore more complex systems are usually of more recent origin than are similar, but simpler, systems.

In this view, viroidlike satellite RNAs and the two self-cleaving viroids are more ancient than are self-splicing introns. Not only are they much smaller than most introns, but sequences required for self-cleaving are much simpler than those necessary for intron self-splicing. Whereas self-splicing group I introns require a minimum of about 200 nucleotides and a complex secondary structure composed of several hairpins for full catalytic activity (Doudna et al., 1991), catalytically active hammerhead ribozymes as small as 13 nucleotides have been constructed (Jeffries and Symons, 1989).

With introns, the RNA-catalyzed reaction has been proposed to represent the ancestral splicing process and the mechanism responsible for the splicing of mRNA precursors to be derived therefrom (Sharp, 1985). Similarly, the self-cleavage mechanism operative with present-day satellite RNAs, ASBVd, and PLMVd may be regarded as the ancestral reaction and the mechanism operative with all other known viroids as derived from it. With introns (Sharp, 1985), as well as with viroids, a major step in the transition would be the appearance of *trans*-acting factors that assist in executing the excision.

Another cogent reason for considering viroids and viroidlike RNAs as more ancient than introns is the fact that, as discussed above, none of the former serve as mRNAs, either *in vivo* or in *in vitro* cell-free protein synthesizing systems, whereas many introns contain expressed open reading frames (Perlman and Butow, 1989). This suggests that some ancestors of small plant pathogenic RNAs may have originated before the evolution of functional translation systems.

4.2 GENOMIC TAGS AND CIRCULAR GENOMES

The discovery of catalytic RNA has helped overcome the long-standing conceptual problem of understanding how, in precellular systems, nucleic acid could be replicated without protein enzymes, or proteins without nucleic acid templates. Many other problems, nevertheless, remain.

As has been pointed out (Weiner and Maizels, 1987; Joyce, 1989), one of the most serious is the need for a primitive RNA polymerase (ribozyme?) to identify a specific terminal site on a nucleic acid template for initiation in order to assure end-to-end copying of a linear template. Without a specific initiation site, replication could initiate anywhere on the template and complete replication would be unlikely. It has been suggested that this problem may have been solved in the RNA world by evolution of a 3'-terminal "genomic-tag" structure which directed the reaction to commence at a specific site (Weiner and Maizels, 1987). I propose that viroids and satellite RNAs have solved the problem of complete replication in a different and simpler manner. The covalently closed circular structure of most of these RNAs and their rolling-circle type synthesis assure complete replication provided that transcription continues at least once around the circle. Thus, no need exists for a "genomic tag". The obvious advantages of circular templates in early evolutionary systems were recognized sometime ago by Reanney and Ralph (1968), who also pointed out that circular molecules are the only ones whose information content remains intact even if the circle is broken --- an event that must have been frequent in primitive replication systems.

Although circular templates, therefore, need not contain defined initiation sites, they require a mechanism for the precise cleavage of monomers out of the oligomeric replication intermediates and an efficient means by which monomers are ligated into circular progeny RNAs. With viroids and satellite RNAs, both of these requirements are fulfilled.

With most present-day viroids, replicative functions have been usurped by proteinaceous enzymes. Thus, PSTVd is replicated by normally DNA-dependent RNA polymerase II (Schindler and Mühlbach, 1992) and, apparently, cleaved and ligated into circular monomers by certain host ribonucleases (Tsagris et al., 1991). With ASBVd, PLMVd, and viroidlike satellite RNAs, however, RNA-catalyzed cleavage is still evident . These self-cleavage reactions generate 5'-OH and 2',3'-cyclic phosphodiester termini, as is the case with ordinary base hydrolysis of RNA, and the only co-factor required is a structure-stabilizing ion, such as magnesium (Buzayan et al., 1986a). Hence, these reactions may represent very primitive RNA processes; they are clearly distinct from the transphosphorylation reactions involved in group I intron splicing, which are more complex (Cech, 1987) and which may represent reactions that have evolved at a later stage in evolution.

The 2',3'-cyclic phosphodiester termini generated during self-cleavage are suitable substrates for RNA ligases (Prody et al., 1986); and most extant self-cleaving satellite RNAs presumably rely on host RNA ligases for circularization of monomers. Spontaneous self-ligation, resulting in covalently closed circular molecules has been demonstrated, however, with the self-cleaved minus strand of sTobRV RNA in the absence of protein (Buzayan et al., 1986b), as well as with a fungal mitochondrial plasmid transcript that combines features of catalytic RNAs and group I introns (Saville and Collins, 1991).

In principle, a ribozyme patterned after the hammerhead structure is capable of performing all of the steps involved in Sharp's (1985) hypothetical RNA self-replication scheme. As compared with intron models, the concept of precellular RNA replication based on viroid and/or satellite RNA precursors requires fewer assumptions. Above all, synthesis in primitive prebiotic systems, of a simple 13- or 19-nucleotide ribozyme and its maintenance is easier to accept than that of a sophisticated 200-nucleotide RNA, as appears necessary with the intron model.

4.3 "POLYPLOIDY" AND CIRCULAR GENOMES

There are still other reasons to consider viroids and satellite RNAs as relics of a stage of precellular evolution earlier than that of introns. Eigen (1971) has pointed out that the fidelity of replication in such systems sets a limit on the length of a master sequence that can be maintained against its distribution of error copies. Also, because of the high stability of G.C basepairs, G+C-rich sequences would have been highly favored (Eigen and Schuster, 1978). With respect to their size, both viroids and satellite RNAs fulfill these requirements. Also, viroids (but not satellite RNAs) are characterized by high G+C ratios and represent the smallest replicating molecules known. In all of these respects, the gap between the characteristics of viroids and satellite RNAs and those expected of primitive self-replicating RNAs in a precellular RNA world is smaller than with self-splicing introns.

Yet, replication and faithful maintenance of 300-400--nucleotide molecules nonenzymatically or by primitive ribozyme systems may appear implausible (Reanney, 1984). Here again, the characteristics of viroids, and particularly their circularity, may help resolve the problem. Initially, the length of oligonucleotides synthesized in these primitive systems must have been very limited. Because circularization would assure complete replication, it would be advantageous for such oligonucleotides to be ligated into small circles. Also, it has been pointed out that the high error rate of prebiotic replication systems (estimated as 10^{-1}-10^{-2} misincorporations per base per generation) (Inoue and Orgel, 1983) would almost certainly favor systems with multiple copies of their genomes (Reanney, 1984). One effective method of achieving such "polyploidy" would consist in rolling-circle transcription of a circular template, resulting in tandem repeats of the genetic information.

It is probable that with improved fidelity of the copying mechanism the need for multiple copies would diminish, but that the sequences required for cleavage would become more stringently defined. As discussed previously (Diener, 1989), lack of cleavage because of a point mutation in the cleavage cassette might conceivably be one of the mechanisms by which larger genomes could have evolved. Indeed, all but one viroid exhibit structural periodicities (Juhasz et al., 1988) which may indicate that viroids have evolved from still smaller, identical subunits.

4.4 FROM EXTRACELLULAR TO INTRACELLULAR EXISTENCE

Unless the sequence similarities between viroids and group I introns are fortuitous, they suggest that the two types of RNA may be evolutionarily related. Several authors have, indeed, entertained this possibility, usually concluding that viroids have evolved from introns (Crick, 1979; Diener, 1981; Dinter-Gottlieb, 1986). In this view, viroids have been regarded as "escaped" introns. Arguments presented above, however, suggest the opposite, namely that viroids and viroidlike satellite RNAs are more ancient than introns and that introns may have evolved from viroidlike RNAs and not vice versa.

It is conceivable that remnants of the "RNA world", namely "free-living", self-replicating viroidlike RNAs may have coexisted with early cells whose information system had evolved to the DNA ---> RNA ---> protein stage and whose nucleic acid replication was already performed by proteinaceous polymerases. Because of their unique molecular structure, coincidental ingestion of viroidlike RNAs into such cells could have led to their passive intracellular replication, as is the case with extant viroids which, on the basis of inhibitor studies, are believed to be replicated by normally DNA-dependent host RNA polymerase II (Schindler and Mühlbach, 1992). Indeed, it has been demonstrated that purified wheat and tomato RNA polymerases II are capable of synthesizing full-length linear minus viroid copies from plus viroid templates, albeit with much less efficiency than is the case with transcription from DNA templates (Rackwitz et al., 1981). Presumably, the rodlike, quasi double-stranded native viroid structure is sufficiently DNA-like for recognition by the enzyme, thus explaining why viroids, but not cellular RNAs, are replicated upon inoculation into plant cells.

In addition to their unique structural properties, other viroid features that were essential for their survival in a precellular world would also have facilitated transformation from extracellular to intracellular existence. Thus, as pointed out above, their circularity would have rendered the evolution of genomic initiation tags unnecessary. Also, because of their self-cleaving abilities, no host-derived mechanism for the processing of oligomers into monomers would have had to evolve.

Although, admittedly, this scenario is highly speculative, it does provide a mechanism by which primitive, self-replicating extracellular RNAs could have become passively replicated, cell-associated, components ---- a crucial process that has received scant attention in speculations on the RNA world.

Aside from their apparent inability to replicate extracellularly, extant viroids fulfill all of the postulated steps. Their mechanical introduction into susceptible plants, either in nature or experimentally, leads to their passive replication by a host enzyme (or enzymes), cell-to-cell translocation, and lasting association with the host organism.

Viroids are the only known RNAs that are synthesized in otherwise uninfected cells without being either transcribed from DNA or contributing to their own replication machinery (by, for example, coding for a replicase subunit). This unique property of viroids again suggests their great antiquity and is consistent with the above evolutionary scenario.

4.5 INTRACELLULAR EVOLUTION

Once permanently within a cell and passively replicated, this "urviroid", as well as its interaction with the host cell, would be subject to Darwinian selection. Although extant viroids have come to light because of their pathogenic properties in certain cultivated plants, many wild plants are infected by viroids without showing recognizable disease symptoms (Diener, 1979). These benign relationships between viroids and some of their host plants are reminiscent of the relatively mild reactions observed in many plants endemically infected by certain fungal or bacterial pathogens --- a phenomenon that is common with host-pathogen systems that have coevolved (Buddenhagen, 1983). It is reasonable to conclude, therefore, that those plant species, in which a viroid remains latent or incites only mild symptoms, have been exposed to the viroid for a long time in evolution and some of these may represent ancestral hosts of the particular viroids.

Taking this argument one step further, viroid infection may, on occasion, be beneficial to the host plant's survival. For example, the stunting so often associated with viroid infection undoubtedly reduces transpiration and may help a plant survive drought conditions. Similarly, association of a viroidlike RNA with early cells may have been beneficial to the latter and hence have favored survival of the viroid-host cell association.

It is even conceivable that certain viroidlike RNAs could in time have become obligatory constituents of their host organisms. Such RNAs, for example, could be involved in cell regulation, as has been suggested (Diener, 1971).

4.6 FROM AUTONOMOUS REPLICATION TO HOST CELL DEPENDENCE

According to the evolutionary scheme proposed, newly cell-associated RNAs would initially be self-replicating and cleaving in the manner assumed for RNAs in the RNA world, but coevolution of the RNAs and their host cells would undoubtedly lead in time to the appearance of more intimate relationships between the two entities. For example, as is the case with extant viroids, replication could become dependent on normally DNA-dependent RNA polymerases and, as is the case with all but two viroids, self-cleavage of RNA oligomers could become superseded by host-assisted processes.

With the evolution of reverse transcriptase-like enzymes, such RNAs could become more permanently associated with their host cells by integration into their hosts' genomes. Fascinating glimpses into the possible evolution of reverse transcriptase have been obtained in investigations of the Mauriceville and Varkud mitochondrial plasmids of *Neurospora*, which apparently replicate by a newly recognized mechanism of reverse transcription in which a full-length RNA transcript of the plasmid DNA is transcribed by a reverse transcriptase-like enzyme encoded by the plasmid into a full-length minus-strand DNA. Interestingly, reverse transcription starts at the 3' end of the plasmid transcript which has t-RNA-like characteristics similar to the 3' ends of many plant viral RNAs (Kuiper and Lambowitz, 1988). Thus, if the "genomic tag" hypothesis of Weiner and Maizels (1987) (see above) is correct, the Mauriceville and Varkud mitochondrial plasmids may be related to the earliest DNA elements (Kuiper and Lambowitz, 1988).

These plasmids may, therefore, represent possible evolutionary links between primitive RNA elements that, after the evolution of proteins, may have acquired a capacity to code for RNA-dependent RNA polymerases in the manner of extant plant viruses, and retroviruses, which have substituted exogenously supplied tRNA for the tRNA-like structures at the 3' end of the RNAs. Furthermore, because the mitochondrial plasmids, like viroids and viroidlike satellite RNAs, contain cognates of sequence elements characteristic of group I introns, they may represent evolutionary links to mitochondrial introns as well (Lambowitz, 1989).

REFERENCES

Branch, A. D., and H. D. Robertson, 1984: A replication cycle for viroids and other small infectious RNA's, Science, 223, 450-455.

Buddenhagen, I. W., 1983: Crop improvement in relation to virus diseases and their epidemiology. In R. T. Plumb and J. M. Thresh, (Eds.), Plant Virus Epidemiology, Blackwell, Oxford, 25-37.

Buzayan, J. M., W. L. Gerlach, and G. Bruening, 1986a: Satellite tobacco ringspot virus RNA: A subset of the RNA sequence is sufficient for autolytic processing, Proc. Natl. Acad. Sci. U. S. A., 83, 8859-8862.

Buzayan, J. M., W. L. Gerlach, and G. Bruening, 1986b: Non-enzymatic cleavage and ligation of RNAs complementary to a plant virus satellite RNA, Nature (London), 323, 349- 353.

Candresse, T., T. O. Diener, and R. A. Owens, 1990: The role of the viroid central conserved region in cDNA infectivity, Virology, 175, 232-237.

Cech, T. R., 1986: A model for the RNA-catalyzed replication of RNA, Proc. Natl. Acad. Sci. U. S. A., 83, 4360-4363.

Cech, T. R., 1987: The chemistry of self-splicing RNA and RNA enzymes, Science, 236, 1532-1539.

Crick, F. H. C., 1979: Split genes and RNA splicing, Science, 204, 264-271.

Diener, T. O., 1971: Potato spindle tuber "virus". IV. A replicating, low molecular weight RNA, Virology, 45, 411-428.

Diener, T.O., 1979: Viroids and Viroid Diseases, Wiley and Sons (Wiley Interscience), New York, NY, pp. 1-252.

Diener, T. O., 1981: Are viroids escaped introns ? Proc. Natl. Acad. Sci. U. S. A., 78, 5014-5015.

Diener, T. O., 1986: Viroid processing: A model involving the central conserved region and hairpin I, Proc. Natl. Acad. Sci. U. S. A., 83, 58-62.

Diener, T.O. (ed.), 1987: The Viroids, Plenum Publishing Corp., New York, NY, pp. 344pp.

Diener, T. O., 1989: Circular RNAs: Relics of precellular evolution? Proc. Natl. Acad. Sci. U. S. A., 86, 9370-9374.

Diener, T. O., 1991: Subviral pathogens of plants: viroids and viroidlike satellite RNAs, FASEB J., 5, 2808-2813.

Dinter-Gottlieb, G., 1986: Viroids and virusoids are related to group I introns, Proc. Natl. Acad. Sci. U. S. A., 83, 6250-6254.

Doudna, J. A., S. Couture, and J. W. Szostak, 1991: A multisubunit ribozyme that is a catalyst of and template for complementary strand RNA synthesis, Science, 251, 1605-1608.

Eigen, M., 1971: Self-organisation of matter and the evolution of biological macromolecules, Naturwissenschaften, 58, 465-532.

Eigen, M., and P. Schuster, 1978: The Hypercycle: A principle of natural self-organization. Part C, Naturwissenschaften, 65, 341-369.

Elena, S. F., J. Dopazo, R. Flores, T. O. Diener, and A. Moya, 1991: Phylogeny of viroids, viroidlike satellite RNAs, and the viroidlike domain of hepatitis virus RNA, Proc. Natl. Acad. Sci. U. S. A., 88, 5631-5634.

Feldstein, P. A., J. M. Buzayan, and G. Bruening, 1989: Two sequences participating in the autolytic processing of satellite tobacco ringspot virus complementary RNA, Gene, 82, 53-61.

Forster, A. C., A. C. Jeffries, C. C. Sheldon, and R. H. Symons, 1987: Structural and ionic requirements for self--cleavage of virusoid RNAs and trans self-cleavage of viroid RNA, Cold Spring Harbor Symp. Quantit. Biol., 52, 249-259.

Forster, A. C., and R. H. Symons, 1987: Self-cleavage of plus and minus RNAs of a virusoid and a structural model for the active site, Cell, 49, 211-220.

Francki, R. I. B., 1987: Encapsidated viroidlike RNA. In T. O. Diener, (Ed.) The Viroids, Plenum Publishing Corporation, New York, 205-218.

Guerrier-Takada, C., K. Gardiner, T. Marsh, N. Pace, and S. Altman, 1983: The RNA moiety of ribonuclease P is the catalytic subunit of the enzyme, Cell, 35, 849-857.

Hammond, R. W., T. O. Diener, and R. A. Owens, 1989: Infectivity of chimeric viroid transcripts reveals the presence of alternative processing sites in potato spindle tuber viroid, Virology, 170, 486-495.

Hernández, C., and R. Flores, 1992: Plus and minus RNAs of peach latent mosaic viroid self-cleave in vitro via hammerhead structures, Proc. Natl. Acad. Sci. U. S. A., 89, 3711- 3715.

Hutchins, C. J., P. D. Rathjen, A. C. Forster, and R. H. Symons, 1986: Self-cleavage of plus and minus RNA transcripts of avocado sunblotch viroid, Nucl. Acids Res., 14, 3627-3640.

Inoue, T., and L. E. Orgel, 1983: A nonenzymatic RNA polymerase model, Science, 219, 859-862.

Jeffries, A. C., and R. H. Symons, 1989: A catalytic 13--mer ribozyme, Nucl. Acids Res., 17, 1371-1377.

Juhasz, A., H. Hegyi, and F. Solymosy, 1988: A novel aspect of the information content of viroids, Biochim. Biophys. Acta, 950, 455-458.

Kiefer, M. C., R. A. Owens, and T. O. Diener, 1983: Structural similarities between viroids and transposable genetic elements, Proc. Natl. Acad. Sci. U. S. A., 80, 6234-6238.

Koltunow, A. M., and M. A. Rezaian, 1989: Grapevine viroid 1B, a new member of the apple scar skin viroid group contains the left terminal region of tomato planta macho viroid, Virology, 170, 575-578.

Kruger, K., P. J. Grabowski, A. J. Zaug, J. Sands, D. E. Gottschling, and T. R. Cech, 1982: Self-splicing RNA: Autoexcision and autocyclization of the ribosomal RNA intervening sequence of Tetrahymena, Cell, 31, 147-157.

Kuiper, M. T. R., and A. M. Lambowitz, 1988: A novel reverse transcriptase activity associated with mitochondrial plasmids of Neurospora, Cell, 55, 693-704.

Lambowitz, A. M., 1989: Infectious introns, Cell, 56, 323-326.

Perlman, P. S., and R. A. Butow, 1989: Mobile introns and intron-encoded proteins, Science, 246, 1106-1109.

Prody, G. A., J. T. Bakos, J. M. Buzayan, I. R. Schneider, and G. Bruening, 1986: Autolytic processing of dimeric plant virus satellite RNA, Science, 231, 1577-1580.

Rackwitz, H.-R., W. Rohde, and H. L. Sänger, 1981: DNA-dependent RNA polymerase II of plant origin transcribes viroid RNA into full-length copies, Nature (London), 291, 297-301.

Reanney, D., 1984: Genetic noise in evolution? Nature (London), 307, 318-319.

Reanney, D. C. and R. K. Ralph, 1968: Genetic circularity and evolution, J. Theoret. Biol., 21, 217-228.

Saville, B. J., and R. A. Collins, 1991: RNA-mediated ligation of self-cleavage products of a Neurospora mitochondrial plasmid transcript, Proc. Natl. Acad. Sci. U. S. A., 88, 8826-8830.

Schindler, I.-M., and H.-P. Mühlbach, 1992: Involvement of nuclear DNA-dependent RNA polymerases in potato spindle tuber viroid replication: A reevaluation, Plant Science, 84, 221-229.

Schneider, I. R., 1969: Satellite-like particle of tobacco ringspot virus that resembles tobacco ringspot virus, Science, 166, 1627-1629.

Sharp, P. A., 1985: On the origin of RNA splicing and introns, Cell, 42, 397-400.

Tsagris, M., M. Tabler, H.-P. Mühlbach, and H. L. Sänger, 1987: Linear oligomeric potato spindle tuber viroid (PSTV) RNAs are accurately processed in vitro to the monomeric circular viroid proper when incubated with a nuclear extract from healthy potato cells, EMBO J., 6, 2173-2183.

Tsagris, M., M. Tabler, and H. L. Sänger, 1991: Ribonuclease T1 generates circular RNA molecules from viroid-specific RNA transcripts by cleavage and intramolecular ligation, Nucl. Acids Res., 19, 1605-1612.

Uhlenbeck, O. C., 1987: A small catalytic oligoribonucleotide, Nature (London), 328, 596-600.

Weiner, A. M., and N. Maizels, 1987: tRNA-like structures tag the 3'ends of genomic RNA molecules for replication: Implications for the origin of protein synthesis, Proc. Natl. Acad. Sci. U. S. A., 84, 7383-7387.

THE WEAK FORCE AND THE ORIGIN OF LIFE

Alexandra J. MacDermott

School of Biological and Molecular Sciences,
Oxford Polytechnic, Oxford OX3 0BP, England
and
Physical Chemistry Laboratory, Oxford University,
South Parks Road, Oxford OX1 3QZ, England.

ABSTRACT

Most biomolecules exist in two mirror image forms, but only one form is used in biochemistry: life is based on L-amino acids and D-sugars rather than the enantiomeric D-amino acids and L-sugars. This broken symmetry is now believed to be a feature of fundamental physics - a result of symmetry-breaking by the weak force, which makes one enantiomer very slightly more stable than the other. We have calculated this electroweak stabilization for a wide range of important biomolecules, and in nearly all cases it is indeed the "natural" enantiomer which is the more stable.

1.THE NEED FOR A PRE-BIOTIC SYMMETRY-BREAKER

Scenarios for the origin of life fall into three main categories: (1) nucleic acids first (with RNA preceding DNA), (2) proteins first, (3) clays first. Interactions at surfaces are likely to have been important catalytically, and in (3) they would have been crucial to Cairns-Smith's "genetic takeover", in which the sequence would be clays $\longrightarrow$ proteins $\longrightarrow$ RNA $\longrightarrow$ DNA. Extraterrestrial theories involve the surface of interstellar grains.

Chirality and polymers are central features of most theories of the origin of life. It has been clear since Fischer (1894) that all biomolecules have to be of one hand: this "homochirality" is a hallmark of life, being essential for an efficient biochemistry, just like the universal adoption of right-hand screws in engineering. It has more recently become recognized that polymerization to form the necessary stereoregular biopolymers - in particular poly-D-ribonucleotides (Joyce et al., 1984) and poly-L-peptides (Blair et al., 1981) - proceeds efficiently only in optically pure monomer solutions: in racemic solution, addition of the "wrong" hand to the growing chain tends to terminate the polymerization. A homochiral pre-biotic chemistry is therefore probably a pre-condition for life.

A symmetry-breaker would thus appear necessary to get life started, unless the need for homochirality could be circumvented with a "prochiral" ancestral replicator such as

ISBN 0-937194-31-X

the glycerol-based polymer (essentially RNA with C2 removed), proposed by Bada and Miller (1987) and Schwartz and Orgel (1985), which is based on achiral monomers and owes its chirality purely to its helical conformation. However the choice must still be made between left and right-hand helices. Some clays are totally achiral, so an ancestral clay replicator would completely circumvent the chiral start-up problem, although symmetry-breaking would still be necessary later in the transition to the present polymer-based biochemistry.

Initial selection of one hand in ancestral biomolecules fixes the handedness of the rest of biochemistry through diastereomeric interactions (Fischer, 1894); we therefore need only account for the handedness of ancestral biomolecules. The diastereomeric connection between the L-amino acids and the D-sugars (Wolfrom et al., 1949; and Melcher, 1974) probably precludes an L-amino acid/L-sugar or D-amino acid/D-sugar biochemistry. However D-amino acid/L-sugar "mirror life " should be just as viable as terrestrial L-amino acid/D-sugar life, so we may ask whether selection of the latter was a "frozen accident" (Miller and Orgel, 1974) or, as Pasteur believed (Pasteur, 1894; Pasteur 1922), the result of some universal chiral influence.

2.THE WEAK FORCE AS A UNIVERSAL SYMMETRY-BREAKER

We argue here that the weak force is the only <u>universal</u> truly chiral influence.

Pasteur was famous for his hand resolution of tartaric acid crystals, but he also noticed that the solar system is non-superimposable on its mirror image (Pasteur, 1922), and it is now known that the whole universe is chiral: there is a predominance among spiral galaxies of rotation to the left with respect to the direction of recession (MacGillivray and Dodd, 1985). Pasteur thought a universal chiral influence could be a magnetic field. Later suggestions for a chirally selective influence on biochemistry have included such combinations as the Earth's magnetic and gravitational fields, or the Coriolis force and gravity.

But none of these influences are "truly chiral" according to Barron's new definition (Barron, 1986). A "truly" chiral system is one that exists in two distinct enantiomeric states that are interconverted by parity (space inversion $x \rightarrow -x$, $y \rightarrow -y$, $z \rightarrow -z$) but not by time reversal ($t \rightarrow -t$). The hallmark of "true" chirality is natural optical rotation, which changes sign under space inversion but not under time reversal, i.e. it is parity-odd, time-even, or P_-T_+. Absolute asymmetric synthesis can therefore be induced only by something with this same symmetry, at least under thermodynamic control.

Under kinetic control, irreversibility destroys any

time-reversal symmetry, so that so-called "false chirality" (P_-T_-) may then suffice, which opens up interesting possibilities given that life is a non-equilibrium phenomenon. A magnetic field on its own is no good as it is not even falsely chiral, being parity-even (axial vector), time-odd (imagine it generated by a circular motion of electrons, which changes direction under time reversal). Some suggested falsely chiral combinations are collinear magnetic (P_+T_-) and electric (P_-T_+) fields, and also stirring (P_+T_-) in the Earth's gravitational field (P_-T_+). However gravity is clearly too weak (10^{27} times weaker than the weak force!) to affect chemical reactions, and although magnetic fields do affect certain reactions - those involving radicals (McLauchlan 1991 and 1992) - it is not clear what the mechanism would be for a chiral effect. A truly chiral influence (and preferably a universal one) is probably more likely to produce a reliable effect.

The classic truly chiral influence is of course the circularly polarized photon (P_-T_+), the enantioselective properties of which are well-established in the laboratory. Unfortunately, although sunlight is 0.1% circularly polarized, it is equally and oppositely so at dawn and dusk (Angel et al., 1972), and so could not have determined biomolecular chirality unless life happened to start in a pool on an east or west facing mountain slope - which would be no more than a "frozen accident" explanation.

Most classical "chiral" influences have now been shown to be either not truly chiral or else even-handed on a time or space average, i.e. not universal. But Pasteur was correct in thinking that there is a universal chiral influence, because it is now known that elementary particles themselves have a handedness, felt only by the weak force, which is one of the four forces of nature:

Force	Carried by	Strength
Electromagnetic	photon	1
Weak	W^+, Z^o	10^{-11}
Strong	mesons, gluons	100
Gravity	gravitons	10^{-38}

The electromagnetic and weak forces were unified in 1968 by Salam and Weinberg, and the strong force is now being brought in with "grand unification" theories, but quantum gravity is proving more elusive. The four forces are quite clearly one, because they all operate by the same mechanism: the world is made of fermions, which interact by exchange of virtual bosons. For example, the electostatic repulsion between two electrons is mediated by exchange of virtual photons, while the weak interaction between an electron and a neutron is mediated by virtual Z^o bosons:

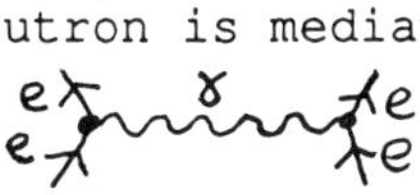

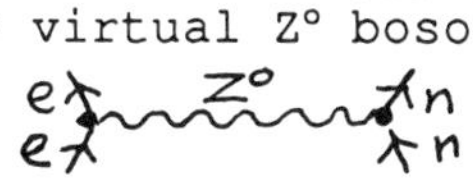

The weak force is the only one of the four forces that is chiral - it can feel the difference between left and right. Fermions exist in two states of opposite chirality, which are interconverted by parity. In a simplified picture, chirality can be taken as the same as helicity: the two helicity states correspond to spin and momentum vectors parallel (right-handed) or anti-parallel (left-handed). There are normally equal numbers of right and left-handed fermions. The two helicity states participate equally in the electromagnetic, strong and gravitational interactions, which therefore do not feel the handedness of fermions, seeing them as "racemic". However, the two helicity states do <u>not</u> participate equally in the weak interaction, which therefore sees fermions not as "racemic" but as more one hand than the other.

Parity is therefore violated, since the weak force treats mirror images unequally: a left-handed electron participates in the weak interaction preferentially compared with a right-handed electron. This suggests that left and right handed electrons are not true mirror images. It turns out, however, that although parity (P) is violated CP (parity + charge conjugation) is conserved, because whereas a left-handed electron is preferred over a right-handed electron in the weak interaction, a right-handed positron is preferred over a left-handed positron to the same degree. So clearly a left-handed electron and a right-handed positron are true mirror-images, and a true mirror is not P, but CP.

The weak force thus sees electrons as mainly left-handed, and positrons as mainly right-handed. The dissymmetry that pervades the Universe is therefore the fact that it is made of matter and not anti-matter. It should be emphasized, however, that electrons are not "really" left-handed: there are normally equal numbers of left and right-handed electrons. But the weak force can feel the left-hand ones better, so it is convenient to regard electrons as left-handed as far as the weak force is concerned.

But if the electrons were produced by a parity-violating weak process there will be an actual excess of left-handed electrons. An example is radioactive β-decay: here a neutron decays to give a proton, an electron, and an anti-neutrino, and there is an excess of left-handed electrons in proportion to v/c, the velocity of the electrons relative to the velocity of light. This raises the possibility of enantioselective β-radiolysis. In the laboratory good differential absorption by chiral molecules of left and right helically polarized electron beams has been observed by Campbell and Farago (1985). Unfortunately there are difficulties in extrapolating this result to enantioselection in nature. β-decay electrons are produced at relativistic velocities, and if they are to have any enantioselective effect they must be decelerated to energies gentle enough not to destroy L and D molecules indiscriminately; however, in nature deceleration occurs through scattering processes, which spoil the preferred helicity. In fact no reproducible bioenantioselective effects have been obtained experimentally from β-radiolysis. This is unfortunate, since in theory the maximum asymmetry obtainable is 10^{-12} (Hegstrom, 1987), which as we shall see later is readily amplifiable to homochirality.

The fact that a true mirror is CP not P has some far-reaching consequences. Firstly, atoms are optically active. The enantiomer of an ordinary atom made of (left-handed) electrons is an anti-matter atom made of (right-handed) positrons. This gives rise to a very small electroweak optical rotation, which has been detected experimentally for the gas-phase heavy metal atoms Tl, Pb, Bi and Cs (Emmons et al., 1983 and 1984). The observed optical rotation is of the expected sign and order of magnitude. It has been suggested that improvements in detection methods could enable atomic parity-violation measurements to distinguish between different unified theories, and provide values of the Weinberg angle (Sandars, 1990), thus giving us "table-top particle physics". Developing improved polarimeters and spectrometers to look for such atomic and molecular "footprints" of fundamental physics would certainly be cheaper than building ever-larger accelerators.

The second consequence of a true mirror being CP not P is that L and D molecules are really diastereoisomers, not enantiomers, owing to the handedness of their constituent elementary particles. The true enantiomer of an L-amino acid is the D-amino acid made of anti-matter. Left and right-handed molecules should therefore differ in many properties, e.g. NMR chemical shifts (Barra et al., 1987), and, most importantly, in energy.

This "parity-violating energy difference" (PVED) between enantiomers arises from weak neutral current interactions, mediated by the Z° boson, between electrons and neutrons. These interactions impart a parity-violating energy shift (PVES), E_{pv}, to the energy of a chiral molecule, and an equal and opposite shift, $-E_{pv}$, to that of its enantiomer, giving a parity-violating energy difference (PVED) of $\Delta E_{pv} = 2E_{pv}$. The magnitude of ΔE_{pv} is typically about 10^{-20} a.u. or 10^{-17} kT. Under thermodynamic control the PVED results, from the Boltzmann distribution, in an enantiomeric excess $|L-D|/L+D = \Delta E_{pv}/kT$ of 10^{-17}. Under kinetic control a similar enantiomeric excess is also obtained because the activation energies of enantiomers are unequal.

Although tiny, the PVED probably represents the most promising possibility for bioenantioselection because it affects all chiral molecules everywhere at all times, providing an everpresent global chiral influence whose effect can amplify over long periods. It is also independent of the detailed molecular mechanisms of the origin of life: it does not assume a photochemical reaction, or require environmental factors such as the presence of fields or β-emitters. And it works under either thermodynamic or kinetic control, whereas most other influences are falsely chiral and so require kinetic control.

3.HOW TO CALCULATE THE PVED

To calculate the PVED (Mason and Tranter, 1985; and MacDermott and Tranter, 1990) we start from the parity-violating Hamiltonian density for the weak neutral current interaction between electrons and nuclei. This is obtained from the Feynman diagram

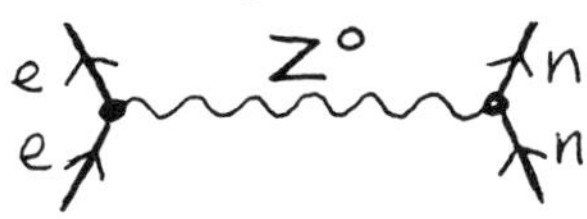

by associating a current J^{μ} with each fermion involved, a coupling constant g_N with each vertex, and a propagator (here $1/M^2_{Z^0}$) with each virtual boson exchanged, giving

$$\mathcal{H}_{PV} = J_{\mu}(e)\; g_N \;\frac{1}{M^2_{Z^0}}\; g_N\; J^{\mu}(n)$$

The weak coupling constant is not small ($g_N \approx e$), but the mass M_{Z^0} of the Z° boson is large, which is why the weak force is weak.

The neutron current $J^{\mu}(n)$ is broken down into currents of the constituent quarks, and the current for each fermion - electron or quark - is broken down into sums of left and right-handed fermion currents. The handedness of the weak force enters through the unequalness of the coefficients of left and right-handed fermion currents, e.g. for electrons $c_L = -1/2 + \sin^2\Theta_W$, $c_R = \sin^2\Theta_W$. Using the theoretical value of 30° for the Weinberg angle Θ_W the Hamiltonian becomes

$$\mathcal{H}_{PV} = -(G_F/2\sqrt{2})\; J_{\mu}(e)\, J^{\mu}(n)$$

where $J_{\mu}(e) = \bar{e}\gamma_{\mu}\gamma_5 e$, $J^{\mu}(n) = \bar{n}\gamma^{\mu}n$. On reduction to non-relativistic quantum mechanics by explicitly multiplying out the γ-matrices, the Hamiltonian density $\mathcal{H}_{PV}$ becomes the parity-violating Hamiltonian

$$\hat{H}_{pv} = -(\Gamma/2) \sum_a \sum_i N_a \left\{ \vec{s}_i.\vec{p}_i\;,\; \delta^3(r_i - r_a) \right\}_+$$

The sums over i and a are over all electrons and nuclei respectively in the molecule, and N_a is the neutron number of nucleus a. This beautifully elegant expression summarizes the physical origin of the PVES. The term in $\vec{s}.\vec{p}$ represents the projection of the spin onto the direction of momentum, thus touching directly on the left-handedness of the electron. $\hat{H}_{pv}$ is of opposite sign for enantiomers: $\vec{p}$ changes sign under parity, being a polar vector, while $\vec{s}$ remains the same, being an angular momentum and therefore an axial vector. The delta function expresses the contact nature of the weak force. The smallness of the PVES arises from the smallness of the constant Γ, which contains the very small weak coupling constant G_F.

One evaluates the PVES E_{pv} by taking matrix elements of $\hat{H}_{pv}$ over the ground state wave functions. In the absence of spin-orbit coupling one can assume separability of the wavefunction into spin and orbital parts, which results in E_{pv} being identically zero because $\vec{p}$ is imaginary. If the ground state wavefunction is corrected for the effect of spin-orbit coupling, the PVES no longer vanishes, and we obtain

$$\Delta E_{pv} = 2 \sum_j^{o} \sum_k^{u} P_{jk} \, (\varepsilon_j - \varepsilon_k)^{-1}$$

where $$P_{jk} = \mathrm{Re} \, \langle \psi_j | \hat{V}_{pv} | \psi_k \rangle . \langle \psi_k | \hat{V}_{so} | \psi_j \rangle$$

is the "parity-violating strength". The sums are over all occupied MOs j and all unoccupied MOs k, ε_j and ε_k are the energies of the MOs, $\hat{V}_{pv}$ is a one-electron version of the parity-violating Hamiltonian, and $\hat{V}_{so}$ is a spin-orbit coupling operator.

The parity-violating strength P_{jk} is closely analogous to the rotational strength

$$R_{OA} = \mathrm{Im} \, \langle \psi_o | \vec{\mu} | \psi_A \rangle . \langle \psi_A | \vec{m} | \psi_o \rangle$$

in optical activity. $\hat{V}_{pv}$ and $\vec{\mu}$ are both parity-odd (being polar vectors), while $\hat{V}_{so}$ and $\vec{m}$ are both parity-even (being axial vectors), with the result that both P_{jk} and R_{OA} are oppositely signed for enantiomers.

The PVED can be calculated by ab initio methods (Mason and Tranter, 1985; MacDermott and Tranter 1989 and 1990), using the LCAO approximation

$$|\psi_j\rangle = \sum_c \sum_\gamma C^j_{c\gamma} \, |\phi_{c\gamma}\rangle$$

to express the molecular orbitals $|\psi_j\rangle$ as linear combinations of atomic orbitals $|\phi_{c\gamma}\rangle$ of type γ on nucleus c. To evaluate the PVED, we need the MO coefficients $C^j_{c\gamma}$ and the MO energies ε_j (the matrix elements of $\hat{V}_{pv}$ and $\hat{V}_{so}$ over the atomic orbitals are pre-evaluated for the required elements using a STO-6-31G basis set). The computation is therefore divided into two stages. First we obtain the MO coefficients and energies from GAUSSIAN 90. Second, we use our own program PVED 84 to combine the MO coefficients and matrix elements to give ΔE_{pv}.

4. AMPLIFICATION OF THE PVED

The calculated PVEDs are of the order of 10^{-20} a.u., about 10^{-17} kT at room temperature, for typical biomolecules, giving an enantiomeric excess of 10^{-17}. These small excesses need amplifying to produce the observed homochirality, and the best-known amplification mechanisms fall into two classes: (a) the Yamagata cumulative mechanism, applicable to crystallizations or polymerizations of optically labile or achiral monomers, and soundly based in equilibrium thermodynamics; and (b) the Kondepudi catastrophic mechanism, applicable to optically non-labile molecules, and rather more speculatively based on non-equilibruim statistical mechanics. More recently a third mechanism has been proposed by Salam (1991) based on quantum mechanical tunnelling.

An example of the Yamagata cumulative mechanism (Yamagata, 1966) is the crystallization of quartz, which consists of helical crystals made of achiral silica units. During crystal growth an achiral unit A may add on to a growing crystal of either hand:

$$L_{n-1} + A \rightleftharpoons L_n \qquad \Delta G_L \neq \Delta G_D$$
$$D_{n-1} + A \rightleftharpoons D_n$$

But owing to the PVED the corresponding free energy changes are unequal, which means that A will add preferentially to one hand of crystal rather than the other, resulting in a fractional excess

$$f_L = \frac{L_n/L_{n-1}}{D_n/D_{n-1}} = e^{-(\Delta G_L - \Delta G_D)/RT}$$

which is operative at each of N stages of crystallization and leads cumulatively to an excess of one hand, given by

$$\frac{L_n}{D_N} = \frac{L_1}{D_1} f_L^{N-1} = (1 + (\Delta G_L - \Delta G_D)/RT)^N$$
$$\approx 1 + N(\Delta G_L - \Delta G_D)/RT$$

This is equivalent to the PVED of the crystal being N times the PVED of the individual units within the crystal:

$$\Delta E_{pv}\ \text{(N-unit crystal)} = N\Delta E_{pv}\ \text{(one unit)}$$

The Kondepudi catastrophic mechanism (Kondepudi, 1987) is based on a kinetic scheme involving autocatalysis and enantiomeric antagonism, i.e. the presence of one enantiomer encourages production of itself but inhibits production of its mirror image. Many polymerization reactions essential to life have precisely these characteristics, and some enantiomeric enrichment can be demonstrated in the laboratory (Brack and Spach, 1980). Kondepudi envisages an open-flow reactor system, such as a lake, fed by an input of achiral substrates A and B, which react both directly (k_1) and autocatalytically (k_2) to form chiral products X_L and X_D:

$$A + B \underset{k_{-1}}{\overset{k_1}{\rightleftharpoons}} X_{L(D)}$$

$$X_{L(D)} + A + B \underset{k_{-2}}{\overset{k_2}{\rightleftharpoons}} 2X_{L(D)} \qquad \text{autocatalysis}$$

$$X_L + X_D \xrightarrow{k_3} P \qquad \text{enantiomeric antagonism}$$

The reaction scheme can accommodate unequal reaction rates for enantiomers, and can be extended to include racemization, thermal fluctuations, etc.

Using the methods of non-equilibrium statistical thermodynamics, Kondepudi showed that for $\Delta E_{pv} \geqslant 10^{-17}$ kT, amplification to homochirality will eventually occur, but for smaller ΔE_{pv} the amplification effect will be overcome by thermal fluctuations. The amplification time for a PVED of 10^{-17} kT is 10^4 years, a very short time on an evolutionary timescale. This figure is based on a small lake of 4×10^9 L (1km x 1km x 4m), with modest concentrations of order 10^{-3} M and a realistic reaction rate of 10^{-10} mol L^{-1} s^{-1}. The amplification can withstand a racemization half-life as low as 10^2 - 10^3 years (cf. typical values of 10^5 - 10^6 years for most amino acids).

The amplification time is very sensitive to the magnitude of the PVED and the lake size. If ΔE_{pv} increases by <u>one</u> order of magnitude, the amplification time decreases by <u>four</u> orders of magnitude, so that whereas a PVED of 10^{-17} kT takes 10^4 years, a PVED of 10^{-16} kT takes only 1 year. Alternatively, if ΔE_{pv} increases by <u>one</u> order of magnitude, one can decrease the lake volume by <u>two</u> orders of magnitude for the same amplification time. So whereas amplification from the enantiomeric excess of 10^{-17} from the PVED takes 10^4 years in 4×10^9 L, amplification from the enantiomeric excess of 10^{-12} theoretically obtainable from β-radiolysis takes 1 year in 40 L, which could be verified on the laboratory scale.

5. COMPUTATIONS OF THE PVED

We have studied the PVEDs of molecules from all three

classes of possible ancestral replicators - nucleic acids, amino acids, and quartz.

The natural L-amino acids L-alanine, L-valine, L-serine and L-aspartate were all found to be more stable than their D-enantiomers, in both the solution and α-helix and β-sheet conformations, by 10^{-17} kT (Mason and Tranter, 1984 and 1985), just enough for Kondepudi amplification. The aldose sugar series $(H_2CO)_n$ was investigated for n=3 to 5, but the results were not so clear-cut as for the amino acids. For glyceraldehyde (n=3), the parent of the higher sugars, the D form is PVED-stabilized by about 10^{-17} kT (Tranter, 1986); D-deoxyribose is also PVED-stabilized in the C2-endo form found in DNA; but D-ribose in the C3-endo form found in RNA is <u>less</u> stable than its enantiomer (Tranter and MacDermott, 1992). These latter results are explained by the PVED of the basic furanose skeleton, tetrahydrofuran (THF): the C2-endo ring conformation of THF is more stable than the enantiomeric C3-endo conformation by about 10^{-17} kT (Tranter and MacDermott, 1986).

Thus the most important biomolecules - amino acids and sugars - have PVEDs of 10^{-17} kT, which is only just amplifiable by the Kondepudi mechanism. If the PVED were only very slightly larger, Kondepudi amplification would be very much easier: 10^{-17} KT takes 10^4 years, while 10^{-16} kT takes only 1 year. It turns out that the PVED is proportional to the sixth power of the atomic number (largely due to the importance of spin-orbit coupling), so attention was turned next to molecules incorporating second-row heavy atoms such as P, Si, S, in the hope of finding larger PVEDs (MacDermott and Tranter, 1989b).

We studied fragments of the sugar-phosphate backbone of DNA without the bases (which are anyway planar and therefore in themselves achiral). The contributions to the PVES from the phosphate and sugar were found to be largely independent and additive, so it was possible to estimate the PVES of DNA by looking firstly at a THF ring between a 3′ and 5′ phosphate (to get the PVES of the sugar in the DNA environment) and secondly at a phosphate link between two THF rings (to get the PVES of the phosphate in the DNA environment). These fragments were studied in the right-hand helical conformation of the DNA double helix, and both sugar and phosphate were found to be PVED-stabilized. It had been hoped that larger PVEDs would be obtained with the heavy phosphorus atom, but the phosphate PVED was still only 10^{-17} kT: this rather disappointing result was traced (MacDermott and Tranter, 1989b) to the fact that the phosphorus atom is very electropositive in phosphates due to the electron-withdrawing effect of the four oxygens, resulting in too little electron density on the phosphorus to feel the potentially larger parity-violating effect.

We saw earlier that although the D-deoxyribose part of

DNA is PVED-stabilized, the D-ribose part of RNA is not. So D-ribose cannot have been selected by its own PVED (since L-ribose is more stable). But the D-sugar series may have been selected either by the PVED of its parent D-glyceraldehyde, or alternatively by the diastereomeric connection with the PVED-stabilized L-amino acids. Ribose is in fact the only molecule so far studied in which the "natural" enantiomer is the less stable, and this may reflect the fact that ribose, being difficult to synthesize, is probably not pre-biotic anyway but a later biological evolution after the first self-replicating systems appeared, in which case its own PVED is irrelevant. We therefore turned our attention to more primitive replicators than DNA or RNA, in particular the glycerol-based polymer, for which the right-hand helix was again found to be PVED-stabilized, by 10^{-17} kT per sugar-phosphate unit.

This raises the interesting question of whether the D-sugars fixed or followed the right-hand double helix. It is usually assumed that the D-sugars fixed the right-handedness of the DNA helix by diastereomeric preference. But it could be that the right-hand helix was PVED-selected in some earlier, simpler replicator (perhaps resembling our glycerol-based double helix), in which case steric considerations in later ring-closure to give ribose would direct selection of the D-sugars.

We had expected the bases to contribute little to the PVED because they are planar and therefore achiral in themselves. But preliminary calculations give them a significant positive PVES which partially cancells the negative PVES of the sugar-phosphate backbone: the PVED of the backbone + bases is about 1.5 x 10^{-17} kT per nucleotide, as compared with about 3 x 10^{-17} kT for the backbone alone. The reason for the unexpectedly large PVED of the bases is probably their aromaticity: the electrons delocalize and sample the highly chiral environment.

Another important system which contains both helices and second-row heavy atoms is quartz: the l-form consists of right-hand 3-fold helices of silica tetrahedra. As a chiral mineral made of achiral units it can undergo Yamagata amplification, and indeed a 1.4% excess of l-quartz has been reported in a large collection of crystals from all over the world (Palache et al., 1942). Our calculations (MacDermott and Tranter, 1989a and 1989b) showed that l-quartz was PVED-stabilized by 10^{-17} kT per SiO_2 unit. This was again disappointingly small for the heavy Si atom: but silicates have the same problem as phosphates in that the electron-withdrawing effect of the oxygens leaves the Si atoms very electropositive. However the PVED does not need to be larger than 10^{-17} kT to account for the 1.4% excess of l-quartz: according to the Yamagata mechanism, ΔE_{pv}(crystal) = N ΔE_{pv}(unit), so ΔE_{pv}(crystal) = 10^{-2} kT (corresponding to a 1% enantiomeric excess) can be obtained from ΔE_{pv}(unit) = 10^{-17} kT if N = 10^{15}, which corresponds (Tranter, 1985) to a realistic small crystal of side 0.1 mm.

These results for quartz are quite a triumph for the idea of electroweak enantioselection, because they predict almost exactly the observed 1% excess of l-quartz, which can now for the first time be regarded as evidence for the global symmetry-breaking effects of the weak force: whereas biomolecular chirality could be the result of a chance effect in a single ancestor, chiral bias in unconnected mineral deposits must be the result of a global influence. Even if quartz-like clay minerals were not ancestral replicators, they could alternatively have played a role in the origin of life through pre-biotic surface catalysis, transferring their chiral bias to biology in the process. L-quartz adsorbs L-amino acids preferentially from a racemic mixture (for purely diastereomeric reasons), with a 1% enantioselectivity (Kavasmaneck and Bonner, 1977); combining this asymmetry of 10^{-2} in adsorption with the 10^{-2} asymmetry in the quartz crystals themselves gives an overall electroweak enantioselectivity of 10^{-4}, which is much greater than the 10^{-17} from the PVED of individual molecules, and could be amplified to homochirality much more easily. Realistic chiral mineral catalysts are not quartz itself, but aluminosilicate quartz analogues, possibly containing heavy metal cations which might produce larger PVEDs.

We have seen that second-row heavy atoms should produce larger PVEDs because of the Z^6 dependence, but phosphates and silicates are too electropositive so we should be looking to more electronegative heavy atoms. To this end we have recently been studying DNA fragments with novel thio-substituted linkages (MacDermott et al., 1992). Here the normal $-O-PO_2^{-}-O-$ linkages are modified to $-O-POS^{-}-O-$ or $-O-PS_2^{-}-O-$ by thiosubstitution of one or both of the side oxygens, or to $-S-S-CH_2-CH_2-$ by thiosubstitution in the helix backbone itself. Thiosubstitution of the side oxygens produced for the first time PVEDs of 10^{-16} kT, and the natural right-hand helix was the more stable. Thiosubstitution in the helix itself resulted in the enormous PVED-stabilization of 0.7 x 10^{-14} kT for the $-S-S-CH_2-$ linkage in the B DNA right-hand helix! This is our most exciting result yet: we saw that a PVED of 10^{-17} kT is only just amplifiable by the Kondepudi mechanism, taking 10^4 years in a 4 x 10^9 L lake, while our new PVEDs of 10^{-14} kT are amplifiable in 1 year in a small 4 x 10^5 L pond.

Thiosubstituted links may have large PVEDs, but can they actually form a viable double helix, and are they relevant to the origin of life? They certainly do form a base-paired double helix, as shown by their recent use as "antisense inhibitors" (Cohen, 1989). If an oncogene, for example, results in an undesirable gene product, one can use a piece of DNA of complementary base sequence ("antisense") to bind to the corresponding m-RNA transcript to form a DNA-

RNA hybrid double helix; this prevents that piece of m-RNA from being translated into undesirable proteins. But if one tries to introduce a normal antisense oligonucleotide into a cell, nucleases break it down, so it is necessary to alter the DNA backbone to make it nuclease-resistant. A thiosubstituted backbone is found to be ideal because it forms a double helix with little change of geometry.

Thiosubstituted DNA analogues could also be relevant to the origin of life because one of the problems is that the Miller-Urey experiment assumed a reducing atmosphere, while the primitive Earth probably had a neutral atmosphere; but deep-sea volcanic vents do provide a reducing environment, and the sulphurous gases present could make a thiosubstituted replicator highly plausible. Moreover, incorporation of phosphate is also a problem in the origin of life, making a purely sulphur-linked ancestral replicator even more likely.

Chiral aromatic systems may be expected to show a large PVED (cf. the huge optical rotation of hexahelicene, which forms a giant chiral chromophore) and fascinating new examples may be afforded by Buckminsterfullerenes, or "buckyballs": these are closed balls of graphite-like sheets with some hexagons substituted by pentagons to provide the necessary positive curvature. The first buckyball was the highly symmetrical C_{60} (Kroto, 1991), but many are chiral, e.g. C_{28}, the smallest chiral buckyball, and C_{76}, which contains helical ribbons of six hexagons and has been isolated (Ettl et al., 1991) but not yet resolved. If heptagons are used instead of pentagons, negative curvature is introduced and we obtain a "buckygym" (Vanderbilt and Tersoff, 1992) - a diamond-type lattice of chiral tetrahedral units which have chiral propellor-like structures on each face of the tetrahedron. It is believed (Kroto 1991) that fullerenes could form in large quantities in outer space in the outflow from carbon stars. There they could act as surface catalysts for prebiotic reactions. One would expect enormous PVEDs for chiral fullerenes as they are giant aromatic chromophores. This would lead to a substantial excess of one hand of fullerenes and hence enantioselectivity in the reactions their surfaces catalyzed. The effect is likely to be best for buckygyms, which form a huge chiral surface with molecules trapped in the cavities, etc..

We have only looked at two buckyballs so far, but surprisingly their PVEDs are very disappointing, at 10^{-19} kT for C_{28} and 4×10^{-17} kT for C_{76}. It could be (and has been suggested by some) that buckyballs are not in fact all that aromatic after all, since the curvature reduces π-overlap. But it may be that these particular examples are not the most chiral: some others, e.g C_{84}, have longer helical ribbons of hexagons at sharper helical angles. Clearly more buckyballs should be tried. Less optimistically, however, it could be that because each carbon atom in a buckyball has almost no charge (homonuclear bonds are non-polar), the atoms cannot "feel" the surrounding chirality, which is normally felt as a chiral disposition of neighbouring charges. Both possible reasons for the small buckyball PVEDs - lack of aromaticity or lack of charge - would tie in with the surprisingly large PVED for the DNA bases, which are both aromatic and with highly polar heteronuclear bonds. To differentiate between the two possibilities, one could try a buckyball with some C=C bonds replaced by the isoelectronic B=N, which would introduce polarity but leave the aromaticity intact - but such hetero-buckyballs are unlikely to have any relevance to the origin of life!

In conclusion, the weak force appears to predict the correct hand whatever came first. For nucleic acids first, D-glyceraldehyde is more stable, so that the D-sugars would be selected, and the right-hand helical backbone of DNA is also PVED-stabilized. For proteins first, the L-amino acids are more stable. For clays first we find that the weak force correctly predicts the reported 1% excess of l-quartz (which preferentially adsorbs L-amino acids) - but it remains to be seen whether this natural bias towards the PVED-stabilized form extends to the aluminosilicates which represent more realistic candidates for Cairns-Smith's clay replicators.

If pre-biotic symmetry-breaking is a pre-condition for life, it is reasonable to ask which start to life provides the easiest symmetry-breaking. For a polymer start, thiosubstituted DNA analogues are clearly best, with their much larger PVEDs. But a crystal start is arguably better still because (1) it uses abundant pre-existing material and requires no special conditions, (2) an achiral replicator is possible, (3) the chirality is easily amplified, once it appears, by the Yamagata mechanism, (4) the chiral bias in crystals is easily transferred to polymers on their surface.

We note that symmetry-breaking is altogether easier if a surface is involved: the weak force acting on the crystal surface leads to a 10^{-4} excess in the polymer, while the weak force acting on the polymer itself produces only a 10^{-17} or 10^{-14} excess. Crystals - aluminosilicate quartz analogues and perhaps buckygyms - are clearly the way forward for electroweak enantioselection.

References

Angel, J.R.P., Illing, R., and Martin, P.G., 1972: Nature 238 389.
Bada, J.L., and Miller, S.L., 1987: BioSystems 20 21.
Barra, A.L., Robert, J.B., and Wiesenfeld, L., 1987: BioSystems 20 57.
Barron, L.D., 1986: Chem. Soc. Rev. 15 189.
Blair, N.E., Dirbas, F.M., and Bonner, W.A., 1981:

Tetrahedron 37 27.
Brack, A., and Spach, G., 1980: J. Mol. Evol. 15 231.
Campbell, D.M., and Farago, P.S., 1985: Nature 318 52.
Cohen, J.S. (ed), 1989: "Oligodeoxynucleotides: antisense inhibitors of gene expression", Topics in Mol. and Struct. Bio. 12.
Emmons, T.P., Reeves, J.M., and Fortson, E.N., 1983: Phys. Rev. Lett. 51 2089.
Emmons, T.P., Reeves, J.M., and Fortson, E.N., 1984: Phys. Rev. Lett. 52 86.
Ettl, R., Chao, I., Diederich, F., and Whetten, R.L., 1991: Nature 353 149-153.
Fischer, E., 1894: Chem. Ber. 27 2985 and 3189.
Hegstrom, R.A., 1987: BioSystems 20 49.
Joyce, G.F., Visser, G.M., van Boeckel, C.A.A., van Boom, J.H., Orgel, L.E., and van Westresen, J., 1984: Nature 310 602.
Kavasmaneck, P.R., and Bonner, W.A., 1977: J. Amer. Chem. Soc. 99 44.
Kondepudi, D.K., 1987: BioSystems 20 75.
Kroto, H.W., 1991: Chem. Rev. 91 1213-1235.
MacDermott, A.J., and Tranter, G.E., 1989a: Croatica Chemica Acta 62 165.
MacDermott, A.J., and Tranter, G.E., 1989b: Chem. Phys. Lett. 163 1-4.
MacDermott, A.J., and Tranter, G.E., 1990: Symmetries in Science IV (B.Gruber and J.H.Yopp, eds), Plenum, NY, p.67.
MacDermott, A.J., Tranter, G.E., and Trainor, S.J., 1992: Chem. Phys. Lett. 194 152-156.
MacGillivray, H.T., and Dodd, R.J., 1985: Astron. Astrophys. 145 269.
McLauchlan, K.A., and Steiner, U.E., 1991: Mol. Phys. 73 241-263.
McLauchlan, K.A., 1992: Phys. World 5 41-45.
Mason, S.F., and Tranter, G.E., 1984: Mol. Phys. 53 1091.
Mason, S.F., and Tranter, G.E., 1985: Proc. R. Soc. Lond. A 397 45.
Melcher, G., 1974: J. Mol. Evol. 3 121.
Miller, S.L., and Orgel, L.E., 1974: "The Origins of Life on the Earth", Prentice Hall, p.171.
Palache, C., Erman, G.B., and Frondel, C., 1962: Dana's System of Mineralogy 7th ed. Vol III, Wiley, New York, p.16.
Pasteur, L., 1894: Rev. Scientifique 7 1.
Pasteur, L., 1922: Oeuvres de Pasteur I (Pasteur Vallery-Radot, ed), Masson et Cie, Paris, p.369.
Salam, A., 1991: J. Mol. Evol. 33 105.
Sandars, P.G.H., 1990: J.Phys. B 23 1655.
Schwartz, A.W., and Orgel, L.E., 1985: Science 228 585.
Tranter, G.E., 1985: Nature 318 172.
Tranter, G.E., 1986: J.Chem. Soc. Chem. Commun. p.60.
Tranter, G.E., MacDermott, A.J., Overill, R.E., and Speers, P.J., 1992: Proc. R. Soc. Lond. A 436 603-615.
Vanderbilt, D., and Tersoff, J., 1992: Phys. Rev. Lett. 68 511-513.
Yamagata, Y., 1966: J. Theor. Biol. 11 495.

THE ORIGIN OF CHIRALITY, THE ROLE OF PHASE TRANSITIONS AND THEIR INDUCTION IN AMINO ACIDS

Abdus Salam
Department of Theoretical Physics, Imperial College,
London, United Kingdom
and
International Centre for Theoretical Physics, Trieste, Italy.

ABSTRACT

We present a brief review of the work in which it was suggested that chirality among the twenty amino acids which make up the proteins may be a consequence of a phase transition which is analogous to that due to BCS superconductivity.

1. THE ROLE OF CHIRALITY IN CHEMICAL EVOLUTION AND THE ORIGIN OF LIFE

The present note summarises the main ideas in two recent publications in which we re–emphasized the role of chirality in the theories which determine chemical evolution and the origin of life – in particular, the fact that almost all amino acids, utilized in living systems, are of L–type (Salam, 1991; 1992).

1.1 One may summarize the presently accepted view of the origin of life as falling into three stages: the cosmic stage; the prebiotic chemical stage; and the biological stage.

1.1.1 The cosmic stage concerns itself with the early history of the universe where the electroweak made a phase transition into two forces, electromagnetic and weak, 10^{-12} secs after the universe was born. The temperature was then 250 GeV and the carriers of the neutral weak force – the Z_0 particle – acquired mass.

1.1.2 Chemistry became important after the planets were formed (some 10 billion years later) though it may have played a role in the *presolar epochs* as well, (long after the quarks of the early cosmic era had condensed into protons and neutrons and much after the recombination with electrons which took place some 10^5 years following the Big Bang). Molecules of future life could thus have formed even before the origin of the Earth itself (Orò *et al.*, 1990).

1.1.3 The biological era which concerns itself with the replication of nucleic acid polymers and protein synthesis. The biological stage may have started some 3.8 billion years ago.

---

* Units $\hbar = c = k_B = 1(1\ °K = 10^{-4}$ eV).

ISBN 0-937194-31-X

1.2. Classically, a chiral molecule and its mirror image (defined by left (L) or right (D) optical/rotatory dispersion) (Mason and Tranter, 1984) have been considered energetically equivalent. However, *the parity–violating weak interactions* give rise to L and D configurations (Mason and Tranter, 1984) and ensure that this equivalence is no longer exact – one of the two molecules, L or D, being energetically stabilized, with energy differences of the order of 3×10^{-19} eV.

In living systems, protein molecules are composed of twenty L–amino acids (though some amino acids of the opposite D–type do occur in cell walls of certain bacteria). The polynucleotides contain sugars in D–configurations only. Clearly, once living processes had selected handedness, the complex machinery of protein synthesis and stereoselectivity of enzymes could have assured that such handedness is perpetuated.

1.3. The most significant of parity violating weak interactions are the weak-neutral, these being mediated by the Z^0 bosons (Harris *et al.*, 1978). These interactions are of exceptionally short range by atomic and strong nuclear physics standards, and to a very good approximation may be taken as contact phenomena. Recent calculations (Mason and Tranter, 1984; Tranter and MacDermott, 1989), indicate that four of the amino acids in aqueous zweitterionic conformation – (essentially all the ones for which these calculations are available – alanine, valine, serine and aspartic acid) – are L–stabilized relative to their 'unnatural' D–mirrors for configurations in *aqueous* media. (A minor triumph, for the sugars – particularly for D–glyceraldehyde – the calculations show that the right–handed variety is the more stable).

This stability affects 1 out of 10^{17} *molecules at room temperatures (since* $10^{-17} \approx \frac{3 \times 10^{-19} ev}{300\,°K}$). It is the smallness of this figure which has prompted many chemists to wonder if this mechanism could indeed be responsible for the ultimate optical asymmetry.

The crucial problem is that of amplification of this electroweak advantage over the course of time so that, for example, the 20 amino acids (which make up the proteins) convert almost entirely from D into L types. This problem has been considered following the seminal ideas of F.C. Franck (Kondepudi and Nelson, 1985).

Consider, for example, quartz crystals which may be taken as *non-equilibrium* statistical mechanical systems at ambient temperatures of the order of T = 300 °K. Using an autocatalytic mechanism and theory of delayed bifurcations, one can show that a lake 1 km sq. and four metres deep would need 10^4 years to produce the necessary electroweak advantage so far as quartz is concerned. Kondepudi and Nelson give a general theory of spontaneous chiral symmetry-breaking in *nonequilibrium* chemical systems and the possible influence of weak-neutral currents in such a process. They conclude that on a long time scale (10^4 years), for reactions occurring in large volumes, such as the oceans, the effects of parity violations due to weak-neutral currents cannot be considered small.

Objections to this work have been voiced by Goldanski (Goldanski, 1987) and following him by Avetisov, Kuz'min and Anikin (Avetisov *et al.*, 1987). For one thing, these authors (Avetisov *et al.*, 1987) (as well as L. Orgel, private communication) maintain that the electroweak advantage takes place by repetitive steps in these calculations and that there are $N = 10^{17}$ steps involved. The earth must therefore have contained $N^2 \approx (10^{17})^2 \approx 10^{34}$ chiral molecules to take proper account of the resulting fluctuations. This, these authors (Goldanski, 1987) find difficult to credit and conclude that "the role of weak neutral currents in the origination of the biomolecular chirality should not be considered essential". Kondepudi (private communication) agrees with this estimate (10^{34}) but maintains that "this number of molecules need not all be reacting at the same time. This is the total number that is *fluxed* through the nonequilibrium flow system in 1.5×10^4 years".

Without necessarily disagreeing with the work of Kondepudi and Nelson (particularly at high temperatures where the equations they use may be considered part of the renormalization group), we *speculate* on an alternative mechanism. The discussion in this note is mainly physical in character. We have discussed the enhancement due to phase transitions (Salam, 1991; 1992).

We would like to treat this as an equilibrium problem quantum mechanically. The quantum mechanical formalism treats the phenomenon as a cooperative one where condensation aspects are emphasized and the transitions D to L are accomplished below a critical temperature T_c.

In general, when global cooperative and condensation phenomena do take place, low temperatures (or high densities, as for example for neutron stars) are necessary. If T_c for the amino acids turns out to be very small, then it would be plausible that the origin-space in which production of chiral amino acids takes place, was larger than the earth. We discuss this problem in Sec.1.8.

1.4. A modern version of the phenomenon of condensation has been described by A. Leggett (Leggett, 1990) who gives a uniform treatment of a superfluid (like liquid helium II), where T_0 – the so–called degeneracy temperature – is ≈ 3 °K, as well as of *metallic superconductivity* for Cooper-paired electrons below a transition temperature T_c. *The latter is more relevant for our purposes as we shall see.*

What is "condensation"?

"Imagine that you are on a mountain-top looking down at a distant city square. The crowd is milling around at random, and each individual is doing something different: Now suppose, however, that it is not market day but the day of a military parade, and the crowd is replaced by a battalion of well drilled soldiers. Every soldier is doing the same thing at the same time, and it is very much easier to see (or hear) from a distance what that is. The physics analogy is that a normal system is like the market day crowd – every atom is doing something different – whereas in a Bose condensed system the atoms (or, more accurately, the fraction of them which is condensed at the temperature in question) are all forced to be in the same quantum state, and therefore resemble the well drilled soldiers: *"every atom must do exactly the same thing at the same time*" (Leggett, 1990).

The analogy of the behaviour of the crowd with racemicity and of the well–drilled soldiers with L-amino acids is apt. The number of particles in any given energy level is fixed as a function of temperature, and as a result the total number of particles occupying the levels cannot be greater than some number $N_{max}(T)$ (which decreases along with T). At some temperature T_0 the quantity N_{max} becomes equal to the total number of particles in the system N, while below T_0 we have $N_{max} < N$. At such temperatures there are simply not enough quantum states available to accommodate all the particles. "The resolution of the problem is remarkably simple: below T_0 the system adjusts by taking all the particles which cannot be accommodated by the distribution formula and putting them in the single quantum state which has the lowest energy (the "ground state".) Since these surplus particles are a finite fraction of the whole (in fact, at zero temperature all of them), we reach the remarkable result that a *macroscopic* number of particles (of order N, which typically is of order say 10^{23}) occupy a *single* quantum state. This phenomenon is known as *"Bose condensation"* (Leggett, 1990).

1.5. The necessary conditions for condensation have been studied by L. Landau (Landau *et al.*, 1980) Landau distinguished between Bose superfluids and Fermi superfluids (even for Fermi superfluids like superconductors the "superfluidity" is produced by bosonic *condensates* like those for Cooper pairs).

"A Fermi gas with attraction between the particles must have the property of superfluidity"; "howsoever weak the attraction is". (Abrikosov, 1987).

1.6. For $T = T_c$ we have the phase transition with all its aspects of suddenness. For $T < T_c$ the superfluid persists reaching its climax at $T = 0$. The question arises as to what happens when $T > T_c$. To answer this question, we consider cooperative phenomena. The main features of a cooperative phenomena can be illustrated by considering with Atkins (Atkins, 1959) the semi-classical treatment of ferromagnetism. At 0°K the electron spins are aligned parallel to one another to give a resultant magnetization even in the absence of an external magnetic field. However at a finite temperature thermal agitation is able to turn over some of the spins and the average magnetic moment in the direction of magnetization is thereby decreased. "As soon as this process starts, an electron chosen at random is likely to have neighbours pointing against the direction of magnetization as well as with it and this reduces the energy needed to reverse the spin of the electron, so that, as the temperature increases and more spins are turned over, it becomes increasingly easier to turn over the remaining spins and the disordering process develops with ever increasing rapidity. At the Curie point the disordering is eventually complete and the spins point equally in both directions" so that only a 50-50 racemic state survives.

We have used the results of this section in Ref.1 (footnote to Appendix B) to Appendix B to obtain one estimate of T_c for most amino acids.

1.7. How important are the effects of the parity violating weak interacitons for specific amino acids? This question may be answered by setting (Harris *et al.*, 1978) $1 - 4\sin^2\theta \approx 0$; the resulting elegant expression for the Hamiltonian has led to *molecular–conformation–dependent values* for parity violation which for Ala, Val, Ser and Asp have been calculated to give (Mason and Tranter, 1984; Tranter and MacDermott, 1989) -3.0, -6.2, -2.3 and -4.8, in units of 10^{-19} eV.

Energy values may be expressed in atomic units (1 atomic unit = 1 Hartree = 27.2 eV) corresponding to the L–configuration of the *aqueous zwetterionic* amino acids (Mason and Tranter, 1984; Tranter and MacDermott, 1989). (The simplest amino acid, glycine (residue R = H) is not resolvable into optical isomers. On account of the further symmetry implied by R = H, parity violation when averaged over a complete rotation of the carboxylate group through all configuration angles, turns out to be zero, in agreement with experiment. For the sugar glyceraldehyde the value is $+1.8 \times 10^{-19}$ eV corresponding to its *D*–configuration (Mason and Tranter, 1984; Tranter and MacDermott, 1989)).

The classical chemist has hitherto used the electromagnetic force as the only fundamental force which can produce chemical effects. He has not considered the electroweak force, and in particular its Z^0 component, since the effects due to Z^0 are supposedly very small at low temperatures. We shall now show that the electroweak interactions (for which there is nothing comparable in classical chemistry) can give rise to a phase transition.

Like all phase transitions, this one will have a behaviour like $(T_c - T)^\alpha, T \to T_c$. This makes the amplitude (or derivatives thereof) an infinite quantity.

To compute T_c exactly is difficult (for one thing, on account of the incompleteness of the standard model of elementary particles), but we can without difficulty measure T_c. *In the sequel we shall assume T_c does exist and that it has been measured for each amino acid.*

1.8. Since the ambient temperature of the Earth's surface is $\approx$ 300 °K (the maximum temperature on the surface being $\approx$ 350 °K), if $T_c \ll 300$ °K, the present formalism may not apply to the Earth. (The fact that the prebiotic temperatures may be less than 300 °K was the content of a 1966 work by Sanchez, Ferris and Orgel (Sanchez *et al.*, 1966).)

In this case, there could be several possible scenarios:

a) Pre–solar contributions may be necessary to get low enough temperatures;

b) Or major contributions come from the more distant and the cooler parts of the solar system with ambient temperatures less than T_c. In this context it is good to remember that the *maximum* surface temperatures have been estimated as 135 °K for Jupiter, 120 °K for Saturn, 85 °K for Uranus, 55 °K for Neptune, and 20 °K for Pluto.

c) If T_c is considerably less than 300 °K, we could entertain the (somewhat odd) notion that the Earth only acted as a junction place where *L*–amino acids came together with *D*-sugars and nucleotides for the replication phenomena to get started for the biotic stage of Sec. 1.2 to be implemented.

In order for the biotic alternative of Sec. 1.2 to prevail, it is necessary to invent a mechanism to deliver organic molecules to the Earth from the cooler locations a) or b). This was first considered by Orò (Orò, 1961; Orò *et al.*, 1990), and in greater detail recently by Chyba, Thomas, Brookshaw and Sagan (Chyba *et al.*, 1990). The latter authors conclude that:

1.8.1 "The Earth did accrete prebiotic organic molecules important for the origins of life from impacts of carbonaceous *asteroids and comets* during the period of heavy bombardment 4.5×10^9 to 3.8×10^9 years ago" (cf. the Chiron comet?).

1.8.2 For plausibly dense (10-bar carbon dioxide) early atmospheres, these authors find that 4.5×10^9 years ago Earth was accreting intact *cometary* organics at a rate of at least $\approx 10^6$ to 10^7 kilograms per year – (a flux that thereafter declined with a half-life of $\sim 10^8$ years). These results may be placed in context by comparison with terrestrial oceanic and total biomasses, $\sim 3 \times 10^{12}$ kilograms and $\sim 6 \times 10^{14}$ kilograms, respectively.

1.8.3 (Contrary to the general impression) the alanine molecules could withstand temperatures as high as $\approx 700\,°$K for one second whereas other amino acids could withstand temperatures in the range of 600° to 800°K for a like period of time and remain stable and intact upon impact with the Earth.

It is tempting to assume further that this material ($\approx 10^6$ to 10^7 kilograms) maintained its chirality at delivery (particularly if T_c had a small value near zero). This follows from the formula $-\frac{\partial\Delta}{\partial T} \approx \frac{3.06}{2} \times \sqrt{\frac{T_c}{T_c - T}}$ if the formula applies to the case of amino acids. At this juncture life could have started and the chirality perpetuated in accordance with the biotic picture in Sec.1.2.

2. PHASE TRANSITIONS AND THEIR INDUCTION IN AMINO ACIDS

2.1. The T_c for BCS superconductivity for metals is of the form $\omega \exp(-\frac{2}{g_{eff}\sigma(0)})$. We conjectured that a similar formula may hold for the case of amino acids chains. In the present section we shall explore this further and show, following Lee and Drell (Lee and Drell, 1991), that a crucial form for the transition temperature T_c involves dynamical symmetry breaking. The t-quarks or supersymmetry (or something similar which ensures a heavy mass) appear to be essential if such mechanisms are to hold.

First, let us review the subject of BCS superconductivity. The best treatment, which I know of has been given by Sakita (Sakita, 1985), following on the work of Ginzburg and Landau ((Ginzburg and Landau, 1950) and of Gorkov (Gorkov, 1958).

The idea is to start from the Feynman Lagrangian methodology of writing down the BCS theory for the superconducting electronic system. One tries to write down the equivalent Ginzburg-Landau equation. From this equation is deduced the value of temperature T_c.

2.2. The BCS *Hamiltonian* is given by

$$
\begin{aligned}
H &= H_0 + H_1 \\
H_0 &= \sum_{s=\uparrow\downarrow} \int d\vec{x}\psi_s^+(\vec{x})(-\nabla\frac{2}{2m} - \mu)\psi_s(\vec{x}) \\
H_1 &= -g_{eff} \int d\vec{x}\psi_\uparrow^+(\vec{x})\psi_\downarrow^+(\vec{x})\psi_\downarrow(\vec{x})\psi_\uparrow(\vec{x})
\end{aligned}
\tag{2.2.1}
$$

where g_{eff} is an attractive coupling constant between spin up (↑) and spin down (↓) electrons and antielectrons is the chemical potential. *The sign of g_{eff} is part of the assumption* of the Hamiltonian which signifies an attractive force between Cooper-paired systems of electrons consist of one of the particles being replaced by its antiparticle with a factor of two which appears in the mass term.

We can introduce electro-magnetic interaction in a gauge-invariant way by the minimal substitution

$$\vec{\nabla} \to \vec{\nabla} - ie\vec{A}$$

and treat the vector potential $\vec{A}(\vec{x})$ as an external source. Then, in general, the Hamiltonian equals

$$H = \sum_{s=\uparrow\downarrow} \int d\vec{x}\psi_s^+(\vec{x})[-\frac{1}{2m}(\vec{\nabla} - ie\vec{A})^2 - \mu]\psi_s(\vec{x})$$

$$-g_{eff} \int d\vec{x}\psi_\uparrow^+(\vec{x})\psi_\downarrow^+(\vec{x})\psi_\downarrow(\vec{x})\psi_\uparrow(\vec{x}) \qquad (2.2.2)$$

which is invariant under gauge transformations: –

$$\psi_s(\vec{x}) \to e^{ie\Lambda(\vec{x})}\psi_s(\vec{x}), \quad \vec{A}(\vec{x}) \to \vec{A}(\vec{x}) + \vec{\nabla}\Lambda(\vec{x}) \, . \qquad (2.2.3)$$

The partition function based on the appropriate Lagrangian corresponding to the above Hamiltonian is given by

$$Z = \int D\psi D\bar{\psi} \exp\left[-\int_0^\beta d\tau \int_V d\vec{x}\bar{\psi}_s \left(\frac{\partial}{\partial\tau} - \frac{1}{2m}(\vec{\nabla} - ie\vec{A})^2 - \mu\right)\psi_s\right]$$

where we have written a non-relativistic equation for the fermion (electron) ψ in the theory.

$$\exp\left[g_{eff}\int_0^\beta d\tau \int_V d\vec{x}\bar{\psi}_\uparrow(\vec{x})\bar{\psi}_\downarrow(\vec{x})\psi_\downarrow(\vec{x})\psi_\uparrow(\vec{x})\right] \qquad (2.2.4)$$

Note that the sign before g_{eff} has changed to plus g_{eff} instead of minus g_{eff} in the Hamiltonian formulation. This is because $\Delta H = -\Delta L$ for the case of potentials which do not contain time derivatives of the fields themselves. The 4-fermion interaction can be expressed in terms of a complex auxiliary Higgs scalar field ϕ:

$$\frac{1}{C}\int D\phi D\phi^* \exp[+\kappa^2 \int d^4x\phi^*\phi + g_{eff}^{1/2}\kappa \int d^4x(\bar{\psi}_\uparrow\bar{\psi}_\downarrow\phi + \psi_\downarrow\psi_\uparrow\phi^*)] \quad (2.2.5)$$

where we have used the notation

$$\int d^4x \equiv \int_0^\beta d\tau \int_V d^3\vec{x} \qquad (2.2.6)$$

The constant C is given by

$$C = \int D\phi D\phi^* \exp[+\kappa^2 \int d^4x\phi^*\phi] \qquad (2.2.7)$$

where κ is a constant with dimension of mass. We shall introduce a source for the ϕ field in the form

$$Z[j,j^*] = \frac{1}{C}\int D\psi D\bar{\psi} D\phi D\phi^* \exp\left[-\int d^4x\bar{\psi}_s\left(\frac{\partial}{\partial\tau} - \frac{(\vec{\nabla}-ie\vec{A})^2}{2m} - \mu\right)\psi_s\right]$$

$$\times \exp[+\kappa^2\int \phi^*\phi d^4x + g_{eff}^{1/2}\kappa\int d^4x(\bar{\psi}_\uparrow\bar{\psi}_\downarrow\phi + \psi_\downarrow\psi_\uparrow\phi^* + j^*\phi + \phi^* j)] \quad (2.2.8)$$

The partition function is given by

$$Z = Z[j,j^*]|_{j=j^*=0} \quad (2.2.9)$$

The action which has appeared in (2.2.8) is $Ld\tau$ where

$$L = L_0 + L_1 \quad (2.2.10)$$

$$L_0 = \int d\vec{x}\left[\bar{\psi}_s\left(\frac{\partial}{\partial\tau} - \frac{\vec{\nabla}^2}{2m} - \mu\right)\psi_s + \kappa^2\phi^*\phi\right] \quad (2.2.11)$$

$$L_1 = \int d\vec{x}\Big[-\frac{ie}{2m}(\bar{\psi}_s\vec{\nabla}\psi_s - (\vec{\nabla}\bar{\psi}_s)\psi_s)\vec{A} + \frac{e^2}{2m}\bar{\psi}_s\psi_s\vec{A}^2 +$$

$$+g_{eff}^{1/2}\kappa(\bar{\psi}_\uparrow\bar{\psi}_\downarrow\phi + \psi_\downarrow\psi^\uparrow\phi^*)\Big] \quad (2.2.12)$$

Using the temperature dependent Fourier decomposition

$$\psi_s(\vec{x},\tau) = (\beta V)^{-1/2}\sum_k \psi_s(\vec{k},n)e^{i(\vec{k}\cdot\vec{x}-\xi_n\tau)} \quad (2.2.13)$$

we obtain the free action

$$\int_0^\beta d\tau L_0 = \sum_{\vec{k},n}\left[\bar{\psi}_s(\vec{k},n)(-i\xi_n + \omega_k)\bar{\psi}_s(\vec{k},n) + \kappa^2\bar{\phi}^*(\vec{k},n)\bar{\phi}(\vec{k},n)\right] \quad (2.2.14)$$

Therefore, the propagator of the electron with four momentum $(\vec{k},n)$ is given by

$$\frac{1}{\omega_k - i\xi_n}, \quad \text{where} \quad \omega_k = \frac{\vec{k}^2}{2m} - \mu, \quad \xi_n = \frac{\pi}{\beta}(2n+1) \quad (2.2.15)$$

and the propagator of ϕ by

$$\frac{1}{\kappa^2} \quad (2.2.16)$$

Notice that the propagator of ϕ to this order does not depend on four momentum, because the Lagrangian (2.2.11) does not contain the kinetic energy term.

2.3. Calculation of the loop diagram:

We shall compute 1-loop diagram with 2 external ϕ-lines. The (temperature dependent) leading terms are

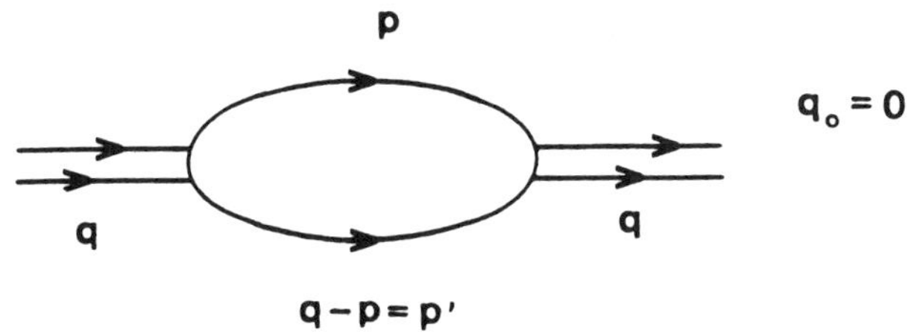

$$= \frac{g_{eff}\kappa^2}{\beta V} \sum_{n,\vec{p}} \frac{1}{(i\xi_n - \omega_p)(-i\xi_n - \omega_{p'})}$$

$$= A + B\vec{q}^{\,2} + \ldots \qquad (2.3.1)$$

where

$$A = \frac{g_{eff}\kappa^2}{\beta V} \sum_{n,\vec{p}} \frac{1}{\xi_n^2 + \omega_p^2} = \sum_n \frac{g_{eff}\kappa^2}{\beta} \frac{1}{(2\pi)^3} \int d^3\vec{p} \frac{1}{\xi_n^2 + \omega_p^2} \qquad (2.3.2)$$

Introducing density of states $\sigma(\omega)$ from the ansatz

$$\frac{1}{(2\pi)^3} \int d^3\vec{p} \equiv \int d\omega_p \sigma(\omega_p) \qquad (2.3.3)$$

we get

$$A = \sum_n \frac{g_{eff}\kappa^2}{\beta} \int_{-\infty}^{\infty} \frac{d\omega\sigma(\omega)}{\xi_n^2 + \omega^2} \approx \sum_n \frac{g_{eff}\kappa^2}{\beta} \sigma(0) \int_{-\infty}^{\infty} \frac{d\omega}{\xi_n^2 + \omega^2} \qquad (2.3.4)$$

where $\omega_p = \frac{p^2}{2m} - \mu = 0$ is the Fermi surface and we have approximated by using density of states at the Fermi surface by putting a cut-off n_{max}

$$\omega_D = k_B T \pi (2 n_{\max} + 1) \qquad (2.3.5)$$

Thus

$$A = g_{eff}\kappa^2 \sigma(0) \ln \frac{4\gamma\omega_D}{\pi k_B T}, \quad \text{where } \gamma = \text{Euler constant} \qquad (2.3.6)$$

The coefficient of $|\phi_c|^2$ terms $\Gamma[\phi_c, \phi_c^*]$ is then

$$\kappa^2 \left(1 - g_{eff}\sigma(0) \ln \frac{4\gamma\omega_D}{\pi k_B T}\right) \phi_c^* \phi_c \qquad (2.3.7)$$

The expression within the brackets gives the critical temperature as

$$T_c = \frac{4\gamma\omega_D}{\pi k_B} \exp\left(-\frac{1}{g_{eff}\sigma(0)}\right) \tag{2.3.8}$$

Note that the terms appearing before the logarithm in (2.3.7) appear in the exponent while terms which are arguments for the log appear as $A_D\omega$ in the formula for T_c.

Since the Higgs mechanism is the relativistic version of the above, one may write the entire calculation of one-loop diagrams as well as the calculation of T_c in the form of corrections to the calculation of mass of the Higgs field. We shall adopt this procedure with slight variations.

2.4. The point we wish to make is summarized by (2.3.8) which has the form $A\omega_D \exp\left(-\frac{2}{g_{eff}\sigma(0)}\right)$. In order that exp factor does not present too much of a restriction $\frac{1}{g_{eff}\sigma(0)}$ should be of the order of unity. This gives $T_c \approx \frac{\gamma\omega_D}{\pi k_B} \times \exp -2$.

Before proceeding, we shall make some remarks about the amino acids. The crystalline structure of amino acids is characterized by the graph of the crystalline structure of amino acids shown in figure 1 of Salam (1992).

The peptide bond which gives rise to proteins is formed, for example, where the molecules of water are consistently expelled and the lattice structure is reduced to the simple form

The hydrogens give up their loose electrons and act as metallic hydrogens. (Superconductivity has been established for a similar case by W.J. Carr (Carr, 1986)).

2.5. Helicity and chirality of the matter fields.

Let us review the definition of chirality, which is the eigenvalue of γ_5, with $\gamma_5 = +1$ corresponding to right-handedness, and $\gamma_5 = -1$ to left-handedness:

$$\gamma_5 R = R, \quad \bar{R}\gamma_5 = -\bar{R},$$

$$\gamma_5 L = -L, \quad \bar{L}\gamma_5 = \bar{L}, \tag{2.5.1}$$

where R, L are Dirac spinors with only two independent components. They may be obtained from a 4-component Dirac spinor ψ by the following projections:

$$R = \frac{1}{2}(1+\gamma_5)\psi, \quad \bar{R} = \frac{1}{2}\bar{\psi}(1-\gamma_5),$$

$$L = \frac{1}{2}(1-\gamma_5)\psi, \quad \bar{L} = \frac{1}{2}\bar{\psi}(1+\gamma_5). \tag{2.5.2}$$

For later applications, it is important to note that

$$\bar{\psi}\psi = \bar{L}R + \bar{R}L \,,$$

$$\bar{\psi}\gamma^\mu\psi = \bar{L}\gamma^\mu L + \bar{R}\gamma^\mu R \,. \tag{2.5.3}$$

written as

$$(\alpha \cdot \mathbf{p} + \beta m)\psi = E\psi, \quad E = (p^2 + m^2)^{1/2} \,. \tag{2.5.4}$$

Using the identity $\alpha = \gamma_5\sigma$, and the fact that γ_5 and σ commute, we can rewrite this in the form

$$\sigma \cdot \hat{\mathbf{p}} R = \frac{E}{p} R - \frac{m}{p} \beta L \,, \quad p = |\mathbf{p}| \,,$$

$$\sigma \cdot \hat{\mathbf{p}} L = -\frac{E}{p} L - \frac{m}{p} \beta R \,. \tag{2.5.5}$$

These equations become decoupled if $m = 0$:

$$\sigma \cdot \hat{\mathbf{p}} R = R$$

$$\sigma \cdot \hat{\mathbf{p}} L = -L \,. \tag{2.5.6}$$

Therefore, for massless Dirac particles, chirality is the same as helicity, for antiparticles, chirality is the opposite of helicity. [An antiparticle has the same chirality as the particle, by definition; but it has the opposite helicity due to a change in the sign of E in (2.5.4)].

A conventional mechanical mass term in the Lagrangian density cannot be invariant under $SU(2)$, because it is proportional to $\bar{\psi}\psi = \bar{L}R + \bar{R}L$. Therefore, in this theory the electron mass can arise only by virtue of a spontaneous breakdown of SU(2). A convenient way to implement this is to introduce a doublet Higgs field

$$\phi = \begin{pmatrix} \phi_+ \\ \phi_0 \end{pmatrix} \,, \tag{2.5.7}$$

where the subscripts refer to the electric charges, write the mass term as

$$\mathcal{L}^{\text{mass}} = \vartheta(\bar{L}\phi R + \bar{R}\phi^\dagger L(\frac{\rho}{\rho_0}) \,, \tag{2.5.8}$$

where $\bar{L}\phi$ is an $SU(2)$ singlet, and a Dirac spinor. If ϕ has non-zero vecuum value, then for low excitations and $\rho = \rho_0$, (2.5.8) is indistinguishable from a conventional mass term.

The Weinberg-Salam model is obtained by gauging $SU(2) \times U(1)$, generated by weak hypercharge.

With these, the covariant derivative can be written in the form

$$D^\mu = \partial^\mu + ig(W_1^\mu t_1 + W_2^\mu t_2) + ieQA^\mu + ieQ'Z_\mu \,, \tag{2.5.9}$$

Electric charge = eQ, $Q = \tau_3 + \tau_0$ where the neutral charge matrix Q' is defined by

$$eQ' = \tau_3 \cot \theta_w - \tau_0 \tan \theta_w . \tag{2.5.10}$$

Here τ_0 is the singlet matrix among $\tau_0, \tau_1, \tau_2, \tau_3$, while $Q = \tau_3 + \tau_0$.

To study the masses of the gauge fields, it is convenient to go to unitary gauge, in which

$$\phi = \begin{pmatrix} 0 \\ \frac{\rho}{\rho_0} \end{pmatrix} \tag{2.5.11}$$

where $\frac{\rho}{\rho_0}$ is a real field.

In terms of fields in the unitary gauge, the Lagrangian density is

$$\begin{aligned} \mathcal{L} = & -\frac{1}{4}(\mathbf{G} \cdot \mathbf{G} + \mathbf{H} \cdot \mathbf{H}) + \frac{1}{4} g^2 \rho^2 \left(W_1^2 + W_2^2 + \frac{Z^2}{\cos^2 \theta_w} \right) \\ & + \partial\rho \cdot \partial\rho \cdots \lambda(\rho^2 - \rho_0^2)^2 , \end{aligned} \tag{2.5.12}$$

Rewrite in the unitary gauge, where η is the real Higgs field in unitary gauge, in which $\phi(x)$ has the form (5.12) and

$$\rho(x) = \rho_0 + \eta(x) . \tag{2.5.13}$$

2.6. To compute any further, we must guarantee the g_{eff} in the effective Lagrangian is a positive number corresponding to attraction. Further, the proton left behind migrates to the nitrogen so that the Z^0 meson before being annihilated gives rise to an interaction between the quarks contained inside the proton and inside the nitrogen. (That it has to be nitrogen rather than the carbon (next door) is empirically guaranteed by the "right" configuration which is imparted to the corresponding molecule which contains sugars and no nitrogens).

To consider Z^0 containing part of the Lagrangian, write it as

$$L_{int}(Z^0) = \frac{eZ^0}{\sin\theta \cos\theta}[(T_{3L} - \sin^2\theta J_{em\ L+R})] . \tag{2.6.1}$$

Here J_{em} is the electromagnetic current, T_{3L} is the left-handed third component of the weak isospin consisting of the (anomaly-free) combination of the proton and neutron (p, n), (or the quarks inside the proton or the neutron). Z^0 here requires a mass because of the sponeaneous symmetry breaking implied by the last equation.

$$L^{\text{mass}} = g_{eff} \times (T_L^3 - (4 \sin^2\theta J_{em\ L+R})) \times L_{int} \times \frac{\rho}{\rho_0} \tag{2.6.2}$$

Here g_{eff} is proportional to $\frac{(e^2)}{16 \sin^2\theta \cos^2\theta} \times \frac{\eta_{\mu\nu}}{k^2 - m_Z^2}$ where $\eta_{\mu\nu} = 1, -1, -1, -1$. This is positive provided $m_Z^2 \gg k^2$ and we consider the 3 space-rotation invariant part of the expression above.

We shall take $(1 - 4\sin^2\theta) \approx \frac{1}{13}$ with the present *empirical value* of the parameter $\sin^2\theta \approx .231$. We shall not take this quantity to equal zero as has been done in (Tranter and MacDermott, 1989) and (Mason and Tranter, 1984) in the hope that the renormalization group will give the "exact" value as $\frac{1}{4}$ (personal communication). We do not believe this will ever happen.

For the mass term we take the ansatz, as the symmetry breaking which is given by Equation (2.2.4), which follows on from (2.2.5) and (2.2.7) where κ is a constant with dimension of mass. We shall introduce a source for the ϕ field in the form of Equation 2.2.8. In terms of the quantity ($\langle\varphi\rangle^0 = 250\,GeV$) the electron mass M_e turns out to be a very tiny number $\approx 2 \times 10^{-6}$. Such a number is large only for the top quark if its mass is in excess of 100 GeV. Thus the term which gives rise to this, looks like $f\varphi\bar{t}_L t_R$ where $f \geq \frac{2}{5}$. Some physicists (like Y. Nambu) take this as the defining property of the field φ, i.e. φ is considered as a $t\bar{t}$ composite.

According to Lee and Drell , a proposal is considered according to which the masses of the fermions in the standard model are determined by dynamical symmetry breaking rather than being introduced as arbitrary parameters in the Lagrangian, they are determined self-consistently by the requirement that the proper self-energy vanish the fermion mass shell. They find that in the one-loop approximation it is possible to generate a heavy top quark mass dynamically while the other fermions remain massless (Lee and Drell, 1991). We find that nonzero solutions for m_t do exist and they are always greater than 70 GeV for all values of cutoffs $\Lambda = 5m_z, 10m_z, 20m_z$ and is the only Higgs mass explored in the problem.

One may thus be led to conclude that, while a heavy top quark mass can be generated dynamically, the observed masses of the light fermion cannot be generated in this approach.

Recently, Ruiz-Altaba, Gonzalez and Vargas (Ruiz–Altaba *et al.*, 1989) extended this to a two-loop calculation predicting $m_t = 124$ GeV, $m_H = 234$ GeV and the weak mixing angle $\sin^2\theta_W \simeq .24$.

Our physical motivation is to remove the fermion masses arbitrary parameters of the Standard Model and treat them as parameters determined by the dynamics in the sub-TeV region. Rather than cancel the divergences, they impose and interpret the Standard Model $\mathcal{L}$ as an effective Lagrangian in order to determine the physical masses. Thus using Sakita's formulation, we get the result.

$$T_c = \frac{\langle\varphi\rangle}{10^3}\exp(-2/g_{eff}\sigma(1 - 4\sin^2\theta)) \approx 2.5 \times 10^2\ {}^\circ K \qquad (2.6.3)$$

The exponential factor gives $\exp -26 \approx 10^{-10}$. Assuming that $\sigma(0) \approx m_Z^2$ we obtain $g_{eff}\sigma(0) \approx 1$.

2.7. Now we go on to the supersymmetry case. The symmetry resolves the dilemma of heirarchy. In a theory possessing fermion–boson symmetry because of the presence of both fermionic and bosonic radiative loops, the radiative shifts for scalar fields are not quadratically divergent. In the limit of broken supersymmetry, the shifts in v^2 only depend logarithmically on the cutoff Λ.

To guarantee that $g_{eff}\sigma(0) \approx 1$, we must take some mass in the theory which is large enough. The simplest mass is obtained by taking supersymmetry such that

$$\phi_0 \rightarrow \phi_0 + \alpha[\Delta M^2]\, \ell n \frac{\Lambda}{\phi_0}, \qquad (2.7.1)$$

where ΔM^2 is the mass splitting between supersymmetry fermions and is of order 1 TeV (Amaldi *et al.*., 1991a; 1991b) which gives $\frac{\Delta M^2}{92 m_z^2} \approx 1$ and thus $g_{eff}\sigma(0)$ of the order of unity once again.

3. DISCUSSIONS AND CONCLUSIONS

3.1 We wonder if the laboratory testing of the central hypothesis of this note is feasible. This can be tested by taking a 50–50 racemic mixture of *crystalline* L and D-amino acids and lowering the temperature.

The crystalline conformations of L and D amino acids do not differ much from the zweitterionic forms except for the absence of water molecules inside the lattice cell. Barring for alanine, there are no definite calculations which show that it is indeed the L-configuration which prevails for these crystals. However this may not affect the laboratory testing of the hypothesis of this paper because lowering of the temperature below the transition temperature would eventually convert *all* amino acids into that particular configuration which *only* partially predominated before. *Conversely, a raising of temperature through T_c would show that a pure configuration below T_c can be converted into a racemic mixture for $T > T_c$.*

To conclude, note that ideally we should be able to compute the values of T_c when electroweak interaction theory is fully worked out. The numerical value of this quantity could lie anywhere, around 2.7°K (the ambient temperature of outer space) or beyond 350°K (but below the dissociation temperatures). *However, the best way to determine T_c for a given amino acid is at present* **experiment**. The analogy of the "superfluidity" exhibited by amino acids and sugars is to "superfluidity" in superconductors and not to the liquid helium. In the case of superconductivity, one has to apply an *external* magnetic field and look for the Meissner effect to determine T_c. Likewise, the "superfluidity" of amino acids (or sugars) is measured by shining on these *external* light sources.

One direct way to test for evidence of the hypothesis (regarding the existence of such a phase transition), is to lower the temperature while measuring the optical activity when polarized light is shone upon a particular amino-acid. If the polarization vector gets rotated, one may be sure that the appropriate phase transition has taken place.

The process itself could perhaps be detected by optical rotatory dispersion (ORD) or circular dichroism (CD) . An alternative means of detecting the process may be by measuring differences of specific heats and looking for anomalies in the curve $C = \gamma T + \beta T^3 + \ldots$ like what has been recently done for the non-amino acids like melanins and tumour melanosomes (Mizutani *et al.*, 1976).

3.2 This paper started with the possibility of defining a science of life based on atomic physics. What we have shown is that if the model is Z_0 it is possible to have the quark instead of the nucleons. The mass of Z_0 is $\approx$ 100 GeV, while the mass of the quarks is $\approx$ 200 GeV so that if Z_0 loops are considered the $t\bar{t}$ quarks go through these loops. The g_{eff} which is proportional to m_Z^{-2} then is of the right magnitude to be able to let these quarks make composites which are manifested through their passage through these loops.

3.3 We have shown that chirality may provide a boundary condition for theories of chemical evolution and the origin of life and that Z^0 interactions – as well as what comes beyond the standard model of fundamental interactions (CP violation, for example) – ought to play a central role in this story. In this context, the following quotation from L. Pasteur (who did not even know of Z^0 particles) is perhaps prophetic: "Life as manifested to us is a function of the asymmetry of the Universe and of the consequences of this fact. The Universe is asymmetrical. Life is dominated by asymmetrical actions. I can even imagine that all living species are primordially in their structure, in their external forms a function of cosmic asymmetry." (Pasteur, 1860).

REFERENCES

Abrikosov, A.A. 1987: Fundamentals of the Theory of Metals, North Holland, pp.345–347.

Ali, A. and Hoodbhoy, P. (Eds.), 1991: M.A.B. Beg Memorial Volume, World Scientific, Singapore.

Amaldi, U., de Boes, W. and Furstenau, H., 1991a: Comparison of grand unified theories with electroweak and strong coupling constants measured at LEP, Phys. Rev., B260, 447–455.

Amaldi, U., de Boes, W. and Furstenau, H., 1991b: Supersymmetry, Science, 253, 1075–1076.

Arnold, P. and McLerran, L., 1987: Sphalerons, small fluctuations and baryon–number violation in electroweak theory, Phys.Rev., D36, 581–595.

Atkins, K.R., 1959: Liquid Helium, Cambrige University Press, p.42.

Avetisov, V.A., Kus'min, V.V. and Anikin, S.A., 1987: Sensitivity of chemical chiral systems t weak asymmetric factors, Chem. Phys., 112, 179-187.

Carr Jr., W.J., 1986: Theory of superconductivity based on direct electron–phonon coupling – I, Phys. Rev., B33, 1585–1600.

Chyba, C.F., Thomas, P.J., Brookshaw, L. and Sagan, C., 1990: Cometary delivery of organic molecules to the early Earth, Science, 249, 366–373.

Ginzburg, V.L. and Landau, L.D., 1950: On the theory of superconductivity, Soviet Phys.JETP, 20, 1064. Reprinted in (Ter Haar, 1965).

Goldanski, V.I., 1987 referred to in the paper by V.A. Avetisov et al. (cf., Avetisov et al., 1987).

Goodstein, D.L., 1985: States of Matter, Dover Publications, New York, pp. 386-391.

Gorkov, L.P., 1958: On the energy spectrum of superconductors, Soviet Phys. JETP, 7, 505–508.

Gorkov, L.P., 1959: Microscopic derivation of the Ginzburg–Landau equations in the theory of superconductivity, Soviet Phys. JETP, 9, 1364–1367.

Harris, M.J., Loving, C.E. and Sandars, P.G.H., 1978: Atomic sheilding and PNC optical rotation in bismuth. J. Phys., B11, L749–L753.

Jack, I. and Jones, D.R.T., 1990: Naturalness without supersymmetry?, Phys. Lett., B234, 321–323.

Kondepudi K.D. and Nelson, C.W., 1985): Weak neutral currents and the origin of biomolecular chirality, Nature, 314, 438–441.

Landau, L.D., Lifschitz, E.M. and Pitaevskii, L.P., 1980: Statistical Physics, Part II, Volume 9, Pergamon Press.

Lee, I–Hsiu and Drell, Sidney D., 1991: Fermion masses in the standard model, in Ali and Hoodbhoy pp.13–20.

Leggett, A., 1990: The New Physics Ed. P. Davies, Cambridge University Press, p. 276.

Mason, S.F. and Tranter, G.E., 1984: The parity–violating energy difference between enantiomeric molecules, Molec. Phys., 53, 1091–1111.

Mizutani, U., Massalski, T.B., McGiness, J.E. and Corry, P.M., 1976: Low temperature specific heat anomalies in melanins and tunor melanosomes, Nature, 259, 505-507.

Oró, J., 1961: Comets and the formation of biochemical compounds on the primitive Earth, Nature, 190, 589.

Orò, J., Miller, S.L. and Lazcano, A., 1990: The origin and early evolution of life on Earth, Annu. Rev. Earth Planet Sci., 18, 317–356.

Orò, J., Squyres, S.W., Reynolds, R.T. and Mills, T.M., 1990: Europa: Prospects for an ocean and exobiological implications, NASA Spec. Publ., (in press, 1990).

Pasteur, L., 1860: Researches on the Molecular Asymmetry of Natural Organic Products, Alembic Club Reprint, No. 14, Livingston, Edinburgh and London.

Ruíz–Altaba, Gonzalez, B. and Vargas, M., 1989: CERN–TH 5558/89, UGVA–DPT 89/11–638 (unpublished); see also discussion in (Jack and Jones, 1990).

Sakita, B., 1985: , Quantum Theory of Many-Variable Systems and Fields, World Scientific, Singapore.

Salam, Abdus, 1991: The role of chirality in the origin of life, J. Mol. Evol., 33, 105–113.

Salam, Abdus, 1992: Chirality, phase transitions, and their induction in amino acids, Phys.Lett., B288 153–160.

Sanchez, R., Ferris, J. and Orgel, L.E., 1966: Conditions for purine synthesis: Did prebiotic synthesis occur at low temperatures?, Science, 153, 72–73.

Ter Haar, D. (Ed.), 1965: Collected papers of L.D. Landau Pergamon Press, Oxford, pp.546–568.

Tranter, G.E. and MacDermott, A.J., 1989: Electroweak Bioenantioselection, Croatica Chem. Acta, 62, (2A) 165–187.

SPONTANEOUS REGULATING MECHANISMS THAT MAY HAVE LED TO THE ORIGIN OF LIFE

Julian Chela-Flores
International Centre for Theoretical Physics,
Miramare P.O. Box 586; 34100 Trieste, Italy
and
Instituto Internacional de Estudios Avanzados,
Apartado 17606 Parque Central, Caracas 1015A,Venezuela.

ABSTRACT

According to Salam condensation may be relevant in biochemistry, as a factor contributing to the homochirality of amino acids. We attempt to show that DNA packaging may be modelled, by interpreting chromatin as a form of soft matter, in which a phase transition has induced chromatin into a condensed mode. In the context of the origin of life we show the relevance of simultaneous discussion of DNA packaging, transcription, and DNA replication. Beyond a certain critical length of the protogenome (RNA), physical properties of inert condensed matter may have given rise to a spontaneous regulating mechanism of certain significance for the evolution of life on Earth.

1. INTRODUCTION

In attempting to describe plausible scenarios for the origin and early evolution of life, several stages have been considered amongst them we may mention the following:

-The origin of the eukaryotic cell
-The origin of natural selection.

In the present work we comment on the second and the last of these aspects of the origin of life. While the former is, according to Margulis and coworkers, presumably of rather recent occurrence the latter is perhaps one of the first difficulties that had to be overcome at the interface of biological and chemical evolution. A common (genetic) theme shall be applied to the discussion of both problems

Similarly, in chemical evolution many stages have been discussed, amongst them we would like to mention only:

-The origin of chirality of amino acids

We restrict our comments to the origin of chirality of amino acids, in view of the conjectures that have been formulated recently by Salam (1990; 1991; 1992). We feel that the main novelty of Salam's work in theoretical

ISBN 0-937194-31-X

biology is his appeal to physical condensation in the sense of Bose and Einstein .

The possibility that phenomena analogous to physical condensation might occur in living systems had been anticipated by Delbrück (1963) and Fröhlich (1969). While Delbrück conjectured the relevance of condensation in general terms, both Fröhlich and Salam share their interest in biochemistry, in the cell membrane and amino acids, respectively. The essence of the analogy used by Salam is that the 'superfluidity' exhibited by amino acids is to 'superfluidity' in superconductors and not to liquid helium. In view of the considerable theoretical difficulties still to be overcome, perhaps the best way to determine T_C may be by means of experiments.

The novelty of Salam's approach lies in the fact that he has been led to conjecture that Bose-like condensation -*a cooperative phenomenon* - is a further factor contributing to biochirality. In spite of some difficulties still to be overcome in Salam's rationalization of the first principle of biochemistry, the proposal of molecular cooperative phenomena in a biological context deserves further consideration (Chela-Flores, 1991).

We feel that the analogy between Bose condensation and biological phenomena may also be fruitfully discussed, in particular in the context of genetics, and from the point of view of the origin of life. This is the subject that shall occupy our attention in this work.

However, it may be worthwhile to point out that in all of these approaches to chemical evolution and the origin of life, we have to face a major difficulty:

Studies of sedimentary rocks may not provide direct evidence of the origin and earliest evolution of life. In fact, we cannot rule out the existence of traces of life from the oldest specimens now available (Schildowski, 1988). More precisely (Schopf *et al*, 1983), we may recall that it is widely accepted that living organisms dating 3.5 Gigayears before the present (Gyrbp) have left fossils in the Warrawoona Group of Western Australia, as well as in the sediments of the Overwacht Group in South Africa. This situation induces us to search for earlier evidence of life in the even older Isua Supracrustal Belt of Western Greenland. The sedimentary rocks retrieved from this region date from the interface of the Hadean/ Early Archean suberas, about 3.8 Gyrbp.

The data indicates that the molecular structure of the Isua kerogen (a mixture of organic substances lacking regular chemical structure) has been severely altered, and that the carbonates (minerals such as dolomite and limestone) have been metamorphosed, rendering impossible a clear estimate of their initial isotopic composition. To sum up, it is not yet possible to exclude either of the following two options:

(a) Even if the isotopic composition of carbonates was due to their biotic origin, their source may have been from autotrophic or heterotrophic ecosystems.

(b) The Isua kerogen may have been of abiotic origin.

1.1 ALL LIVING SYSTEMS CONTAIN WITHIN THEM EVIDENCE OF THEIR HERITAGE.

Nevertheless, the biochemical and molecular genetics of all living systems contain within them evidence of their heritage. Thus, comparative metabolic, biochemical, and genetic studies of extant organisms may provide important insights into the nature of the early stages of evolution, which are beyond the reach of geologist's research (Gest and Schopf, 1983). This argument invites studies on biochemistry and genetics as a possible means of studying life's origins. Many clear examples of biological properties that may have been preserved over significant time periods may be cited, amongst them:

(i) The first is provided by the molecular mechanisms of transcriptional regulation in yeast. The mechanism of transcriptional initiation, in particular, is substantially conserved throughout the eukarya domain. Important illustrations are provided, for instance, by vertebrate transcription factors, which are known to fuction in yeast cells, and by the fact that yeast and mammalian TATA factors are functionally interchangeable (Struhl, 1989).

(ii) A second example of a mechanism preserved over geologic time is the remarkable homologies observed in heat-shock proteins extending from bacteria to man (Lindquist, 1986).

A third example will now be considered more carefully, since it will serve a useful purpose in our scenario of life's origin.

1.2 THE RELATIONSHIP BETWEEN TRANSCRIPTION, DNA PACKAGING, AND DNA REPLICATION.

We have some insights into the underlying mechanisms of gene silencing, which is observed in mammals. Since the most frequent mode of gene control is at the transcriptional level (Darnell, 1982), the important phenomenon of X chromosome inactivation is probably due to inhibition of RNA synthesis (transcription). Thus the remarkable structural alterations in the heterochromatic X chromosome may be directly related to the regulation of transcription (Chela-Flores, 1987).

The above relationships in gene expression should be seen together with an intimate relationship that is observed between transcription and DNA replication. Remarkable examples of such coupling are known in molecular genetics (DePamphilis, 1988).Thus we have been led to study the consequences of assuming:

1.2.2 Hypothesis 1

The packaging of DNA with structural proteins is correlated with the way in which DNA can interact with and, hence, respond to the strongly coupled transcriptional and replication signals.

On the strength of this working hypothesis, based on an earlier one by Sperling and Weiss (1980), we have made a study of the *chromatin repeat length* (λ), which is the average amount of DNA wound around a core of histones.

What is new in the above assumption is the hypothetical exclusion of factors other than DNA replication and transcription in the correlated function of gene expression and replication with DNA packaging. One important possibility that our hypothesis leaves out is a further coupling with the nuclear membrane (lamin proteins and other factors).

To be more precise, the λ length parameter consists of two segments:

(i) The first one (*L*) consists of two full turns of DNA wound about the histone core, contributing a fixed number of base pairs to the overall value of λ.

(ii) The second segment is called the linker DNA (λ). It is a variable length of DNA; its average length varies with species, cell type and developmental stage. It links adjacent nucleosomes, contributing to the flexibility of the chromatin fiber.

We may summarize the above information in terms of a simple formula:

$$\lambda = L + l$$

On the basis of Hypothesis 1 we have been able to obtain some insight into experimental results (Jaegger and Kuenzle, 1982) that show variations of λ for the same type of cells (neurons) at different stages in ontogeny (Chela-Flores, 1992a):

The temporal evolution of the repeat length is a decreasing function of time in mammalian cortical neurons, whereas it is an increasing function of time for granule cells around the time of birth.

A certain consensus is now emerging in the sense that *chromatin is an essential part of the transcriptional mechanism* (Felsenfeld, 1992): Instead of the once held opinion that the chromatin most prominent DNA-binding proteins (the histones) played mainly structural roles, the view is now emerging that nucleosomes are not merely inducing DNA bending, but rather they are biochemically active complexes. Likewise, λ should be envisaged as a dynamic parameter to be understood in its intimate relationship with the transcription process. This implies that a relationship between chromatin structure at the nucleosome level and transcription seems possible.

The support for this approach is based on a selection of values of λ given in table 1 for a wide range of eukaryotes, from sac fungi and true slime molds to cells from a variety of organisms ranging from protists to echinoderms and chordates (Chambon, 1978; Godde and Widom, 1992):

TABLE 1. THE CHROMATIN REPEAT LENGTH FOR A SAMPLE OF CELLS

CELL TYPE	λ IN BASE PAIRS
Schizosaccharomyces pombe	156
Aspergillus nidulans	158
Neurospora	170
Physarum	171, 173
Tetrahymena micronucleus	175
Rat fetal liver	193
Rat liver	196, 198
Syrian hamster liver	196
Chick oviduct	202
Tetrahymena macronucleus	202
Rabbit cerebellar neuron	200
Chicken erythrocyte	207, 212
Sea urchin gastrula	218
Sea urchin sperm	241

In other words, this parameter ranges from 154 to 241 bps. However, what is more interesting to notice is that for a given cell, λ also changes at different stages of ontogeny (cf., the examples of the rat liver cells).

1.3. GENETIC STUDIES OF EXTANT ORGANISMS MAY PROVIDE INSIGHTS INTO THE EARLY STAGES OF EVOLUTION

General genetic relationships have been identified in extant organisms. Some examples are the ones we have just described above in the case of eukaryotic cells (i.e., gene silencing and gene expression in the neurons of the embryonic brain). These relationships, if extrapolated back to the early stages of evolution, may provide important insights into some aspects of the origin of life which, as we have seen, are beyond the reach of geological research.

To sum up, we are led to:

1.3.3 Hypothesis 2

We conjecture that the proposed coupling of DNA packaging, transcription, and DNA replication may be extrapolated back to the early stages of the evolution of life on Earth.

2. CAN GENETICS BE MODELLED WITH PHYSICAL ANALOGIES?

As stated above, an important insight into basic molecular mechanisms of genetics was anticipated by Delbrück by suggesting that both, biological systems as well as simpler condensed systems (such as inert condensed matter), may exhibit the following property:

In spite of being exposed to a supply of external energy, a physical system itself will not respond (i.e., the energy that has been supplied to it is either stored or dissipated). Such system will only respond when the energy available is larger than a certain threshold, but the system may respond to subthreshold energies if some external effect contributes to destroy some underlying order or structural property. Physicists (quantum liquid specialists) and biologists (geneticists) are familiar with such phenomena:

Quantum liquid theorists call these phenomena superconductivity, or superfluidity. For the sake of brevity we limit ourselves to superconductivity. Electrons may not be excited from their ground state with arbitrarily small energies (at least in metals and alloys at low temperatures, where the microscopic mechanisms are well understood). This situation changes when the excitation energy is larger than a threshold value, which is called an 'energy gap'. However, the presence of certain magnetic fields may disrupt the order of motion that has originally given rise to the phenomenon.

Geneticists, as we have seen above, call such phenomena gene silencing. Free energy is supplied by metabolism. Normally, in eukaryotic cells wastes from fermented food molecules in the cytoplasm enter the mitochondria and pass through a cycle of reactions, which produce most of the ATP for both the mitochondria and the rest of the cell, in particular for the chromosomes inside the nuclear membrane. In spite of free energy being supplied by metabolism, a whole X chromosome remains silent (Lyons, 1961). Thus, in female mammals two identical X chromosomes respond differently to metabolic energy; the Barr body normally does not respond and remains silent, but this situation is reversible as follows from recent experiments of Lyon and coworkers in aging female cats (Wareham *et al.*, 1987).

Also very early in embryogenesis the silencing of the Barr body is reversible (Migeon *et al*, 1986). In the case of genetics the analogous concept is called the *chromatin packaging* or, alternatively, *chromatin condensation*, but we prefer to reserve the term condensation to the physical phenomenon studied in physics by Bose and Einstein. We may conclude that analogies such as those of Delbrück, Fröhlich, and Salam may have some heuristic value in genetics.

2.1 CHROMATIN IS A FORM OF CONDENSED MATTER, COMPARABLE IN DENSITY WITH SOME LIQUIDS AND SOLIDS

Chromatin packaging may be studied in terms of the variable η parameter. Tight packing of chromatin yields DNA concentrations ρ within localized regions of **interphase nuclei** given by (Olins and Olins, 1974)

$$\rho \approx 200 \ mg / ml$$

Significantly higher values are expected for **metaphase chromosomes**, since the η parameter may be one or two orders of magnitude larger. This value is of the same order of magnitude as the upper bound for the densitiy of **liquid helium four** (London, 1954a):

$$\rho \approx 145 \ mg / ml$$

where the phenomenon of condensation is known to occur. Another example is **liquid helium three**, where the phenomenon of superfluidity also occurs at low temperatures at even a lower value of ρ (80 mg / ml; London, 1954b). On the other hand, the packing ratio η (degree of condensation) of chromatin is given by the formula (Finch and Klug, 1976):

$$\eta = L_1 / L_2$$

where L_1 denotes DNA length in the fully extended state, and L_2 denotes the length of the coiled, or folded state, achieved at any state of condensation. In the **nucleosome** we have 165 base pairs (bps) of DNA rolled over a histone core; in this case,

$$L_1 \approx 600 \ \text{Å} \qquad \text{and} \qquad L_2 \approx 55 \ \text{Å}$$

Thus,

$$\eta \ (\text{nucleosome}) \approx 10$$

2.2 CAN GENE EXPRESSION BE MODELLED USING THE ANALOGY WITH PHYSICAL CONDENSATION?

In a search for some quantitative bases for the problem of the origin of natural selection, we may turn to a study of genetics in which we considered some regularities that have been observed in the rate of propagation of DNA polymerase (r_f) and RNA polymerase (r_t). We have remarked that some data suggests the linear relation (Chela-Flores, 1992b):

$$r_f = \mu r_t$$

where μ is a constant. We have shown that interpreting Hypothesis 1 in terms of the physical analogy suggested earlier (in different contexts) by Delbrück, Fröhlich, and Salam, allows us to study the linear formula in a thermodynamic formulation.

This reinforces our arguments in favour of the analogy between genetics and condensed matter, which may be seen in this light as a valuable working hypothesis for approaching quantitative aspects of DNA packaging, and gene expression, in the early Earth.

3. ON THE ORIGIN OF NATURAL SELECTION

Ultraviolet (UV) radiation played a dual role at the origin of life. In the first place it provided ample free energy to drive a variety of chemical reactions that were needed in order to generate the earliest replicators. The effect of UV radiation on the proto-genome may be discussed in two stages:

(i) As we have discussed above, in the second half of the Early Archean, some 3,5 Gyrbp, prokaryotic cells definitely existed (cf., for example the dome-like calcareous structures, the stromatolites, discovered three decades ago in the hypersaline waters of Western Australia, an environment which is hostile to its predators). Such organisms may have used light in an anoxygenic atmosphere. Seeking protection from UV rays by burying themselves in the mud was no solution, since they required light. The solution may have been to seek filters, or UV defense mechanisms (UVDMs).

For instance, they may have lived in solutions rich in certain salts, so that UV radiation may have been absorbed, but visible light may have been let in. Another strategy may have been to develop mechanisms for repairing sun-damaged DNA (i.e., thymine dimers formed by the UV effect on DNA). Such repair enzymes occur in contemporary organisms, removing the unnecessary chemical units that UV energy may generate.

(ii) Besides from the rigorous discussion of Isua kerogen and carbonates, we cannot exclude the existence of life during the latter part of the Hadean subera (earlier than 3.9 Gyrbp). Whatever conditions we conjecture for the Hadean environment cannot be confirmed by geologic data, since sedimentary rocks from this earliest subera have not been preserved.In order to address the question of how to rationalize the UVDM in terms of molecular relics, rather than geological data, we reconsider DNA packaging

3.1 THE UV PARADOX

Candidates for UVDMs have to be found for the protection of the very products of the UV-driven reactions (the early replicators). Otherwise, fast mutation rates would be unavoidable, due to the large doses of UV rays delivered by the early Sun (less than one billion years old) to the surface of the Earth. In such conditions it is not possible to have slow mutation rates, which are needed for natural selection to take place. This is a statement of the UV paradox as formulated by Sagan (1965). One hint as to how to rationalize this paradox may be found in a contemporary genome, namely that of a **viroid:**

This is a pathogenic stretch of nucleic acid which succeeds to be resistant to an UV-rich environment. In fact, a viroid is a nucleic acid species without capsid: Its molecular weight is relatively small, in fact, smaller than prokaryotic RNA polymerase ($1.1 - 1.3x10^5$). Its length can be as small as 240 bps. They can be isolated from certain organisms affected by disease. Viroids cannot be detected in healthy organisms of the same species. They have been extensively studied by Diener (1979).

Nevertheless, viroids may be injected into an organism; for instance, into a dicotyledonous plant by means of the purpule bacterium Agrobacterium tumefaciens. Then autonomous replication of the nucleic acid follows, giving rise to a characteristic syndrome. A specific example is the potato viroid (potato spindle tuber viroid) PSTV, which is a firm double helix structure arising from base pairing amongst its 359 bps. With the resulting degree of packing this viroid presents an elusive target for inactivation by means of UV radiation. From Hypothesis 2, we may conclude that by modulating DNA packaging the proto-gene may have been sufficiently short, to have turned them into elusive targets to UV rays.

3.2 AN ILLUSTRATION

A simple order-of-magnitude estimate may clarify the above arguments on the PSTV resistance to the UV radiation. The length of a typical prokaryotic gene, for instance rDNA of *Escherichia coli,* is given by:

$$L_1 = 6 \ kb,$$

If η assumes the value typical of a contemporary 30 nm fiber

$$250 < \eta < 400,$$

then the length of the coiled state would satisfy the following inequality:

$$L_2 < 150 \ nm$$

However, as we know, during the Hadean subera the far UV window

$$150 \ nm < \lambda_{UV} < 300 \ nm$$

was the source of 3,400 cal/cm^2/yr, as opposed to 2 cal/cm^2/yr for wavelenghths smaller than 150nm (Sagan, 1965). We have denoted the UV wavelength as λ_{UV}.

In other words, we have considered a plausible prokaryotic gene with an equally plausible packing density (Chela-Flores, 1985). This protogene presents a target for UV light which is in the range of wavelengths corresponding to the primordial Sun exposure. We have considered particularly the most damaging UV window, where the most abundant flux of energy was being delivered through the early anoxygenic atmosphere.

Hence, just as in the case of the potato viroid (PSTV), which is known today to present an elusive target to UV radiation, the protogene may have escaped fast mutation rates *purely on structural basis, independent of its environment.* On this basis, natural selection may have been the plausible trigger to the early evolution of a protogenome, provided the total length was somewhat longer than the above example, and the packing density was somewhat higher than the corresponding value considered.

To sum up, Hypothesis 2 suggests that the origin of natural selection - independent of any given pattern of transcription and replication - may be understood in terms of DNA packaging of a short (viroid-like) stretch of nucleic acid (either DNA or RNA). In this scenario the phenomenon of condensation in nucleic acids may have provided a rationale for the UV paradox. Hypothesis 1 and 2 encourage us to extend their implications to other problems confronting our understanding of the origin of life.

4. THE SPONTANEOUS ORIGIN OF THE MAIN REGULATING MECHANISMS OF GENE EXPRESSION

Our next priority is to rationalize in terms of standard chemistry and physics **the spontaneous origin of the main regulating mechanisms of gene expression:** DNA replication, transcription and DNA packaging. Subsequently, the question of the origin of translation and the genetic code must also be addressed, but this has been done adequately by other researchers. In this work we have considered physical condensation, in the same sense as it was used by Salam in the context of chemical evolution.

We are suggesting that condensation may have played the role of a regulating mechanism during the course of biological evolution, and that understanding such a mechanism may be relevant at the macromolecular level to both natural selection and symbiosis.

Some modest progress in the direction of clarifying the origin of regulatory mechanisms is possible by adopting the view that cells are *environments* for macromolecules which may migrate into a host cell from a wide variety of sources.

Foreign DNA is capable of crossing the membrane of the host cell. For instance, DNA may invade either prokaryotic cells (such a *Daptobacter*), or eukaryotic cells (such as the ancestral cryptomonad) in the form of plasmids or transposons. Once the holobiont is formed, regulating mechanisms may be set in motion. These may eventually control the interactions between the associated bionts.

One extreme example of such interaction is to be found in insects: the leafhopper *Euscelidius* is a genera of large homopterans. Schwemmler has studied the symbiogenesis in insects as models for cell differentiation, morphogenesis, and speciation (1991). He found that if the symbionts of *Euscelidius* of one generation were prevented from infecting the eggs, only head-thorax embryos are produced (i.e., no abdomen is produced).

4.1 A GENERAL HYPOTHESIS

To sum up, at the present stage of our understanding of the interconnections of chromatin structure and function, suggested by our hypothesis 1, we cannot make much progress in some important problems concerning symbiosis such as pinning down the genetic mechanisms that may predispose organisms for symbiosis, but a general hypothesis nevertheless may be useful as a frame for our discussions:

4.1.1 Hypothesis 3.

In eukaryotic organisms that succeed to establish symbioses with each other, the host and its cytobiont alter their metabolism in such a way that the consequent alterations in gene expression are characterized by λ-parameter changes in both the endosymbiont's DNA-, as well as host's native DNA-packaging.

In the protozoan *Tetrahymena* nuclei, which presumably may have been the result of early symbioses, have different λ parameters. This characterizes different patterns of gene expression for both its macronucleus and the micronuclei. Symbiosis between two eukaryotes are known to occur in a species of freshwater alga called cryptomonad - the flagellated protist *Cryptomonas* ϕ (Douglas *et al* , 1991).

Cryptomonads contain an unusual membranous structure, which is sometimes also called a "nucleosome". This organelle contains both DNA and RNA. The DNA is more similar to that of red algae than that of cryptomonad nucleus. Indeed the ancestral cryptomonad may have become photosynthetic by the incorporation of red algae containing chloroplasts.

We may recall that the accepted view is that chloroplasts were originally photosynthetic bacteria. The recent work of Douglas and coworkers leads us to the view that some eukaryotes obtained their chloroplasts from other eukaryotes: An ancestral eukaryote may have been the host of another - a red algae. The holobiont is an organism containing several genomes, but the novelty is that two of them should be complexes of DNA and proteins. In fact, in deciding the origin of cryptomonad, the standard approach of comparing homologies between the genomes of different eukaryotes may, for the above reasons, be supplemented with a further test involving the λ parameter. However, in so doing we are not binding ourselves to a given gene (rRNA) of the eukaryotic nucleus in order to address questions of phylogeny, but rather we are binding ourselves to global properties - expressed by the λ parameter - of both genomes (endosymbiont and host), in order to discuss questions of phylogeny.

5. DISCUSSION AND CONCLUSIONS

We have illustrated with chromatin repeat lengths of the mammalian brain that global statements of the structure of the genome may be rationalized, at least in some cases when transcription and replication can be separated. We are suggesting, in agreement with Hypothesis 1 that the λ parameter may be correlated with the changing patterns of gene expression for different cells. New experiments on the repeat lengths of both nuclear genomes of cryptomonad algae are, therefore, suggested. The pattern of λ parameter variations should be characteristic of the symbiotic partners, before they entered into mutualistic dependence.

It seems premature to investigate the qualitative and quantitative implications of the genetic mechanisms that seem to underlie symbiosis; for that purpose it may be necessary to model DNA packaging, as we have illustrated above, in the as yet unresolved problem of polymerase dynamics. This remains to be done.

We have aimed at clarifying a few questions:

(i) We have restricted ourselves to the genetic level and have included the possibility of *molecular* cooperation (i.e., collective phenomena) as in the example studied by Salam. In this manner we have shown the relevance in the context of the origin of life of simultaneous discussion of DNA packaging, transcription, and DNA replication.

(ii) Physical properties of inert condensed matter, such as condensation, may have provided a spontaneous regulating mechanism for the evolution of life on Earth, which may be tested in contemporary genomes, for instance, in models such as the one we have discussed in relation with polymerase dynamics. The present work may be interpreted as some support for Salam's conjecture in his chemical evolution context.

(iii) For the discussion of questions of phylogeny in organisms which are the result of endosymbiosis of two eukaryotes, we are not necessarily restricting ourselves to a given nuclear gene (rDNA), but rather we have indicated in this context the relevance of global properties of both genomes (endosymbiont and host).

ACKNOWLEDGMENTS

The author would like to thank Professor Abdus Salam for discussions on the origin of chirality.

REFERENCES

Chambon, P., 1978: Summary: The Molecular Biology of the Eukaryotic Genome is Coming of Age, Cold Spring Harbor Symp. Quant. Biol., 42, 1209-1234.

Chela-Flores, J., 1985: Evolution as a collective phenomenon, J. Theor. Biol., 117, 107-118.

Chela-Flores, J., 1987: Towards a collective biology of the gene, J. Theor. Biol., 126, 127-136.

Chela-Flores, J., 1991: Comments on a novel approach to the role of chirality in the origin of life, Chirality, 3, 389-392

Chela-Flores, J., 1992a: Influence of Chromatin Molecular Changes on RNA Synthesis during Embryonic Development, Acta Biotheoretica 40,41-49.

Chela-Flores, J., 1992b: Towards the Molecular Basis of Polymerase Dynamics, J. Theor. Biol., 154, 519-539, and Erratum J. Theor. Biol., 157, 269.

Darnell, J.E., 1982: Variety in the level of gene control in eukaryotic cells, Nature, 297, 365-371.

Delbrück, M., 1963: In Conmemoration of the 50th. Anniversary of Niels Bohr's Papers on Atomic Constitution, Session on Cosmos and Life. Institute for Theoretical Physics, Copenhaguen, 41-67.

DePamphilis, M.L., 1988: Transcriptional Elements as Components of Eukaryotic Origins of DNA Replication, Cell, 52, 635-638.

Diener, T.O., 1979: Viroids: Structure and function, Science, 205, 859-866.

Douglas, S.E., C.A. Murphy, D.F. Spenser, and M.W. Gray, 1991: Cryptomonad algae are evolutionary chimaeras of two phylogenetically distinct unicellular eukaryotes, Nature, 350, 148-151.

G. Felsenfeld, 1992: Chromatin as an essential part of the transcriptional mechanism, Nature, 355, 219-224

Finch, J.T., and A. Klug, A., 1976: Solenoidal model for superstructure in chromatin, Proc. Natl. Acad. Sci. USA, 73, 1897-1901.

Fröhlich, H., 1977: Long-range coherence in biological systems, Rivista del Nuovo Cimento, 7, 399-418.

Gest, H., and J.W. Schopf, 1983: Biochemical evolution of anaerobic energy conversion: The transition from fermentation to anoxygenic photosynthesis. In Schopf, 1983, 135-148.

Godde, J.S. and Widom, J., 1992: Chromatin Structure of *Schizosaccharomyces pombe* A Nucleomatosome Repeat Length that is Shorter than the Chromatosomal DNA Length, J. Mol. Biol. 226, 1009-1025.

Jaegger, A.W., and C.C. Kuenzle, 1982: The Chromatin Repeat Length of Brain Cortex and Cerebellar Neurons Changes Concomitant with Terminal Differentiation, EMBO J., 1, 811-816.

Lindquist, S., 1986 : The heat-shock response, Ann. Rev. Biochem., 55, 1151-1191.

London, F., 1954a: Superfluids, 2. Dover,New York, 2.

London, F., 1954b: Superfluids, 2. Dover, New York, 172.

Lyon, M.F., 1961: Gene action in the X-chromosome of the mouse (*Mus musculus L.*), Nature, 190, 372-373.

Migeon, B.R., M. Schmidt, J. Axelman, and C.R. Cullen, 1986: Complete reactivation of X chromosomes from human chorionic villi with a switch to early DNA replication, Proc. Natl. Acad. Sci. USA, 83, 2182-2186.

Olins, A.L.& Olins, D.E., 1974: Spheroid chromatin units (ν bodies), Science, 183, 330-332.

Sagan, C., 1965: In S.W. Fox (Ed.), The origins of prebiological systems, Academic Press, New York, 207.

Salam, A., 1990: Unification of the fundamental forces. Cambridge: Cambridge University Press. pp. 56-58.

Salam, A., 1991: The role of chirality in the origin of life, J. Mol. Evol., 33, 105-113.

Salam, A., 1992: Chirality, phase transitions, and their induction in amino acids, Phys. Lett. B288, 153-160.

Schidlowski, M., 1988: A 3,800-million-year isotopic record of life from carbon in sedimentary rocks, Nature, 333, 313-318.

Schopf, J.W. (Ed.), 1983: Earth's earliest biosphere its origin and evolution. Princeton University Press, Princeton.

Schopf, J.W., J.M. Hayes, and M.R. Walter, 1983: Evolution of Earth's Earliest Ecosystems: Recent Progress and Unsolved Problems. In Schopf (ed.), 1983, 361-384.

Schwemmler, W., 1991: Symbiogenesis in Insects as a Model for Morphogenesis, Cell Differentiation, and Speciation. In L.Margulis and R. Fester (Eds.), Symbiosis as a source of Evolutionary Innovation, The MIT Press, Cambridge, Mass, 178-204.

Sperling, L., and M.C. Weiss, 1980: Chromatin Repeat Length Correlates With Phenotypic Expression In Hepatoma Cells, their Didifferentiated Variants, and Somatic Hybrids, Proc. Natl. Sci. Acad. USA , 77, 3412-3416.

Struhl, K., 1989: Molecular mechanisms of transcriptional regulation in yeast, Ann. Rev. Biochem., 58, 1051-1077.

Wareham, K.A., M.F. Lyon, P.H. Glenister, and E.D. Williams, 1987: Age-related reactivation of an X-linked gene, Nature, 327, 725-727.

CHIRALITY AND THE ORIGINS OF LIFE[1]

R. Navarro-González[2], R.K. Khanna and C. Ponnamperuma

Laboratory of Chemical Evolution
Department of Chemistry and Biochemistry
University of Maryland, College Park, Maryland 20742 U.S.A.

ABSTRACT

The origins of life and of chirality are integrally related. Numerous theoretical and experimental attempts have been made over the years to understand this issue and basically two different hypotheses on the origin of chirality have originated: Abiotic and Biotic. A recent quantum mechanical treatment of the parity violating weak interactions predicts a spontaneous phase transition from *d*- to *l*- form of amino acids and, therefore, supports the abiotic hypothesis. In an effort to gain insight into these problems several analytical, computational and synthetic studies have been carried out. This report presents summaries of the results obtained in our laboratories and elsewhere and their implications on the origin of chirality.

1. INTRODUCTION

One important feature of all living organisms is that they are built of asymmetric building block molecules: *l*-amino acids (making up the proteins and enzymes) and *d*-sugars (making up the back bone of DNA and RNA). The origin of chirality in the biosphere must be integrally related to the origin of life itself (Ponnamperuma, 1974); however, it is not yet

---

[1]This work was partly supported by a NASA grant NAGW-637.

[2]Instituto de Ciencias Nucleares, Universidad Nacional Autónoma de México, Circuito Exterior, C.U., Apartado Postal 70-543, México D.F. 04510, México.

ISBN 0-937194-31-X

clear how chirality in the biosphere arose and what the driving force was to push the decision in just one direction. The origin of chirality has been a fundamental problem since the period of Pasteur. Numerous experimental and theoretical attempts have been made over the years to understand this issue (for reviews see Bonner, 1972, 1991; Quack, 1989; Ponnamperuma *et al.*, 1990; Thiemman, 1990). Basically two different hypotheses; namely, "Abiotic" and "Biotic" have been put forward.

1.1 ABIOTIC MECHANISMS

It is postulated that molecular homochirality arose before the origin of life by either accidental or deterministic mechanisms (for a comprehensive review see Bonner, 1991). Under the accidental mechanisms it is proposed that, even though, symmetry-breaking processes at the molecular level have equal probabilities of producing an excess of either enantiomer, a chance event resulted in preferential incorporation of *l*-amino acids and *d*-sugars in the life essential molecules. The suggested mechanisms include:

(a) Autocatalytic processes. In a racemic system, an optically active compound may be both a catalyst for its own production and an anti-catalyst for the production of its enantiomer (Frank, 1953; Calvin, 1969).

(b) Chiral crystallization. Crystallization from a supersaturated solution of a racemic compound may produce racemic crystals or less commonly conglomerates (mixtures of pure crystals of each enantiomer). In this mechanism it is suggested that one enantiomer may crystallize preferentially (Bonner, 1972).

(c) Asymmetric synthesis in chiral crystals. A racemic reactant, in which enantiomers are interconvertible in the liquid state, crystallizes as conglomerates with crystals of each enantiomer. These crystals then undergo changes to form a nonracemizable chiral product which leads to additional product of the same chirality (Green and Heller, 1974).

(d) Asymmetric adsorption on quartz. Quartz, often found in nature as well defined enantiomorphic crystals, may lead to asymmetric adsorption in racemic solutions (Karagounis and Coumoulos, 1938; Bonner *et al.*, 1975).

(e) Asymmetric adsorption on clays. Clays have no chirality associated with their crystal structures and thus should not promote stereoselective interactions with chiral molecules; nevertheless, asymmetric effects have been claimed (Degens *et al.*, 1970; Bondy and Harrington, 1979; Julg, 1990).

Deterministic mechanisms suggest that some non-random physical forces interacting with racemic or prochiral substances leading to asymmetric syntheses or degradations resulted in chiral products. The deterministic mechanisms include:

(f) Regional and/or temporal processes. These involve external forces whose intrinsic asymmetry may alter the direction or magnitude at different locations on the Earth or at different epochs. Among the forces that have been studied include electric, magnetic and gravitational fields and circularly polarized light (Bonner, 1991).

(g) Universal processes. These involve external (parity violation) and internal (weak interactions) forces that persisted in the Universe in an invariant manner during its entire history (Vester *et al.*, 1959; Salam, 1991).

1.2 BIOTIC MECHANISMS

It is postulated that life originated in all possible chiral configurations by chemical and prebiological evolution in a racemic primordial soup. Chiral selection and chiral homogeneity were assumed to be the inevitable consequences of the evolution of life (Morozov, 1979). This hypothesis has generally been ignored.

2. EXPERIMENTAL AND THEORETICAL STUDIES

In our laboratory we have been interested in understanding the origins of life and of chirality, and have used analytical, computational and synthetic approaches to gain insight into these problems. In the analytical approach, we have been able to go back in time and look at the records of ancient rocks and/or sediments from the Earth and from extraterrestrial bodies such as the Moon, Mars, and meteorites. In the computational approach we have investigated the likelihood of comets as carriers of intact homochiral biomolecules from interstellar space. In our synthetic approach we have carried out studies related to (1) synthesis of small molecules by various forms of energy under plausible primitive Earth conditions, (2) interaction among small molecules with inorganic matrices, (3) asymmetric inclusion compounds, (4) search for a *d*- to *l*-phase transition, (5) association between amino acids and nucleotides, and (6) synthesis and properties of asymmetric biomacromolecules. The purpose of this paper is to describe our results and attempts to relate them to other studies which will help in understanding the origins of life and of chirality.

2.1 SEARCH FOR ORGANIC COMPOUNDS IN THE SOLAR SYSTEM

Of paramount importance in the understanding of the processes that led to the beginning of life is the information that can be gathered from the study of organic matter in the solar system. In this regard we have searched for evidence of chemical evolutionary processes in the Moon, Mars, and meteorites.

Analysis of Lunar samples (Chang *et al.*, 1971; Chang and Kvenvolden, 1972; Gehrke *et al.*, 1987) and Martian rigolith (Biemann *et al.*, 1976 and 1977; Horowitz *et al.*, 1976) showed that organic molecules were present in very low abundance. No attempt was made to examine the chirality of the molecules present.

Meteorites are sources of extraterrestrial material amenable to laboratory study on Earth. In 1969 the Murchison meteorite which landed in Australia was analyzed within days. The first analysis established unequivocally that six amino acids commonly found in proteins, and eight which do nor occur in natural proteins, were present in the meteorite (Kvenvolden *et al.*, 1970). Alanine, aspartic acid, glutamic acid, proline, and valine were present in approximately equal amounts of the *d*- and *l*-isomers (Figure 1a). Similar results have now been obtained with other meteorites such as the Murray (Figure 1b, Lawless *et al.*, 1971), the Mighei, the Alan Hills, and the Yamato (Ponnamperuma, 1972).

The presence of non-biological amino acids and almost equal amounts of the *d*- and *l*-enantiomers minimizes the possibility of terrestrial contamination, and suggests a possible extraterrestrial origin. The question immediately arises as to their origin. Two possibilities present themselves:

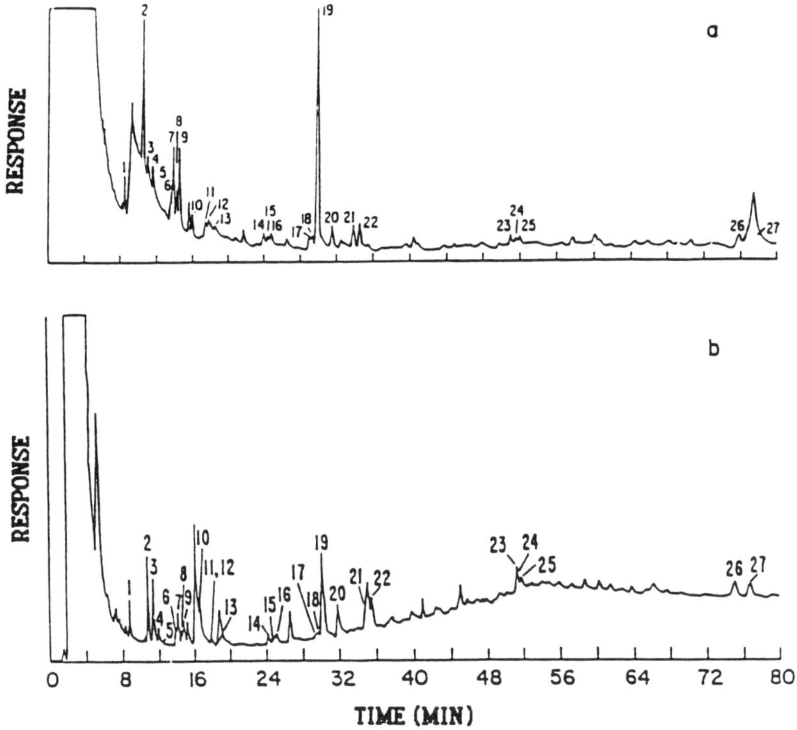

Fig. 1. Gas chromatograms of N-trifluoroacetyl-*d*-2-butyl esters of amino acids in acid-hydrolyzed aqueous extracts of Murchison (a) and Murray (b) meteorites (Ponnamperuma, 1972). Identification for the samples is a follows: (1) (*d,l* ?)-Isovaline; (2) α-Amino-isobutyric acid; (3) *d*-Valine; (4) *l*-Valine; (5) (*d,l* ?)-N-Methylanine (?); (6) *d*-Amino-n-butyric acid; (7) *d*-α-Alanine; (8) *l*-α-Amino-n-butyric acid; (9) *l*-α-Alanine; (10) N-Methylglycine; (11) N-Ethylglycine (?); (12) *d*-Norvaline (?); (13) *l*-Norvaline (?); (14) *d*-β-Aminoisobutyric acid (?); (15) *l*-β-Aminoisobutyric acid (?); (16) (*d,l* ?)-β-Amino-n-butyric acid; (17) *d*-Pipecolic acid; (18) *l*-Pipecolic acid; (19) Glycine; (20) β-Alanine; (21) *d*-Proline; (22) *l*-Proline; (23) γ-Amino-n-butyric acid; (24) *d*-Aspartic acid; (25) *l*-Aspartic acid; (26) *d*-Glutamic acid; (27) *l*-Glutamic acid.

1) these amino acids were existing at some period of time in the meteorite in one stereoisomeric form and were then racemized in the course of time and 2) the two forms were always present in nearly equal amounts (Kvenvolden *et al.*, 1971). Recently it has been reported that *l*-alanine exceeded *d*-alanine by about 18% in the Murchison Meteorite (Engel *et al.*, 1990). The ^{13}C content of each enantiomer indicates an extraterrestrial origin. This excess of the *l*-alanine, and presumably that of other *l*-amino acids reported earlier by Engel and Nagy (1982) imply asymmetry in the distribution of optically active materials in the early Solar system. However, these unique findings have yet to receive decisive independent corroboration.

2.2 COMETS AS CARRIERS OF INTACT HOMOCHIRAL BIOMOLECULES FROM INTERSTELLAR SPACE

It has been suggested that circularly polarized synchrotron ultraviolet radiation from remnants of a supernova explosion interacted with interstellar grains from the presolar nebula producing racemic compounds in the organic mantles of interstellar grains (Bonner and Rubenstein, 1990). Furthermore, it has also been suggested that comets were the carriers of extraterrestrial sources of molecular homochirality from interstellar space to the primitive Earth (Bonner, 1991). Under this scenario it is possible that the ionizing radiation resulting from the decay of radionuclides distributed throughout the comet nuclei may have led to the racemization of the biomolecules possibly present in the interiors of comets.

We have recently developed a computer model that enabled us to estimate the radiation chemical processing of cometary nuclei (Navarro-Gonzâlez *et al.*, 1992). In this study we examined the possible degree of radiation-induced destruction of homochiral amino acids and carboxylic acids. If these compounds were present in comets at abundances comparable to those found in carbonaceous chondrites, 60 and 15 ppm, respectively (Cronin *et al.*, 1988), and if alanine and lactic acid were the only representatives of these classes of organic compounds at such abundances, we predict that the degree of radiation-induced racemization and destruction is expected to be about 0.2%, if only ^{40}K, ^{235}U, ^{238}U and ^{232}Th are considered (Table 1). If we now include the additional contributions of ^{244}Pu, ^{129}I, ^{247}Cm, ^{10}Be, ^{237}Np and ^{26}Al, then the degree of radiation-induced racemization and destruction could increase by as much as 12.6% These figures do not change significantly if the abundance of amino- and hydroxy-acids is increased by two orders of magnitude. Based on these data we conclude that comets are possible carriers of intact homochiral biomolecules from interstellar space.

TABLE 1. % Destruction of alanine and lactic acid in the center of comet nuclei.

Compound	Abundance (ppm)	% Destruction		
		Case 1*	Case 2**	Case 3***
Alanine	60-6000	0.2	2.2	12.6
Lactic acid	15-1500	0.2	2.2	12.6

*Case 1: ^{40}K, ^{235}U, ^{238}U and ^{232}Th.
**Case 2: ^{40}K, ^{235}U, ^{238}U and ^{232}Th, ^{244}Pu, ^{129}I, ^{247}Cm, ^{10}Be, ^{237}Np.
***Case 3: ^{40}K, ^{235}U, ^{238}U and ^{232}Th, ^{244}Pu, ^{129}I, ^{247}Cm, ^{10}Be, ^{237}Np and ^{26}Al.

2.3 PREBIOTIC SYNTHESIS

We have synthesized a number of prebiotically important molecules such as hydrocarbons, aldehydes (Honda, *et al.*, 1989), carboxylic acids (Negrón-Mendoza, *et al.*, 1983; Honda, *et al.*, 1989), amino acids (Kobayashi and Ponnamperuma, 1985), and nucleobases (Ponnamperuma *et al.*, 1963 a,b) in experiments that simulate possible environments of the primitive Earth. All possible isomers are usually produced in prebiotic synthesis, e.g., n-pentane, isopentane and 2-methyl-butane (Ponnamperuma and Woeller, 1964).

Enantiomeric analysis of amino acids produced in electric discharge experiments was reported in the presence or absence of clays (Shimoyama *et al.*, 1978). In this study the relative abundance of enantiomers was determined by their peak areas in the gas chromatograms. The results showed that *d*- and *l*-enantiomers were formed in equal amounts (Figure 2). The ratios varied from 48 % to 52 % for alanine and α-amino-n-butyric acid. This variation was also obtained for the same sample when *d*- and *l*-stationary phases were used. Discrepancies of this magnitude are frequently observed for authentic *d*- and *l*-amino acid samples. Therefore, it is reasonable to conclude that the amino acids are produced in equal

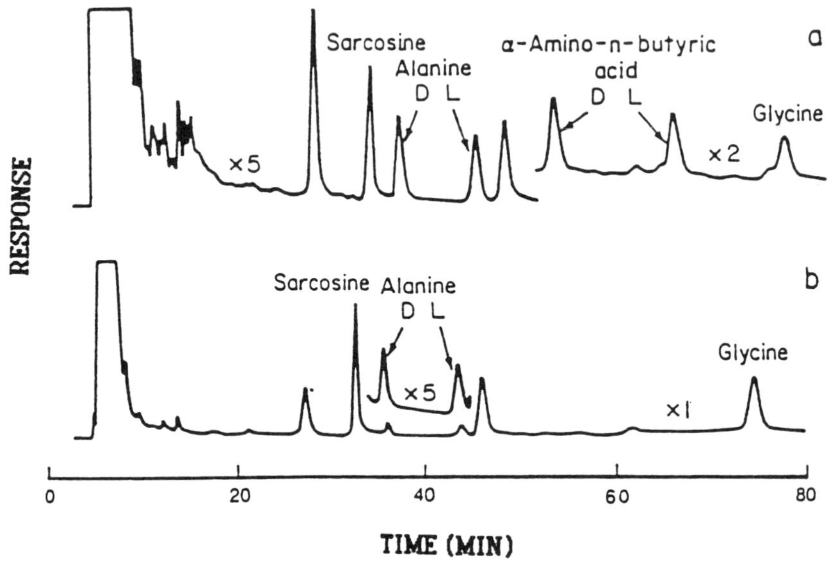

Fig. 2. Gas chromatogram of amino acid enantiomers produced in the electric discharge from a simulated atmosphere-hydrosphere environment in the presence (a) or absence (b) of Na-montmorillonite (shimoyama, et al., 1978).

amounts. The presence or absence of clay in the systems did not affect in the enantiomeric synthesis of amino acids.

Further investigations are needed to extend other chemical systems and energy sources to examine any selective synthesis of *l*- or *d*-enantiomers. Nonetheless, it seems likely that wherever an organic compound with an asymmetric center is formed almost equal amounts of *l*- and *d*-isomers appear to be present in the absence of an asymmetric medium.

2.4 SELECTIVE ADSORPTION OF ENANTIOMERS IN MINERAL MATRICES

One of most interesting possible roles of clays in chemical evolution which was suggested by Bernal in 1951 is the preferential adsorption of certain biological monomers. Many investigators have studied the adsorption of biologically important organic compounds in clays (Ponnamperuma *et al.*, 1982). Selective adsorption of optical isomers by clays has also been of interest in the study of the origin of asymmetric molecules. In order to determine whether clays could have played a part in the selection process of *l*-amino acids over *d*-forms in prebiotic times, adsorption of several amino acids in Na-montmorillonite was studied at various pH's, see Table 2 (Friebele *et al.*, 1980). Gas chromatography analyses showed that the *l*-enantiomeric form of amino acids was in 0.5-2.0% more adsorbed in the clay than the *d*-form. However enantiomeric analyses of the same samples by amino acid analyzer yielded results which were not consistent in all cases with gas chromatographic data (Table 2). The results suggest that apparently clays have no significant ability to sort out *l*- and *d*-amino acids under the experimental conditions. Our findings demonstrate that clays may not have played an important role in the enantiomeric selection of amino acids in the process of chemical evolution.

TABLE 2. *d:l* Ratio of Amino Acids Adsorbed on Na-Montmorillonite as Determined by Gas-Liquid Chromatography*

	pH		
Amino Acid	3	7	10
α-Alanine	1.022	0.976	0.935
		1.001**	0.983**
α-Aminobutyric acid	1.005	0.996	0.925
Norvaline	0.986	0.979	0.986
Valine	0.977	0.996	1.015

*Friebele *et al.*, 1981. Reprinted by permission of Kluwer Academic Publishers.

**Analysis determined by amino acid analyzer.

2.5 UREA AND ASYMMETRIC INCLUSION COMPOUNDS

Inclusion compounds are molecular aggregates wherein one or more "guest" molecules are contained within the open spaces in a crystalline or molecular framework made up of "host" molecules. Urea yields a number of crystalline inclusion complexes with a variety of straight chain aliphatic compounds, provided their molecules contain 6 or more carbon atoms (Bonner, 1972). Such compounds crystallize as hexagonal lattices of urea molecules surrounding parallel channels of about 5-6 Å in diameter, within and along which the included guest molecules reside (see figure 3, Cramer 1955). Note that the urea molecules in the hexagonal lattice (dark circles in figure 3) may be arranged in the chiral sequences of either a right-handed or left-handed helix.

Formation of urea inclusion compound of any suitable guest substance ordinarily affords crystals whose lattices are of either right-handed (R) or left-handed (S) lattices. Which lattice forms initially is determined by chance, but the original crystal nuclei act as seeds for the remainder of the product. If a racemic substance ($\pm$A) is included as guest molecules in a urea lattice, four stereochemically different crystalline products (I-IV) may arise (Bonner, 1972):

(+)A-(R) lattice	(+)A-(S) Lattice	(-)A-(R) Lattice	(-)A-(S) Lattice
I	II	III	IV

These products have properties of enantiomers (pairs I-IV and II-III) or diastereomers (pairs I-III and II-IV). If the inclusion compounds by chance crystallizes in the (R) lattice, the products are limited to I and III. If product I happens to be the least soluble diastereomer it will crystallize preferentially, leaving product III in solution. These principles of novel resolution technique have been used successfully by Schlenk, Jr. (1952a,b), who resolved racemic 2-chlorooctane in 95% yield by repeated formation of its urea complex and using the initial formed nuclei as seed crystals. This undoubtedly represents a possible mechanism for the prebiotic appearance of optical activity on the early Earth (Schlenk, 1952; Bonner, 1972) since urea is readily formed under a variety of prebiotic conditions (Navarro-González *et al.*, 1989). Moreover, It is even more plausible since we have found that urea is extremely stable to degradation under presumed prebiotic conditions (Navarro-González *et al.*,1989); therefore,

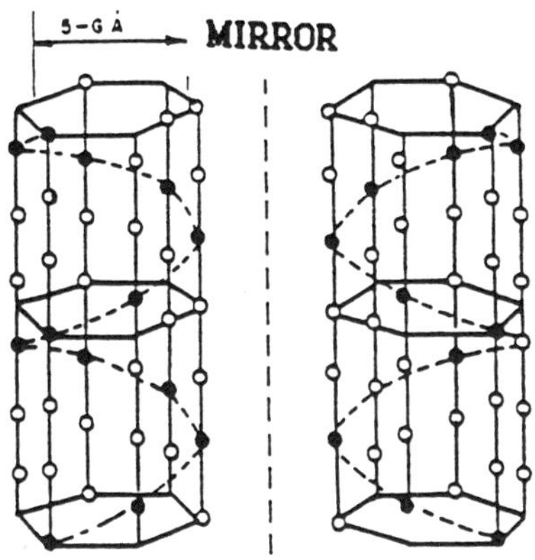

Fig.3 . Crystal lattice in urea inclusion compounds (Bonner, 1972). Only host molecules (○ ●) are shown.

it is likely that urea gradually accumulated in the primitive hydrosphere, where it could have played a role in the chiral selection of biomolecules.

2.6 SEARCH FOR A *d*- TO *l*-PHASE TRANSITION

Salam (1991) has recently suggested that the origin of chirality in nature may have arisen from a second-order phase transition (including *d*- to *l*-transformation in amino acids) below a critical temperature T_c. This prediction is based on an analogy with superconducting phase transition arising due to a quantum mechanical cooperative and condensation phenomena.

This phase transition is suggested to account for the observed amplification of the ground state population of the *l*-amino acids. The difference between these energies of the *d*- and *l*-enantiomers, $\sim 3.1 \times 10^{-19}$ eV, is due to the parity-violating electroweak interaction. The proposed phase transition is suggested to be characterized by a T_c which may be studied theoretically by extending the standard electroweak theory to include either the top quark or supersymmetry or experimentally (Chela-Flores, 1991). In an effort to test for this *d*- to *l*-phase transition, the following experiments were performed:

(1). 10 g of a racemic alanine sample were dissolved in 100 ml water and its optical rotation was measured with a polarimeter (accuracy of ~ 0.0005 deg). The optical rotation of this sample was 0.0005 deg. Other 10 g of the sample from the same batch were sealed in an evacuated quartz tube, immersed in liquid helium and kept at 4.2 K for 5 days. After taking the sample out of the Dewar flask, it was warmed up to room temperature, dissolved in water, and the optical rotation measured.

(2). The above experiment was repeated with the sample except that this time, the sample was heated to 503 K (just below the decomposition temperature of alanine).

(3). A saturated solution of alanine in water was slowly evaporated to obtain single crystals. The idea was to make sure that the impurities were removed to a maximum extent. A collection of these crystals were subjected to the same experiment as in (1).

In all three cases, the optical rotation of the sample after cooling did not change beyond the limits of uncertainty. We can conclude from these results that at temperatures as low as 4.2 K or as high as 503 K no transformation from *d*- to *l*- (to the extent measurable by optical polarimeter- estimated to be less than 1 part in 100,000) was detected. It is quite likely that a temperature much lower than 4.2 K would be required for the transition. Under equilibrium conditions, even at 4.2 K, the fractional excess of *l*- over *d*-molecules is 1 in 10^{16}, on the basis of the energy difference of $\sim 10^{-19}$ eV. A similar situation exists in the difference between the conductor and superconductor states; and a cooperative condensation phenomena results in an observable superconducting phase transition, and a decrease in electrical resistance by 6-7 orders of magnitude is noted. The problem to overcome in optical rotation measurements is the small magnitude of the rotation. Some techniques, for example, circular dichroism, use of intracavity laser for the amplification of the rotation are essential.

2.7 INTERACTION BETWEEN AMINO ACIDS AND NUCLEOTIDES

The origin of nucleic acid directed protein synthesis is one of the principal problems in the study of the origin of living systems. In contemporary organisms, the genetic information recorded in DNA is first transcribed into mRNA, and the nucleotide sequence of mRNA is then translated into the amino acid sequence of proteins. Each group of three consecutive nucleotides defined as a "codon" in the mRNA, is read as a signal for the incorporation of a particular amino acid. To achieve this, amino acids are attached to a special class of adaptor molecules, the tRNAs, each of which carry a special trinucleotide sequence known as the "anticodon" by synthetase. The question to be answered concerns the physical basis by which nucleic acid components could contain information regarding amino acids, thereby initiating a process that led to the evolution of this current elaborated system of coding and translation. Speculations on the origin of the genetic code began even before the code was deciphered (Pauling and Delbruck, 1950; Gamow, 1954; Nirenberg and Mattaei, 1961; Leder and Nirenberg, 1964). At one extreme of these speculations is the direct interaction theory which states that the codon assignments are based on preferential physical and chemical interactions between amino acids and component nucleotides of the genetic code. This is amenable to experimental verification as it may be presumed that traces of those early selective interactions are even now discernible, though they may not play any role in the contemporary system.

Since the deciphering of the genetic code, stereochemical models for the origin of the genetic code have fallen primarily into two classes, that is, amino acid-codon and amino acid-anticodon interaction models (Ponnamperuma and Hobish, 1984). Studies on the association of amino acids and nucleotides have been done in our laboratory to understand the origin of the codon assignments. Measurements of the changes in the chemical shifts by NMR in an amino acid-nucleotide mixture allows quantitative determinations of the binding process.

The association of homocodonic amino acids (*l*-Phe, *d*-Phe, *l*-Lys, *l*-Pro, and Gly) with nucleotide 5'-monophosphates (AMP, UMP, GMP, and CMP) was analyzed. *l*-Homocodonic amino acids have higher association constants with anticodonic nucleotides than with codonic nucleotides. In addition, anticodonic nucleotides have low association constants with *d*-homocodonic amino acids than with *l*-homocodonic amino acids (Senaratne, 1986; Ponnamperuma *et al.*, 1992). These results provide a possible physical and chemical bases for the origin of the genetic code. Further work is needed to look at the interactions between *l*-nucleotides and *d*-amino acids to learn more on the role of asymmetry in the origin of the genetic code of living organisms.

2.8 SYNTHESIS AND PROPERTIES OF ASYMMETRIC BIO-MACROMOLECULES

We have been interested in examining the effect of enantiomeric inversion of amino acid residues on protein structure, and have formed a synthetic fragment complex whose properties provide valuable insights into the relationships between structure and function of polypeptides of biological interest. In this respect, one of most well-characterized systems is bovine pancreatic ribonuclease S, a stable and functionally active complex consisting of S-peptide (residues 1-20) and S-protein (residues 21-124). An all-*l*-model S-peptide, H-Ala-Glu-Ala_4-Lys-Phe-Ala-Arg-Ala-His-Met-Ala_2-OH, have been previously synthesized and shown to form a stable complex with S-protein, exhibiting 36% of the specific activity of fully native ribonuclease S against the substrate, cytidine 2',3'-cyclic monophosphate, *d*-cCMP, (Komoriya and Chaiken, 1982). In order to examine the effect of a defined enantiomeric sequence on protein structure, the all-*d*-model S-peptide was synthesized. In the presence of S-protein, the all-*l*-S-peptide has hydrolytic activity against the substrate *d*-cCMP, whereas the all-*d*-peptide was inactive. Addition of the all-*d*-peptide to mixtures of *l*-peptide and S-protein did not lead to inhibition of enzymatic activity. These results indicate lack of binding of *d*-peptide to S-protein to produce either an active or inactive species (Corigliano-Murphy *et al.*, 1985). To get a better insight into the catalytic activity of an all-*d*-model ribonuclease S, we would need to synthesize an all-*d*-model S-protein and to assay its activity against *l*-cCMP as substrate.

A different study has recently shown that the *d*- and *l*- forms of the enzyme HIV-1 protease (HIV-1P) have reciprocal chiral specificity on peptide substrates (Milton *et al.*, 1992). That is, each enzyme enantiomer cuts only the corresponding substrate enantiomer. These data also imply that the folded forms of the *d*- and *l*-HIV-1P have similar chemical properties and are mirror images of one another in all elements of the three-dimensional structure.

3. CONCLUSIONS

While considering the problem of the origins of life, it is necessary to give special attention to the origin of chirality. Our attempts to understand both problems have been analytical, computational and synthetic approaches. The analysis of meteorites had indicated the presence of racemic mixtures of amino acids, although more recently, there is a report which showed an excess of *l*-alanine from the Murchison meteorite. While this work needs independent corroboration, it suggests that optically active materials were already present in the early phase of the solar system. In our computational approach we have shown that comets could be carriers of intact homochiral biomolecules from interstellar space to the primitive Earth. Our synthetic approach has shown that wherever there is an organic compound with an asymmetric center, it forms in racemates in the absence of an asymmetric medium. Our studies on the selective adsorption of enantiomers by clay minerals indicate that they do have no significant ability to sort out *l*- and *d*-amino acids. Our studies on the stability of urea indicate that such a compound may have gradually accumulated in the early oceans where it may have played a role in the chiral selection of biomolecules by forming asymmetric inclusion compounds. Search for a *d*- to *l*-phase transition has produced negative results in the temperature range between 4.2 to 503 K. Studies on the interaction between amino acids and nucleotides have provided possible physical and chemical basis for the origin of the genetic code. These results indicate that anticodonic nucleotides have higher association constants with *l*-homocodonic amino acids than with the *d*-forms. Further work is, however, needed to investigate the interactions between *l*-nucleotides and *d*-amino acids. We have also examined the effect of enantiomeric inversion of amino acid residues on protein structure and function and found that both enantiomeric polypeptides are similar chemically and conformationally, *e.g.*, they are mirror images of one another in all elements of the three-dimensional structure. Our knowledge of the origin of chirality in nature is still limited and further research is needed before we can uncover the origins of life and of chirality.

REFERENCES

Bernal, J.D., 1951: The Physical Basis of Life, Routledge and Kegan Paul, London, 34.

Biemann, K., Oró, J., Toulmin, P., Orgel, L. E., Nier, A.O., Anderson, D.M., Simmonds, P.G., Flory, D., Diaz, A.V., Rushneck, D.R., and Biller, J.A., 1976: Search for Organic and Volatile Inorganic Compounds in Two Surface Samples from the Chryse Planitia Region of Mars, Science, 194, 72.

Biemann, K., Oró, J., Toulmin, P., 3rd, Orgel. L.E., Nier, A.O., Anderson, D.M., Simmonds, P.G., Flory, D., Diaz, A.V. Rushneck, D.R., Biller, J.E. and Lafleur, A.L., 1977: The Search for Organic Substances and Inorganic Volatiles Compounds in the Surface of Mars., J. Geophys. Res., 82, 4641.

Bondy, S.C. and Harrington, M.E., 1979: *l*-Amino Acids and *d*-Glucose Bind Stereoespecifically to a Colloidal Clay, Science, 203, 1243.

Bonner, W.A., 1972: Origins of Molecular Chirality, in C. Ponnamperuma (ed.), Exobiology, North-Holland, Amsterdam, 170.

Bonner, W.A., 1991: The Origin and Amplification of Biomolecular Chirality, Origins of Life Evol. Biosph., 21, 59.

Bonner, W.A., Kavasmaneck, P.R., Martin, F.S. and Flores, J.J., 1975, Asymmetric Adsorption by Quartz. Model for the Prebiotic Origin of Optical Activity, Origins of Life, 6, 367.

Bonner, W.A., and Rubenstein, E., 1990, Photochemical Origins of Biomolecular Chirality, in C. Ponnamperuma and F.R. Eirich (eds.), Prebiological Self Organization of Matter, A. Deepak Publ. Co, 35.

Calvin, M., 1969, Chemical Evolution. Molecular Evolution Towards the Origin of Living Systems on Earth and Elsewhere, Oxford University Press, New York, 151.

Chang, S. K., Kvenvolden, K., Lawless, J., and Ponnamperuma, C., 1971: Carbon, Carbides, and Methane in an Apollo 12 Sample, Science,

171, 474.

Chang, S. and Kvenvolden, K. A., 1972: Distribution and Significance of Carbon Compounds on the Moon, in C. Ponnamperuma (ed.), Exobiology, North-Holland, Amsterdam, 400.

Chela-Flores, J., 1991, Comments on a Novel Approach to the Role of Chirality in the Origin of Life, Chirality, 3(5) (in press).

Corigliano-Murphy, M.A., Liang, X., Ponnamperuma, C., Dalzoppo, D., Fontana, A., Kanmera, T. and Chaiken, I.M., 1985: Synthesis and Properties of an All-D Model Ribonuclease S-Peptide, Int. J. Peptide. Protein Res., 25, 225.

Cramer, F., 1955: Inclusion Compounds, Rev. Pure Appl. Chem., 5(3), 143.

Cronin, J.R., Pizzarello, S. and Cruikshank, D.P., 1988: Organic Matter in Carbonaceous Chondrites, Planetary Satellites, Asteroids and Comets, in J.F. Kerridge and M.S. Matthews, *Meteorites and the Early Solar System*, The University Press, Tucson, 819.

Degens, E.T., Matheja, J. and Jackson, T., 1970: Template Catalysis: Asymmetric Polymerization of Amino Acids on Clay Minerals, Nature, 227, 492.

Engel, M.H., and Nagy, B., 1982: Distribution and Enantiomeric Composition of Amino Acids in the Murchison Meteorite, Nature, 296, 847.

Engel, M.H. Macko, S.A., and Silfer, J.A., 1990: Carbon Isotope Composition of Individual Amino Acids in the Murchison Meteorite, Nature, 348, 47.

Frank, F.C., 1953: On Spontaneous Asymmetric Synthesis, Biochim. Biophys. Acta, 11, 459.

Friebele, E., Shimoyama, A., and Ponnamperuma, C., 1980: Adsorption of Protein and Non-Protein Amino Acids on a Clay Mineral: A Possible Role of Selection in Chemical Evolution, J. Mol. Evol., 16, 269.

Friebele, E., Shimoyama, A., Hare, P. E., and Ponnamperuma, C., 1981: Adsorption of Amino Acid Enantiomers by Na-Montmorillonite, Origins of Life, 11, 173.

Gamow, G., 1954: Possible Relation between Deoxyribonucleic Acid and Protein Structure, Nature, 173, 318.

Gehrke, C.W., Zumwalt, R.W., Kuo, K., Ponnamperuma, C., and Shimoyama, A., 1987: Search for Amino Acids in Apollo Returned Lunar Soil, in C.W. Gehrke, K.C.T. Kuo, and R.W. Zumwalt (eds.), Amino Acid Analysis by Gas Chromatography, CRC Press, Boca Raton, Florida, Vol. II, 151.

Green, B.S. and Heller, L., 1974: Mechanisms for the Autocatalytic Formation of Optically Active Compounds Under Abiotic Conditions, Science, 185, 525.

Honda, Y., Navarro-González, R. and Ponnamperuma, C., 1989: A Quantitative Assay of Biologically Important Compounds in Simulated Primitive Earth Experiments, Adv. Space Res, 9(6), 63.

Horowitz, N.H., Hobby, G.L., and Hubbard, J.S., 1976: The Viking Carbon Assimilation Experiments: Interim Report, Science, 194, 1321.

Julg, A., 1990, Asymmetrical Adsorption on Kaolinite and the Origin of the *l*-Homochirality of the Amino Acids in the Proteins of Living Organisms, Sciences Géologiques, Mémoire, 85(I), 25.

Karagounis, G. and Coumoulos, G., 1938: A New Method of Resolving a Racemic Compound, Nature, 142, 162.

Kobayashi, K. and Ponnamperuma, C.: 1985, Trace Elements in Chemical Evolution II: Synthesis of Amino Acids under Simulated Primitive Earth Conditions in the Presence of Trace Elements, Origins of Life, 16, 57.

Komoriya, A. and Chaiken, I.M., 1982: Sequence Modeling Using Semisynthetic Ribonuclease S, J. Biol. Chem., 257, 2599.

Kvenvolden, K.A., Lawless, J., Pering, K., Peterson, E., Flores, J., Ponnamperuma, C., Kaplan, I.R., and Moore, C., 1970: Evidence for

Extraterrestrial Amino Acids and Hydrocarbons in the Murchison Meteorite, Nature, 228, 923.

Kvenvolden, K.A., Lawless, J.G., and Ponnamperuma, C., 1971: Nonprotein Amino Acids in Murchison Meteorite, Proc. Natl. Acad. Sci. U.S., 68, 486.

Lawless, J.G., Kvenvolden, K.A., Peterson, E., Ponnamperuma, C., and Moore, C., 1972, Amino Acids Indigenous to the Murray Meteorite, Science, 173, 626.

Leder, P. and Nirenberg. M.W., 1964: RNA Code words and Protein Synthesis, II. Nucleotide Sequence of a Valine RNA Codeword, Proc. Natl. Acad. Sci. U.S., 52, 420.

Milton, R.C. de L., Milton, S.C.F., and Kent, S.B.H., 1992: Total Chemical Synthesis of a *d*-Enzyme: The Enantiomers of the HIV-1 Protease Show Demonstration of Reciprocal Chiral Substrate Specificity, Science, 256, 1445.

Morozov, L., 1979: Mirror Symmetry Breaking in Biochemical Evolution, Origins of Life, 9, 187.

Navarro-González, R., Negrón-Mendoza, A., and Chacón, E., 1989: The γ-Irradiation of Aqueous Solutions of Urea. Implications for Chemical Evolution, Origins of Life Evol. Biosph., 19, 109.

Navarro-González, R., Ponnamperuma, C. and Khanna, R.K., 1992: Computational Study of the Radiation Chemical Processing in Comet Nuclei, Origins of Life Evol. Biosph. 21, 359.

Negrón-Mendoza, A., Draganic, Z.D., Navarro-González, R., and Draganic, I.G., 1983: Aldehydes, Ketones, and Carboxylic Acids Formed Radiolytically in Aqueous Solutions of Cyanides and Simple Nitriles, Rad. Res. 95, 248.

Nirenberg, M. W. and Mattaei, J. H., 1961: The Dependence of Cell-Free Protein Synthesis in E. Coli upon Naturally Occurring or Synthetic Polyribonucleotides, Proc. Natl. Acad. Sci. U.S., 47, 1588x.

Pauling, L. and Delbruck, M., 1940: Nature of the Intermolecular Forces Operative in Biological Processes, Science, 92, 77.

Ponnamperuma, C., 1972: Organic Compounds in the Murchison Meteorite, Ann. N.Y. Acad. Sci., 194, 56.

Ponnamperuma, C., 1974: Optical Activity and Origin of Life, in W. Thiemann (ed.), International Symposium on Generation and Amplification of Asymmetry in Chemical Systems, Kernforschungsanlage Jülich, West Germany, 131.

Ponnamperuma, C., and Hobish, M. K., 1984: The Stereochemical Approach to Studies of the Origin of the Genetic Code, in K. Matsuno, K. Dose, K. Harada, and L. Rohlfing (eds.), Molecular Evolution and Protobiology, Plenum Publishing Co., New York, 295.

Ponnamperuma, C., Hobish, M. K., and Senaratne, N., 1992, A Stereochemical Basis for the Origin of the Genetic Code (F3.5-M.1.06), The World Space Congress, Washington, D.C., August 28-September, 5, 588.

Ponnamperuma, C., Honda, Y. and Navarro-González, R., 1990: Asymmetry and the Origin of Life, in B. Gruber and J.H. Yopp (eds.), Symmetries in Science IV, Plenum Press, New York, 193.

Ponnamperuma, C., Lemmon, R. M., Mariner, R., and Calvin, M., 1963a: Formation of Adenine by Electron Irradiation of Methane, Ammonia, and Water, Proc. Natl. Acad. Sci. U.S., 49, 737.

Ponnamperuma, C., Shimoyama, A., and Friebele, E.: 1982, Clay and the Origin of Life, Origins of Life, 12, 9.

Ponnamperuma, C. and Woeller, F., 1964: Differences in Character of C_6 to C_9 Hydrocarbons from Gaseous Methane in Low-Frequency Electric Discharges, Nature, 203, 272.

Ponnamperuma, C., Young, R. S., and Muñoz, E., 1963b: The Formation of Guanine during the Thermal Polymerization of Amino Acids, Fed. Proc., 22, 479.

Quack, V.M., 1989: Structure and Dynamics of Chiral Molecules, Angew. Chem., 101, 588.

Salam, A., 1991: The Role of Chirality in the Origin of Life, J. Mol. Evol., 33, 105.

Schlenk, Jr., W., 1952a: Separation of optical antipodes by complex formation without utilization of an asymmetric molecule, Experientia, 8, 337.

Schlenk, Jr., W., 1952b: The fractionation of ozocerites and the Separation of optical Isomers, Analyst, 77, 867.

Senaratne, S.M.D.N., 1986: Direct Interactions between Amino Acids and Nucleotides as a Possible Explanation for the Origin of the Genetic Code, Doctoral dissertation, University of Maryland.

Shimoyama, A., Blair, N., and Ponnamperuma, C., 1978: Synthesis of Amino Acids under Primitive Earth Conditions in Presence of Clay, in H. Noda (ed.), Origin of Life, Center for Academic Publications, Tokyo, Japan, 95.

Thiemann, W., 1990: Some Ideas on the Evolution of Symmetry Breaking of Biosphere on Earth, in B. Gruber and J.H. Yopp (eds.), Symmetries in Science IV, Plenum Press, New York, 205.

Vester, F., Ulbright, T.L.V. and Krauch, H., 1959: Optische Aktivitat und Die Paritatsverletzung im b- Zerfall, Naturwissenschaften, 46, 68.

SEARCH FOR PHASE TRANSITIONS CHANGING MOLECULAR CHIRALITY

A. Figureau, E. Duval[1], A. Boukenter[1]
Institut de Physique Nucléaire de Lyon
IN2P3-CNRS and Université Claude Bernard
F-69622 Villeurbanne Cedex, France

ABSTRACT

Since Pasteur discovered in 1848 that biological molecules possess a rotatory power, the origin of the chiral purity in living organisms has been a constant preoccupation in biology, but the problem is not solved yet. In particular, the appeal to weak interactions, a fundamental physical process which is known to violate parity, has not permitted so far to establish any firm relation between parity nonconservation and the complete dissymmetry between mirror image biological molecules. The main difficulty resides in the weakness of the physical forces, and can be overcome only when some amplification process can be proved to be at work. Recently such a mechanism was proposed, which does not seem to ask for any adhoc new concept : due to the attractive character of the parity violating force in electro-weak interactions, a phase transition leading eventually to enantiomeric purity is predicted. Phase transitions at low temperatures have already been detected in biological materials but no signature concerning the parity aspect was obtained. We undertook this year in Lyon a series of experiments to measure the rotatory power of solutions containing organic dissymmetric molecules, in order to observe if it varies with temperature. Our first measures involved cystine, which possesses a high rotatory power. No variation of this quantity was observed down to 0.6 K. Lower temperatures will be attained in a next step.

[1] Permanent address : Laboratoire de Physicochimie des Matériaux Luminescents URA 442 CNRS, Université Claude Bernard F-69622 Villeurbanne Cedex

ISBN 0-937194-31-X

1. INTRODUCTION

In 1848, Pasteur observed that crystals formed from a salt of tartaric acid were of two types, each one a mirror image of the other. He also discovered that solutions of each type of these crystals rotated polarized light in opposite directions. He interpreted these experimental facts as the manifestation of two different types of the same molecule. More than a century later, this is not a surprise to us, who know the structure of matter down to the sub-nuclear scale : the two types of crystals correspond to mirror-symmetric arrangements of atoms in the molecule.

Nine years later, Pasteur made another fundamental discovery, the profound understanding of which stays as a challenge for our modern science : micro-organisms could change an optically inactive solution to an optically active one. He also realized that this was only one example of a more general phenomenon, namely that the whole chemistry of life had a preferred handedness. In our present view, there is no mystery in this selective functionning : since most biochemical reactions are strongly dependent on a stereospecific recognition between reactants, it is quite natural that homochirality of molecules perdures as an essential property, and in such an absolute manner as to be considered as a characteristic of the living systems structure. A problem arises, however, when one considers the origins of life : up to now, there was no convincing proposition for a process that would allow to constitute a homochiral medium from a racemic mixture. For sure, there exist many physical processes which might give rise to some asymmetry between two molecules which are mirror-symmetric of each other, but all the suggested mechanisms are far from accounting for the chiral purity necessary for the good-working of any self-replicating organism. Furthermore, gradual enrichments are faced with the difficult fact that known living systems do not tolerate racemization in their delicate mechanisms. A complete review of the subject was published recently and we will refer to it for information not directly related to our experiment (Bonner, 1991).

The violation of parity discovered in weak interactions (Lee and Yang, 1956) is probably the most promising physical effect for inducing effects on a macroscopic scale : its constant intensity has been there for eons, and is not subject to modification by any circumstantial event. On the contrary its effects should be very weak, and in fact experiments trying to demonstrate a relation between parity violation in weak interactions and chiral purity in biological matter are generally considered as negative (Bonner, 1991).

Soon after the unification of weak and electromagnetic interactions (Glashow, 1980, Salam, 1980, and Weinberg, 1980), it was realized that different enantiomers would possess different physical properties, due to the newly proved existence of neutral weak currents. In particular energy differences would arise to modify the classical strict equality of masses for two molecules which are mirror symmetric of each other. From thermodynamic considerations, the state with minimal energy must then be favored as compared to the other : a certain dissymmetry between the populations of two enantiomers is then expected. Ab initio calculations were carried out for various amino-acids in aqueous solutions (Mason and Tranter, 1984), showing parity violating energy differences of the order of 10^{-17} kT ; this corresponds to an excess of approximately 10^{-17} for the molecules of lower energy, a very small number indeed. Some orders of magnitude can be gained if one considers instead some other (heavier) biomolecules (MacDermott and Tranter, 1989), but an immense gap is still to be bridged before a connection is established with biomolecular homochirality.

A remarkable and encouraging result of these calculations, however, is the prediction that amino-acids which produce optical rotation of polarized light to the left are favored, and not the reverse. This is exactly what is observed from biological materials, though on a very different scale. Moreover, the prediction for sugars is a preference for the right-handed variety, in accordance with experimental facts. So, weak interactions provide a fundamental and natural force which might participate in the homochirality properties of life discovered by Pasteur. Obviously, some amplification process must be at work to transform this tiny advantage into a plausible mechanism. In this respect however, all the other proposed sources of asymmetry do not fare much better, though the amplification factor needed for weak interactions effects may seen desperately larger.

A recent theoretical work (Salam, 1991) has introduced in this context new ideas which might provide a natural solution to the amplification problem. It is briefly described in Section 2. There we also discuss in more detail some points which are of interest for the definition of an experimental test for this theory. Of the predicted parity-violating phase transition, we only attempted to observe the "global" signature, mimicking in the crudest possible way the processes which might have been at the origin of chiral purity in prebiotic material. This exploratory experiment is described in Section 3. It brought no evidence for the phenomenon, but this negative result only restricts the domain of the phase transition to temperatures lower than our minimal one (0.6 K). Lower temperatures and different conditions should be explored. In Section 4, we analyse briefly the different consequences of our experiment that can be derived for physics and for biology.

2. A SIMPLE TEST FOR A NEW TYPE OF PHASE TRANSITION

We do not intend to explore here in details the theoretical work at the origin of this experiment, and we refer to Salam (1991) and Chela-Flores (1991) for a complete account of the ideas which are to be tested. We simply note that weak and electromagnetic interactions have been unified in a single model, called the Standard Model (Glashow, 1980, Salam, 1980 and Weinberg, 1980), which allows to describe quite precisely the whole set of experimental data produced in these domains of physics. Up to now, there has been no need to go beyond this model, though its limits have been perseverantly searched for. Its direct predictions are then on relatively solid grounds. This is the case for the recent observation

(Salam, 1991) that the exchange of a Z° boson gives rise to an attractive force between certain pairs of electrons. Since it is a weak interaction phenomenon, the force is minuscule, but there is little doubt about its existence. A second and essential point is that, under some critical temperature T_c at which one should observe a phase transition, the electrons in matter must be sensitive to the attraction and be the seat of a condensation phenomenon. Such a collective process is the equivalent of superconductivity which results from the Cooper pairs interactions. This analogy provides the basis for the amplification of the minute effects classically expected from the weak interaction, since a macroscopic number of particles are then allowed to participate coherently.

The estimate of the critical temperature is much less sure than the existence of the collective effect, and the prudent conclusions in the original publication (Salam, 1991) will be here our credo : "The best way to determine T_c for a given amino-acid is at present experiment". The fact is that no detailed mechanism exists, which could help the numerical evaluation of the parameter describing the phase transition. The uncertainty is even so large as to cast doubts on the nature itself of the phase transition, and consequently on its observability by our method. This point will be discussed in Section 4. It seems reasonable however, considering the importance of the prediction for physics as well as for biology, to look for the predicted new type of ordering of matter.

For the time being we will then suppose that the phase transition consists as proposed by Salam (1991) in the transformation of the configurations of mirror nonsymmetric biomolecules, from the L type to the D type or the inverse, depending on the precise energy difference between them. The best signature for the phase transition would then be the corresponding inversion of the optical activity when the temperature is lowered under T_c, as already suggested by Salam (1991). We remark further that such a measurement would also supply the most direct and interpretation-free proof for the homochirality problem : the quantity to be measured is precisely the rotatory power which is considered since Pasteur a characteristic signal for life, and in our times as a necessary condition for its emergence (Gol'danskii and Kuz'min, 1989).

The ideal experiment consists then in measuring the optical rotation produced by a racemic mixture of chiral molecules on a polarized beam of light : as the temperature is slowly lowered, the rotation should abruptly grow at some time from the initial zero value to a sizeable fraction of the nominal rotatory power for the homochiral equivalent preparation. In practice, such a measurement is not easy and we found more convenient to measure the optical rotation outside the cryogenic vessel. We were then faced with the rapid return of the sample to warmer temperature, and eventually the disparition of the expected effect. We devised then our experiment so as to bypass this difficulty : two samples with the same molecular composition but with a different chirality (generally one racemic and one non racemic in our experiments) are simultaneously brought to low temperatures. If the critical temperature is attained, both samples should become identical, possessing now in principle the same degree of homochirality. When heated back to room temperature, both should follow the same path and eventually become racemic at the passage through the critical temperature, so that in the end at least one of the two samples should have a chirality different from its inital value, whatever happens when going back above T_c, and provided that the phase transition consists in the predicted chirality change. Let us note finally that our argument is not modified if the homochiral phase exists only in a limited range of temperatures : the only condition to be fulfilled is that T_c is attained and that the transition has time to develop. Anticipating over the short discussion of Section 4, we believe this is at the moment matter for experiment.

3. EXPERIMENT

A natural choice for our samples would be one of the 4 amino-acids for which there exists a parity violating energy difference estimation (Mason and Tranter, 1984). However the rotatory power of these substances is not very high, and we preferred to use a more sensitive detector of chirality. A priori, most complex molecules are mirror asymmetric and can then constitute a good target for our experiment, since the weak interaction parity violating energy difference has no reason to be zero, except accidentally. In particular, biomolecules are not expected to present any advantage in this context. We chose nevertheless cystine, because of its similarity with the amino-acid cystein, its high rotatory power ($[\alpha] = 218°$ at 589.3 nm and 20° C) and its availability in the pure homochiral states. Two samples were used simultaneously in our experiments : one was racemic and the other left-handed.

The experimental procedure was the following. Polycristalline cystine powders were brought by pairs (one racemic, one left-handed) to low temperature in a cryostat and kept cold during several days to allow for an eventually slow transition. After that period, the two samples were taken out of the cryostat, dissolved (separately) at room temperature in the 2N HCl solution, and placed for the polarization rotation measurement in transparent cells set between polarizers. They were then illuminated with a He-Ne laser at 632.8 nm. At this wavelength, the specific rotatory power of cystine is $[\alpha] = 203°$.

Three different cooling conditions were successively experimented : **1)** Three days at 77 K, **2)** three days at 4 K, and **3)** four days at 0.6 K. For all three racemic cystine specimens, absolutely no rotation was observed after cooling. On the other hand, no change in the polarization rotation was measured for the left-handed samples : with a concentration of 46 g/l of this homochiral cystine, and a light path length of 8 cm, the measured rotation was 7° after as well as before the days inside the cryostat. This value corresponds to a rotatory power $[\alpha] = 190°$, in good agreement with the specific value 203° if one takes into account our experimental uncertainty, which is about 30' for the measure of the polarization rotation.

In another experiment, we used recrystallized right-handed and left-handed samples. The size of the crystals was close to 5 mm, a much larger dimension than the one of the first received samples. (On the contrary, it was impossible to increase the crystal size of our racemic specimens so that right-handed cystine was used in place of the racemic mixture). After four days at 0.6 K we did not detect any change in the two opposite polarization rotations. This points to a relative independance of our conclusions on the detailed state of the macroscopic cystine material.

4. DISCUSSION

In this first exploration, our measures have shown no evidence for a sizeable transition from one state of the molecule cystine to its mirror-symmetric. On the contrary, the existence of some phase transition due to the parity violating sector of weak interactions is on relatively firm grounds. It is then interesting to compare more in detail what is observed and what is expected.

Our most obvious concern is the temperature range which is to be explored. Estimations, drawn from the analogy with superconductivity, give critical temperatures as high as 300 K, but this number could be varied very easily in any direction (Salam, 1989). On the contrary, an order based on energy differences around 10^{-18}eV would be destroyed by thermal motion as soon as the temperature is higher than 10^{-14} K, i.e. the phase transition would be unattainable before long. Such energy differences should not be considered as characteristic ones, however : the potential barrier between one state and its mirror symmetric is more representative of the transition which produces the chirality change in molecules. A typical energy would then be the dissociation energy, corresponding to temperatures of the order of 5-600 K, but intermediate steps depending strongly on the detailed mechanism could bring T_c to a much lower value : in the absence of a model, experiment stays as the best means of exploration. So in a next step, our measurements will be performed after an exposition of samples down to the temperature 0.02 K.

The second question arising from our analysis concerns the choice of cystine. This biomolecule is the association of two cysteins linked by a disulfide bridge. No theoretical calculation is available for the amino-acid cystein, and even less for cystine, but the generality of the argument for the existence of a phase transition implies that cystine is a priori as good a candidate for a test as amino-acids. The presence of sulfur atoms in this molecule is even an advantage, a priori, because of the Z^6 dependence of the parity-violating energy difference (MacDermott and Tranter, 1989). Our choice might be unlucky, however : the structure of the molecule as an arrangement of two identical pieces may lead in this case to a particular and unpredictable lack of sensitivity of our measures. A check of our results with another molecule will then be our objective in the near future.

These uncertainties are mainly due to our ignorance of the mechanism which might transfer the parity violation from the particle level to the molecular one. However, it has been know experimentally for more than 10 years now that the weak force, and more particularly the Z° exchange, is felt measurably by the outermost electrons in atoms (Bouchiat and Pottier, 1984). Thus, the problem reduces to know if the collectivity of electrons in the medium is sufficient to produce a phase transition involving energies as high as the potential barrier between left and right-handed configurations. This is a complex matter from which arise various questions : is the dependance with Z of the parity violation effect valid for atoms still applicable for complex molecules ? What is the importance of the medium density and of the crystal structure ? What can be the intermediate states between the two enantiomers, and consequently how high is the effective potential barrier ? Finally, what is the characteristic time necessary for the establishment of the phenomenon ? None of these is really easy to answer at present or even to formulate so as to lead to clear experimental tests. The problem is nevertheless so important, and the recently provided answer so elegant, that more work is needed along these lines both on the theoretical and on the experimental sides.

One natural extension of our present work concerns the complex molecules formed in outer space. In effect, the main consequence of the existence of the phase transition at some low or very low temperature would be to displace the origins of life, or at least some of its primordial steps, from the earth to extraterrestrial space - with obvious consequences for all prebiotic chemistry researches. Amino acids have been found on meteorites and their chiral purity analyzed : no dissymmetry between the left and right-handed molecular populations was observed * but such measures only prove that T_c if it exists is lower than earth-like temperatures and that the unknown time characteristic for the disparition of the pure enantiomeric phase is shorter than the duration between entry in the atmosphere and the measurement in a laboratory. Extraterrestrial space on the contrary furnishes favourable conditions for an unbiased experiment : in addition to the existence of amino-acids which proves that some step at least of prebiotic chemistry do occur, the temperature is generally very low (though limited by the lower value 2.7 K of the cosmic background radiation), and the duration of this natural cryostat is incommensurably longer that any plausible human experiment. It is then of utmost interest to search in space itself for a signature of nonracemic mixtures of molecules, whether organic or not since a priori most types of complex chemicals can be the seat of the predicted phase transition.

ACKNOWLEDGMENTS

It is a pleasure to thank Dr J. Delorme for many enlightening discussions and Prof. M. Ericson for warm encouragements. We also are grateful to Prof. I. Berkes for his hospitality in his low temperature laboratory and for having

* Note added in proof : If confirmed from various extraterrestrial samples, the recent observation of a small D/L value for alanine, and also for other amino acids (Engel, Macko and Silfer, 1990) would constitute a first indication for the existence of a non-racemic chemistry outside our planet which might take its origin in the phase transition that we were looking for in this experiment.

performed the very low temperature heat treatments and to J.P. Hadjout for his technical assistance.

REFERENCES

Bonner W.A., 1991 : The origin and amplification of biomolecular chirality, Origins of life 21, 59-111

Bouchiat M.A. and Pottier L., 1984 : An atomic preference between left and right, Sci. Amer., 250 (6), 76-85

Chela-Flores J., 1991 : Comments on a novel approach to the role of chirality in the origin of life, Chirality 3, 389-392

Engel M.H., Macko S.A. and Silfer J.A., 1990 : Carbon isotope composition of individual amino acids in the Murchison meteorite, Nature 348, 47-49

Glashow S.L., 1980 : Towards a unified theory : threads in a tapestry, Rev. of Mod. Phys. 52, 539-543

Gol'danskii V.I. and Kuz'min V.V., 1989 : Spontaneous breaking of mirror symmetry in nature and the origins of life, Sov. Phys. Usp 32, 1-29

Lee T.D. and Yang C.N., 1956 : Question of parity conservation in weak interactions, Phys. Rev. 104, 254-258

Mason S.F. and Tranter G.E., 1984 : The parity-violating energy difference between enantiomeric molecules, Mol. Phys. 53, 1091-1111

Salam A., 1980 : Gauge unification of fundamental forces, Rev. of Mod. Phys. 52, 525-538

Salam A., 1991 : The role of chirality in the origin of life, J. Mol. Evol. 33, 105-113

MacDermott A.J. and Tranter G.E., 1989 : Electroweak bioenantioselection, Croatica Chem. Acta 62 (2A), 165-187

Weinberg S., 1980 : Conceptual foundations of the unified theory of weak and electromagnetic interactions, Rev. Mod. Phys. 52, 515-523

THEORETICAL AND EXPERIMENTAL STUDIES ON THE POSSIBILITY OF CHIRALITY DEPENDENT TIME DIRECTION IN MOLECULES

A. S. Garay

Department of Biochemistry and Biophysics,
Texas A and M University, College Station, TX 77843[1]
and
Institute of Biophysics, Biological Research Centre,
Hungarian Academy of Sciences, P.O.Box 521,
H-6701 Szeged, Hungary

ABSTRACT

The modern time concept varies considerably in different branches of physics. It fails to incorporate a dynamic element which is obvious in symmetry violations and in the evolution of the universe and life. Chirality on the other hand is a dynamic concept related to time direction. Since chirality is a general phenomenon the mixed time direction state may not be confined to the K_L^0 meson; it may exist in molecules and at higher levels of organization. Attempts were made to measure time direction in two liquid enantiomeric pairs by the determination of the half life of ^{32}P and ^{22}Na dissolved in them. This was successful. Chirality related differences were found, yet it is premature to draw any quantitative conclusion.

1. INTRODUCTION, SHORT REVIEW AND THE OBJECTIVES OF THIS PAPER

1.1 INTRODUCTION

When I first heard that this conference will discuss the origin of chirality in several invited presentations, the famous astronomist, Hoyle came to my mind. In his novel "The Black Cloud" a non scientist says of scientists: "I can not understand what makes them tick. They are always wrong and they always go on." This is a very true and positive statement about us, the participants of this conference. According to many our field of research, the origin of life, and particularly the origin of chirality present a desolate picture. Yet we are here to discuss the problems, and to design and report new experiments. Thanks are due to Dr. Salam and Dr. Ponnamperuma for making this possible. In my contribution I will make an attempt to discuss the nature of time in relation to the origin of life and particularly the origin of biochirality. I will also present experimental data in support or in falsification of my thesis.

1.2 THE PRESENT STATUS OF CHIRALITY RESEARCH

In 1991 two comprehensive reviews were published on molecular chirality (Bonner, 1991; Barron, 1991). Bonner reviewed about 500 papers. He discusses chance and determinate mechanisms, such as fluctuations, circularly polarized light, violation of parity, weak neutral current etc. It has been claimed that any of these can lead to the homochirality of life from racemic mixtures. His conclusion is that none of the recommended mechanisms could ever be effective or have reasonable

---

[1]Address to which correspondence should be sent

ISBN 0-937194-31-X

capability of producing the uniquely one handed molecules of the cell, because the effects are minuscule. E.g. it is a striking statement that every atomic nuclei and every non chiral molecules are left handed due to the interaction of the omnipresent parity violating weak nuclear current (PVNC) between the nucleus and the electron. On the other hand it is equally striking that 10^{10} times magnification is needed to visualize the left helical electron probability current density in atoms (Hegström et al. 1988). The contribution of binding energy by the PVNC is even less, 10^{17} times magnification is needed. Therefore Bonner concludes "that the source of terrestrial chirality must have been extra terrestrial and furthermore it must have been capable of providing an ongoing influx of chiral molecules having uniform chirality". This is supported by the majority of experimental data.

However, quite a few researchers reported conspicuously larger effects, but these are either neglected or deemed as faulty because they exceed the theoretically predicted range (e.g. Akaboshi et al. 1990, Conte and Pieralice 1987, Garay 1968, Garay and Ahlgren-Beckendorf 1990, Merwitz 1976, Norden et al. 1985, Thiemann and Wagener, 1970).

The second review of the past year by Barron (1991) evaluates 181 papers and puts molecular chirality in a new perspective. He reviews the application of the fundamental symmetry aspects, charge conjugation, space inversion and time reversal to molecules. On that basis he discusses the concepts of "true and false chirality", magnetic monopoles and chiral molecules, CPT theorem, and the significance of the PVNC. However, he barely mentions any biological implications. Maybe this is why, in contrast to Bonner, he considers the research of molecular chirality an exciting and promising field.

1.3 THE OBJECTIVE OF THIS PAPER

My intention is to incorporate the notion of time in the search for the origins and evolution of life. I called attention to this problem several times since 1974, however with not much positive results (Garay 1974, 1978, 1979, 1987a, 1987b). Yet I am convinced that a more complete and comprehensive understanding of the nature of time will have a profound effect on the understanding of biological phenomena. Bronowski (1969) emphasized this by asking the question: "how does it come about that evolution as a whole appears to have a direction of time, namely the direction of greater complexity. If there is any single question to be answered about biology it is: why does biological time have this particular direction?" The question implies that Bronowski, along with many others, is sceptical regarding the assumption of the contemporary biology that the combination of the known laws of physics and chemistry with Darwin's theory can satisfactorily explain evolution. Mendelian inheritance and natural selection undoubtedly played an important role but they still leave room for studying the nature of time in relation to biology. Let us now have a look at the physical notion of time direction and time reversal in this context. The relation to chirality will be pointed out wherever possible.

2. THE NOTION OF TIME, TIME DIRECTION AND TIME REVERSAL IN DIFFERENT BRANCHES OF PHYSICS

2.1 TIME DIRECTION AND TIME REVERSAL IN NEWTONIAN PHYSICS

In physics time reversal means the reversal of the direction of motion with no implication whatsoever to time direction. Yet, in newtonian physics the direction of motion of an ensemble of particles defines time direction, the well known "entropy dictated" irreversible disorder. One may say that a single particle alone has complete freedom to move, and its motion is reversible. However if a particle is really alone, then there is no reference point and it is meaningless to speak about its motion. On the other hand, if there is a reference point, then the particle is not alone because the reference point plus the body of interest constitutes a minimum multitude, and the particle's motion is subjected to the probabilistic direction towards disorder. Thus motion is inseparable from time direction, and it is not justified to use the concept of time reversal "only" for motion and exclude time direction. I will use time reversal for the reversal of the motion of multitudes of particles. It is emphasized again that this is in terms of newtonian physics which claims in Newton's words that "time flows equably without relation to anything external". Of course the speed of particles in a multitude (e.g. the dissolution of a piece of cube sugar in water), does not mean that time "goes" faster. It means though that past and future are distinguishable, time direction points into disorder, a state of translational and reflectional symmetry. This is, however, not a law like but a fact like statement.

2.2 CONCEPT OF TIME IN RELATIVITY, AND QUANTUM THEORY

Relativity theory describes space (x,y,z) and time(t) as a unified four dimensional continuum (t,x,y,z), which essentially means that extension in time implies extension in space. Thus time is not an independent parameter in which something happens. It does not seem to be a serious exaggeration to say: "The objective world simply is, it does not happen" (Weyl 1949). Consequently it is a nonsense to suppose that time moves or has an arrow of intrinsic orientation. Time reversal is a purely mathematical concept. The change of the sign of t into -t has no physical meaning at all simply because all known laws of physics are invariant under this operation. Therefore the notion of spacetime should be extended into all branches of physics. This includes the time dilation of special relativity theory where "time" slows down depending on velocity. It also includes quantum theory because the Hamiltonian operator is invariant under T transformation, even if according the uncertainty principle the wave function extends both into the future and into the past. Asymmetry in time, or time asymmetry, i.e. the historical asymmetry of the world with respect to time does not mean the asymmetry *of* time. The idea of the "flow of time like a river" is a psychological illusion. Many physicists adopt this view of time including Davies (1977) in his famous book "The Physics of Time Asymmetry". He admits, however, that insisting to the above contradiction free view is only possible at the cost of disregarding the time symmetry violation in the decay of the K_L^0 meson. We, of course, can not do this because from the point of view of chirality and biological time direction the CP violation of K_L^0 meson is of utmost importance.

Along this line the idea of time reversal played an important role in the formulation of quantum electro dynamics. In Feynman's (1966) words: "if an electron is moving back from the future, that is equivalent to changing the sign of the charge, and therefore it would act like a positron... Positrons could be simply represented as electrons going from the future to the past".

2.3 TIME DIRECTION AND TIME REVERSAL IN COSMOLOGY

Although cosmology is governed by relativity theory and quantum theory it also relates to the newtonian physics in respect to time concept because of a pervasive global asymmetry in virtue of the expansion of the universe. It has been shown by an astrophysical school (Gal Or 1972) that the second law of thermodynamics can be traced back to this most significant timelike arrow. Because the expansion of the universe becomes a huge "thermodynamic sink" for all the radiation flowing out into empty space. It should be noted that a link with the expansion of the universe and CP violation in the decay of K_L^0 meson has been suggested (Aharony and Ne'eman 1970). Thus different time arrows point to the same direction. According to a recent cosmological model (Gunzig et al. 1987) the specific entropy per baryon of the universe depends only on the mass of black holes and on the universal constants h, C and k, where h is the planck constant, C the speed of light and k the Robertson-Walker factor. Remarkably in this deduction CP violation was omitted because it may "destroy the predictive power of the model".

Although cosmology is dominated by relativity theory, in a somewhat contradictory way, it permits antiparallel past, future arrows in different regions of the universe and also the directional change of these arrows i.e. time reversal in an eventual recontracting universe (Davies 1977). Even the possibility of time travel is heavily discussed in the most recent literature (Allen and Simon 1992). Gott (1991) pointed out that time machines can be built without violating the presently known laws of physics. He proposes to use two cosmic strings, an exotic form of matter, which if moving with $v \approx 0.7C$ velocity in opposite directions can generate a closed timelike curve. Thus it could take a person back into its own past. In other words the cosmic strings are able to bend and distort their internal spacetime into a circle. "The circumference of the circle is less than the euclidean circumference, $2\pi r$, as if a wedge has been cut out of a circular paper and glued together to form a cone. The spacetime outside the curve remains uncurved or flat. If the string is moving to the left, then transporting across the wedge to the right results in a jump backward in time, and transporting across the wedge to the right results in a jump forwards "(Allen and Simon 1992). Because the transporting is not on a circle but on the surface of a cone it seems chirality is incorporated in Gott's time machine. Do cosmic strings exist? Can a person be duplicated by going back into the past? Is cell duplication a consequence of backward pointing time direction in living systems? The duplication of a person certainly shows that backward pointing time direction resulted in a more complex system because the two persons are not exactly the same. One of them was in the future. Does this mean a sort of evolution? These questions and ideas belong partly to science and partly to science fiction. In fact Gott's time machine seems to be doomed (Allen and Simon 1992).

2.4 TIME REVERSAL AND SYMMETRY VIOLATIONS

When I heard the first time that ß decay produces left handed electrons I had the feeling that there is no more equilibrium. Events must have a direction. Consider one of the simplest steps in the origin and evolution of life, the choice of one chiral form over racemic mixture. Thermodynamics do not permit this initial step. They dictate full racemization of all non completely racemic mixtures. In this respect (!) weak nuclear interactions seemingly do not obey the second law. They never produce a full racemic mixture of ß particles. Neither do weak interactions produce a fully homochiral system. In order to produce a homochiral system the nuclei must emit the ß particles with the velocity of light, and this would require infinite energy as relativity theory claims. In order to produce a fully racemic system the nuclei must emit the ß particles with zero energy. This is however, a contradiction because no beta particles can be emitted with zero energy. Therefore, if the second law dictates racemization and weak interactions never permits it, the universality of entropy seems to be broken. In a more general form, the energy imposed restriction means that full disorder and full order are equally unreachable but any transitions between the two extremes are permitted. However entropy is saved at cosmic level because ß-decay in a charge conjugated form gives rise to right handed positrons, i.e. racemic mixture is produced. Under the condition of the earth however, electrons dominate over positrons, thus it seems that entropy broke down locally to a very minuscule extent.

The universal violation of time reversal was observed and firmly corroborated in the charge and chirality balanced decay of K_L^0 meson, (Christenson et al. 1964) as formula shows (Bennet et al. 1967).

$$\frac{R(K_L^o \rightarrow \pi^- e^+ {}_d\nu_\ell)}{R(K_L^o \rightarrow \pi^+ e^- {}_\ell\tilde{\nu}_d)} = 1.003 \pm 0.0004$$

Were R means the decay rate, *d* and *l* right and left handed spin directions, + and - sign charge. Because the ratio is not equal to 1 and we believe in the validity of the CPT theorem, the K_L^0 meson must carry a time direction which is antiparallel to the entropy dictated time asymmetry. It also carries an entropy parallel time direction, but the decay rate is slower when electrons are produced. The entropy parallel time direction is mixed with the entropy antiparallel time orientation. Clearly the K_L^0 meson cannot produce entropy decrease because the entropy increasing decay dominates over the time symmetry violating part of the interaction (Ne'eman 1972). Yet the decay of K_L^0 meson has a privileged time direction, namely the direction towards antimatter, positron instead of towards matter, electron.

It is very essential that all symmetry violations demonstrate an interlocking of handedness with charge and in case of the K_L^0 meson with time as well. In this respect symmetry violations reminds to matter space time interlocking in relativity theory. The new feature is that the interlocking seems to be different for the left handed electron, (e^-,-x,-y,-z,-t) and for right handed positron (e^+,+x,+y,+z,+t).

2.5 THE BIOLOGICAL TIME DIRECTION

The biological time direction towards complexity is explained by the permanent energy input into biological systems. The ultimate energy source, the sunshine is transformed into adenosinetriphosphate (ATP). Which is then used in nonspontaneous entropy decreasing reactions to provide free energy for producing complex structures. This coupling of non spontaneous reactions with free energy producing ATP runs life. A human being makes as much ATP a day as its body weight; a well known statement of modern elementary biochemistry. In this view there is no need to study the notion of time of physics. Mutations and the selective power of the environment combined with steady flux of energy satisfactorily explains life and its evolution, the temporal chronology, the time direction towards complexity.

It is very remarkable that this widely held euphoric view was of no help whatsoever to explain the origin of life, the seemingly very simple step to choose one enantiomer over the other. After many decades of research we are referred to extraterrestrial source of chirality (Bonner 1991), to cometary delivery of organic molecules to the early earth (Christopher 1990), and in fact to extraterrestrial life (Chyba 1990). Maybe this is the solution! I accept it, yet we should not stop to search through all avenues. One of the logical choice was to review the concept of time in physics.

3. CONCLUSIONS AND COMMENTS

As we have seen there is no universal concept of time. The time concept varies considerably in different branches of physics. It is a common place that any conclusion can be drawn from an inconsistent framework of thoughts. Yet I venture to make two statements.

3.1 THE TIME CONCEPT OF QUANTUM THEORY AND RELATIVITY THEORY IS STATIC

The first statement is that both relativity theory and quantum theory describes matter and the universe as a static entity. Their time concept fails to incorporate the dynamic element which is obvious in symmetry violations and in the evolution of the universe and life. May be this is the reason why quantum theory excludes the existence of enantiomeric molecules. Born and Jordan (1930) stated: "The Hamiltonian is always invariant to space inversion, consequently there could not exist any optically active molecules, which contradicts experience."

Dirac (1977) attempted to enrich relativity theory with a dynamic time concept by uniting it with his large number hypothesis. This claims that universal constants are subjected to change as the universe evolves. The most interesting consequence of Dirac's theory is that it predicts symmetry violations i.e. there is no symmetry between a quantity increasing and the same quantity decreasing. Consequently there is no symmetry between negative and positive charge, which is debated today (Maddox 1992). Dirac's theory breaks the charge and time symmetry while preserving their product CT. This is caused by a long range force unlike weak interactions. The theory, however, has not been accepted because as vectors are shifted in space their length changes as in Weyl's (1918) geometry. However, it points into the right direction.

3.2 CHIRALITY IS A DYNAMIC CONCEPT

The second conclusion is that chirality seems to be a dynamic concept, related to time direction and mixed time direction states. The decay of the K_L^0 meson substantiates this conclusion along with the fact that it is involved in the time direction dictating expansion of the universe. The mixed time direction state may not be restricted to the K_L^0 meson, efforts should be made to find it in other systems. Molecules are candidates also. Consider the following example. During excitation of enantiomers a parallel or antiparallel transitional electric and magnetic moments are produced. In this respect enantiomers are time reversed pairs, and not only space inverted bodies. It should be noted though that time reversal operation does not affect the molecular structures, and only quantum theoretical model leads to the conclusion, that they are time reversed forms. Classical treatment fails in this respect, (Garay 1979, 1987a and personal communication, Bela Gazdy, Department of Chemistry, Emory University, Atlanta, Georgia 30322). It should also be noted that energy is pumped permanently into the living cell. Therefore chiral molecules are permanently in excited state. Thus their transitional moments can be considered permanent, i.e. they are time reversed pairs.

Finally I want to mention a very relevant, recent experiment. Milton et al. (1992) synthesized a protease from D amino acids and its enantiomeric pair from L amino acids. They reported that "the two synthetic enzyme molecules were equally active, but showed reciprocal chiral specificity in that the L-enzyme cleaved only the L-substrate whereas the D enzyme cleaved only the corresponding D-substrate". It is clear, that if there is any difference between the L-, and D-form with respect to time direction then Milton's results would be difficult to explain. A careful comparison of their data revealed, however, that the two mirror image enzymes exhibit different kinetics. (Fig. 1. Copyright 1992, by the AAAS. The courtesy of "Science and authors is highly appreciated.)

In short: it is possible that mixed time direction states exist in molecules and even at higher levels of organization such as electrontransport, brain etc.

4. THE PROBLEM OF MEASURING TIME IN MOLECULAR SYSTEMS

One of the best way to measure time direction in molecular systems seems to be the determination of the ß decay constant of radioactive labelled molecules. The exponential decay is a time asymmetric event. It has been used in many experiments to measure the elapsed time e.g. for demonstrating time dilation predicted by special relativity theory. ß decay, however, is not fully insensitive to chemical environmental factors. In Fermi's theory the emission intensity of a ß particle by the nucleus in solids depends on two factors: an outgoing wave from the ß emitter and a wave scattered back to the nucleus from the nearby atoms. It is the constructive or the destructive interference of these components that leads to increased or decreased emission (Rehr 1991). Both theoretical calculations and experimental results indicate that the magnitude of the effect is in 1 KeV region, thus it is only observable in low energy emitters such as tritium, where it results in an oscillatory structure in the ß spectrum (Koonin 1991).

In addition, the ß decay rate was supposed to be sensitive to electron helicity in chiral molecules (Hrasko 1973). The helical electron gas model (Hrasko and Garay 1975), however, is controversial. Experiments

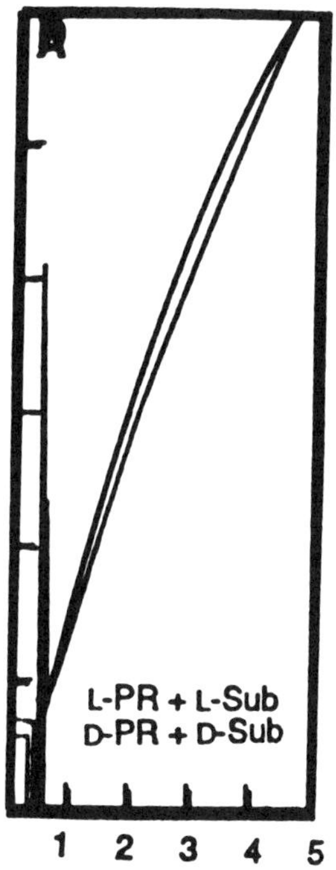

Fig.1. Comparative activity of the HIV protease enantiomers on enantiomers of a chiral fluorogenic substrate according to Milton et al. (1992). Ordinate: fluorescence intensity (arbitrary units). Abscissa: time in minutes. Upper curve L-enzyme with L-substrate. Lower curve: D-enzyme with D-substrate. In contrast to author's conclusion the difference in kinetics between the two combinations is apparent. Figure has been constructed by superimposing author's figures 2A and 2D.

show that electron helicity in chiral molecules is negligible (Soós and Bagyinka 1976, Gidley et al. 1982, Hegström et al. 1982). In view of the above intuition suggests choosing a high energy ß emitter with a relatively long half time. Then there is a chance to measure elapsed time in the system of interest.

In the late seventies I have measured the decay constant of ^{32}P in different molecules. No differences were found. Similar results were published by Soós and Bagyinka (1976). Then in the early eighties liquid enantiomers became available offering a new experimental approach. One can put radioactive material in liquid enantiomers and immediately check the eventual changes in the decay constant either in the form of a sudden increase or a sudden drop. In fact a sudden increase, a spike was observed.

5. EXPERIMENTS

5.1 MATERIALS AND METHODS

Two liquid enantiomers were used, R-, and S-2-phenylbutyric acid (PBA), and R-, and S-diethyltartrate (DET). In order to avoid systematic mistakes both substances were purchased from two different companies: the

PBA from Norse and Aldrich, the DET from Tokyo Casei and Sigma. For purity check absorption, excitation and luminescence spectra were taken. No contaminations were found in PBA samples whereas in DET contaminations were quite common. High pressure liquid chromatography also revealed a few contaminated samples. They were thrown out.

Two different carrier free samples of $H_3{}^{32}PO_4$ and $^{22}NaCl$ in water was purchased from ICN and New England Nuclear Radiochemicals. Since water does not dissolve in PBA, part of the samples were evaporated to dryness and redissolved in spectroscopically pure methanol. The ^{32}P and ^{22}Na concentrations were chosen to be close to 80,000 CPM/10 microliter, in order to avoid overspill or saturation of the photomultiplier of the instrument by using too much radioactive material. A Beckman LS 3801 scintillation counter were used to measure radioactivity in the Cerenkov mode.

Seven hundred microliters of R-, and S-enantiomers or 350+350 microliters for racemic mixtures (M) were placed into 2 millilitre glass vials. Then 10 microliters of ^{32}P or ^{22}Na was added to each vial. This was immediately vigorously vortexed and shaken for 30 seconds to assure homogenous distribution of the nuclids in the enantiomers. The CPM was measured for 5x1 minutes. The average of this measurements served as zero time value. Then the CPM was taken for several hours in 1 to 5 minute intervals. After that, the measurement continued in longer intervals, however in a random way to avoid systematic errors and the influence of the instability of the instrument. The different PBA and DET samples as well as the different ^{32}P and ^{22}Na samples were divided into 3 groups containing 4, 4 and 6 samples each. Thus a total of 14 repetitions were made in random distributions. Two different controls were prepared: 700 microliters water was placed into 28 vials and 10 microliters of the nuclide solutions were added to them. Fourteen samples were evaporated to dryness, the other 14 samples were treated the same way like the enantiomers. The first 14 are called the air control, the second 14 are called the water control.

5.2 RESULTS AND DISCUSSION

Only the initial phase of the decay will be presented and discussed. The long term experiments are still in progress, they are not ready for evaluation. To detect differences in the initial part of the decay is experimentally much easier than to follow the decay for a long time. The long term measurements are now in their fourth year and the difficulties to stabilize the instrument for such a long time are enormous. For a couple of hours the instrument is stable.

Results are shown on figures 2-7. All figures are computer printouts. The controls will be discussed first, then the decay of ^{32}P and ^{22}Na will be compared in both enantiomeric pairs. Figures 2 and 3 show that in case of ^{22}Na data scattering is considerably higher than in case of ^{32}P. Maybe this is due to the more complex decay of ^{22}Na as compared to ^{32}P. ^{22}Na emits high energy γ rays of 1274 KeV, which produce Compton electrons. The annihilation of positrons gives rise to high energy γ-s of 510 KeV too. All these contribute to the photon production measured in the Cerenkov mode.

A more important observation is that both the ^{32}P and the ^{22}Na exhibit a well defined spike in water. This is difficult to understand since the radioactive nuclids were dissolved in water and a further dilution with water should not result in an increased ß emission. Does vortexing have an effect on the structure of the solution? A delayed luminescence of the glass vial is excluded.

No quantitative conclusions can be drawn from the results presented on figures 4-7. A few solid qualitative statements however can be made.

a. All fourteen repetitions gave essentially the same results like the 4 parallel measurements presented on the figures.

b. The initial faster decay rate (spike) was observed in both molecular species, PBA and DET and also in water with both radioactive nuclids, ^{32}P

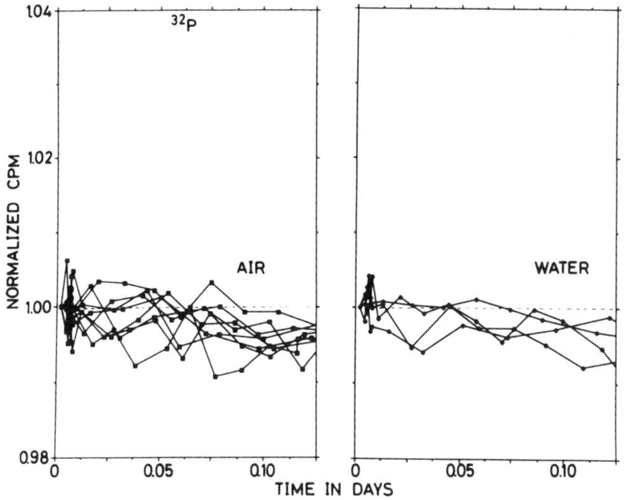

Fig.2. The initial decay of ^{32}P in air and water. Sudden rise of the decay in water is observable.

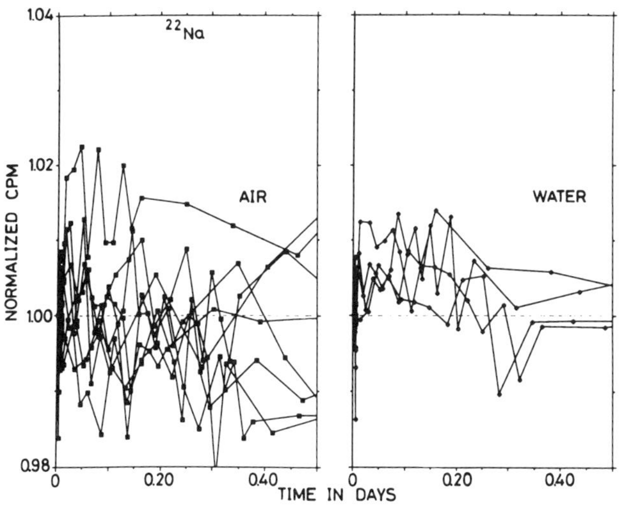

Fig.3. The initial decay of ^{22}Na in air and water. The difference between air and water is not significant, yet it is well observable. There is a small definite spike of the decay in water.

and ^{22}Na. The spike is not only chirality dependent.

c. The hight of the spikes seems to be inversely related to the helicity of ß particles. Compare figures 4 and 6 with figures 5 and 7. The average helicity of ^{22}Na β^+ particles is 0.690, the correspondent hight of the spike is 1.02 CPM. The values for ^{32}P are 0.906 helicity and 1.002 CPM spike hight.

d. Comparing figures 4 and 5 with figures 6 and 7, it is clear that the heights of the spikes are approximately equal in PBA and DET (1.02 CPM for ^{22}Na), yet the strength of the helical potential field is considerably bigger in PBA, $-49.9 \pm 18.8 \times 10^{-4}$ eV, than in DET, $-8.5 \pm 5.06 \times 10^{-4}$ eV. The negative sign means left handed potential field for the R-enantiomers.

e. Both the heights and the lengths of the spikes are directly related to the half life of the nuclids, but no quantitative relation was found. The half life of ^{22}Na is 950.45 days, that of ^{32}P is 14.28 days. Compare figures 5 and 7 with 4 and 6.

f. The length of the spike is longer in the S-enantiomers as compared to the R-enantiomers. This applies for all combinations, DET, PBA, ^{32}P, ^{22}Na. However in case of ^{22}Na in PBA enantiomers the spike lengths could not be determined satisfactorily due to very big scattering.

g. The hight of the ^{22}Na spike is bigger in R-DET than in S-DET. See figure 7.

h. The half life of ^{32}P is significantly shorter in R-PBA than in M-PBA and S-PBA. This has been calculated from 200 days decay curves by fitting an exponential curve on all 14 repetitions. Data are the followings in days: R-PBA 13.52 ± 0.18, M-PBA 13.86 ± 0.24, S-PBA 14.27 ± 30. There are no differences of ^{32}P half life in DET enantiomers.

i. The scattering of the data is always biggest in the racemic mixtures followed by the S-enantiomers. The smallest scattering was observed in the R-enantiomers. This may be important in view of my earlier speculations concerning the relationship between scattering and time arrow (Garay 1978).

5.3 SOME PITFALLS, COMMENTS AND CONCLUSION

It seems that one source of error is the mixing of enantiomers with the radioactive nuclids, especially because both the PBA and the DET are very viscous. Density of PBA and DET is 1.055 and 1.205 respectively. Viscosity data are not known. It is possible that the mixing is incomplete and the ^{32}P and ^{22}Na is surrounded with water or methanol after the mixing. As the mixing becomes gradually complete, they come into direct contact with the PBA and DET. Due to the very high refractive index of PBA (1.515) and DET (1.446), the efficiency of Cerenkov radiation increases gradually and this gives rise to the initial decay spike. This hypothesis however does not explain the observed differences in the R-, and S-enantiomers. Furthermore the eventual thickness of the water coat around the nuclids must be in the Angstrom range whereas the average distance travelled by the β^- particle in R-PBA and R-DET is 1.268 ± 0.017 mm and 1.510 ± 0.020 mm. The β^+ particles travel 0.347 ± 0.012 mm in R-PBA and 0.802 ± 0.011 mm in R-DET. The values for S-PBA and S-DET are somewhat bigger. The negligible thickness of the water coat compared with the travelling distances can not influence the Cerenkov radiation considerably. The contribution of other factors such as electron density seems negligible because it is the same for both enantiomers. Electron density in e^-/cm^3 are in the order of water PBA and DET, 2.94×10^{24}, 3.40×10^{23} and 3.87×10^{23}.

One of the solutions could be to determine the spike and decay constant by γ rays instead ß-s because γ particles are orders of

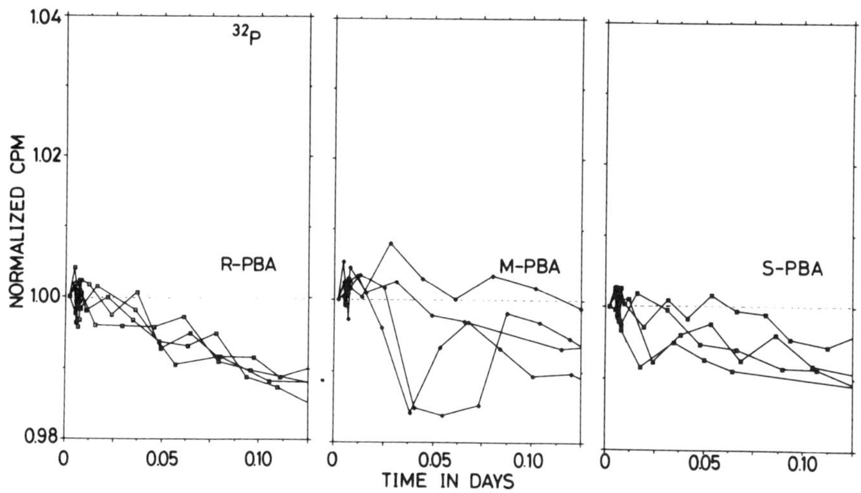

Fig.4. The initial decay of ^{32}P in R-, M-, and S-PBA. The spikes are barely observable because the resolution below 0.01 days is unsatisfactory.

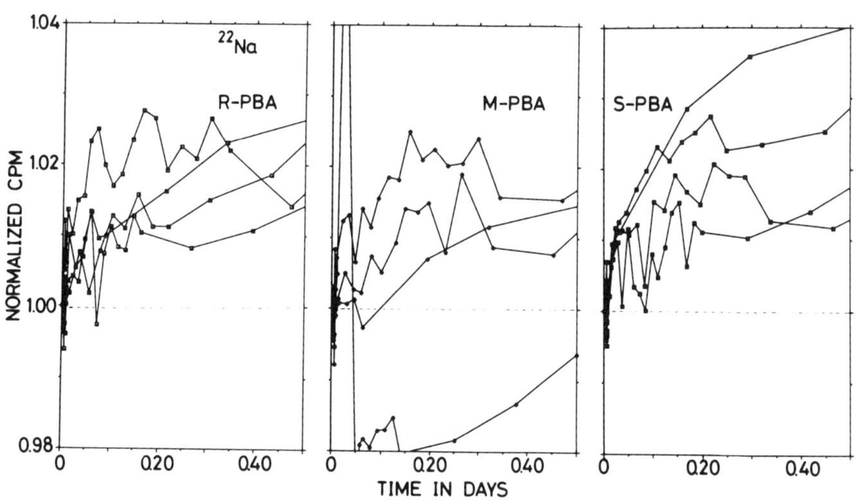

Fig.5. The initial rise of the decay of ^{22}Na in R-, M-, and S-PBA. The spikes are huge.

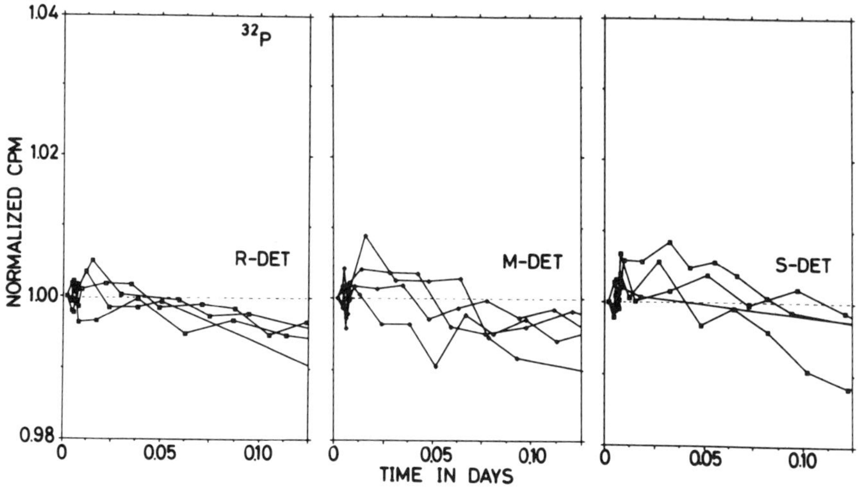

Fig.6. The initial decay of ^{32}P in R-, M-, and S-DET. The lengths of the spikes are longer in S-DET than in R-DET.

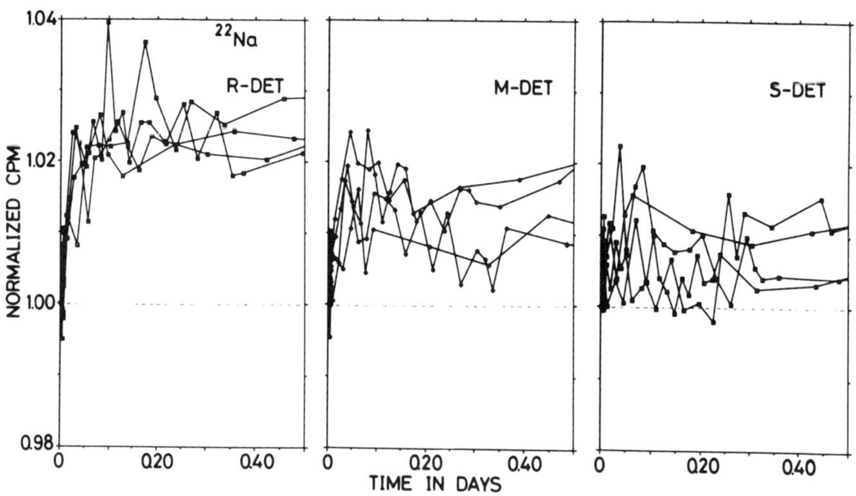

Fig.7. The initial rise of the decay of ^{22}Na in R-, M-, and S-DET. The differences between the enantiomers and racemic mixture are apparent.

magnitude less sensitive to environmental factors than the ß-s. Unfortunately my efforts to measure γ rays resulted in unreliable data because the scattering masked the expected differences between the enantiomers.

In short: this paper neither proved *nor falsified* the hypothesis according which enantiomeric molecules have different time direction states. Further physical, chemical or biological inferences should not drawn from the experiments presented.

ACKNOWLEDGMENTS

The financial helps of the Texas Agricultural Experiment Station and Sir Emil and Lady Clementine Ogden are gratefully acknowledged. Thanks are due to Mrs. Christine Snaples for her excellent technical assistance with the computer evaluation.

REFERENCES

Aharony,A., and Y.Ne'eman, 1970: Time Reversal Symmetry Violation and the Oscillating Universe, Int.J.Theor.Physics., 3, 437-441.

Akaboshi,M., K.Kawai, H.Maki, W.Ehrlich, and Y.Honda, 1990: Selective Decomposition of either Enantiomer of Aspartic Acid Irradiated with ^{60}Co-γ-rays in the Mixed Aqueous Solution with D-, or L-alanine, Origins of Life and Evolution of Biosphere, 20, 111-119.

Allen,B., and J.Simon, 1992: Time Travel on a String, Nature, 357, 19-21.

Barron,L.D., 1991: Fundamental Symmetry Aspects of Molecular Chirality, in P.G.Mezey, (Ed.), New Developments in Molecular Chirality, Kluver Academic Publisher, Dordrecht, 1-55.

Bennet,S., D.Nygren, J.Steinberger, and J.Sunderland, 1967: Measurement of the Charge Asymmetry in the Decay of $K_L^0 \to \pi + e + \nu$., Phys.Rev.Letter, 19, 933-997.

Bonner,W.A., 1991: The Origin and Amplification of Biomolecular Chirality, Origins of Life and Evolution of Biosphere, 21, 57-111.

Born,M., and P.Jordan., 1930: Elementary Quantum Mechanics, Springer, Berlin.

Bronowski,J., 1969: Physics and the Explanation of Life, 136th Meeting of AAAS, Boston, Mass., 40-69, 3-2678 Casette.

Christenson,J.H., J.W.Cronin, V.L.Fitch, and R.Turlay, 1964: Evidence of the 2π Decay of the K_L^0 Meson, Phys.Rev.Letter, 13, 138-140.

Chyba,C.F., 1990: Extraterrestrial Life, Nature, 348, 113-114.

Chyba,C.F., P.J.Thomas, L.Brookshaw, and C.Sagan, 1990: Cometary Delivery of Organic Molecules to the Early Earth, Science, 249, 366-373.

Conte,E., 1989: On the Asymmetrical Interaction of Positrons from ^{22}Na Decay with Homochiral Molecules: a Proof on the Existence of Electrons with Net Helicity in Biomolecules, Nuovo Cim., 101A, 683-689.

Conte,E., and M.Pieralice, 1987: Investigation on the Chirality of Positrons from ^{22}Na Decay and their Asymmetrical Interactions with D-alanines, L-alanines and DL-alanines, Nuovo Cimento D., 9, 283-288.

Davies,P.C.W., 1974: The Physics of Time Asymmetry, Surrey University

Press, London, U.K., 214 pp.

Dirac,P.A.M., 1973: New Ideas of Space and Time. Lindau Nobelpreisträger Tagung in P.T.Landsberg, (Ed.), Enigma of Time, A.Hilger, Bristol, 109-113.

Feynman,R.P., 1966: The Development of the Space Time View of Quantum Electrodynamics, Science, 153, 699-708.

Gal-Or,B., 1972: The Crisis about the Origin and Irreversibility and Time Anisotropy, Science, 176, 11-51.

Garay,A.S., 1974: Comments on the Time Direction of Biological Evolution, Acta Bot.Acad.Sci.Hung., 20, 43-47.

Garay,A.S., 1978: Superweak Interactions and the Biological Time Direction, Origins of Life, 9, 1-5.

Garay,A.S., 1979: Broken Symmetries in Physics and their Relevance in Chemistry and Biology, in D.C.Walker, (Ed.), Origins of Optical Activity in Nature, Elsevier, Amsterdam, 245-257.

Garay,A.S., 1987a: Why is Chirality so Important, BioSystem, 20, 1-6.

Garay,A.S., 1987b: Differential Interactions of ß Particles with Liquid Enantiomers and Internal Timing of Chiral Molecules, BioSystems, 20, 63-74.

Garay,A.S., and J.A.Ahlgren-Beckendorf, 1990: Differential Interaction of Chiral ß Particles with Enantiomers, Nature, 346, 451-453.

Garay,A.S., J.A.Beckendorf, and B.Gazdy, 1992: Absolute Molecular Handedness and Chiral Strength Determined by Matter Wave Circular Dichroism, (Submitted).

Garay,A.S., P.Hrasko, 1975: Neutral Currents in Weak Interactions and Molecular Asymmetry, J.Mol.Evol., 6, 77-89.

Garay,A.S., L.Keszthelyi, I.Demeter, and P.Hrasko, 1973: Differences in the Annihilation of Positrons in Optical Isomers, Chem.Phys.Letter, 23, 549-552.

Gidley,D.W., A.Rich, J.Van House, and P.W.Zitzewitz, 1982: ß Decay and the Origins of Biological Chirality: Experimental Results, Nature, 297, 639-643.

Gunzig,E., J.Géhéniau, and I.Prigogine, 1987: Entropy and Cosmology, Nature, 330, 621-624.

Gott,J.R., 1991: Closed Timelike Curves Produced by Pairs of Moving Cosmic Strings. Exact Solutions, Phys.Rev.Lett., 66, 1126-1129.

Hegstróm,R.A., 1982: ß Decay and the Origins of Biological Chirality: Theoretical Results, Nature, 297, 643-647.

Hegstróm,R.A., J.P.Chamberlain, K.Seto, and R.G.Watson, 1988: Mapping of the Weak Chirality of Atoms, Am.J.Phys., 56, 1086-1092.

Hrasko,P., 1973: ß Decay in Optical Isomers, KFKI, Internal Report, P.O.Box 49, 1525 Budapest, Hungary.

Koonin,S.E., 1991: Environmental Fine Structure in Low Energy ß Particle Spectra, Nature, 354, 468-469.

Maddox,J., 1992: Is Charge Quantization Exact?, Nature, 358, 449.

Merwitz,O., 1976: Selective Radiolysis of Enantiomers, Radiat.Environ.Biophysics, 13, 63-69.

Milton,deL.R.C., S.C.F.Milton, and S.B.H.Kent, 1992: Total Chemical Synthesis of a D-enzyme: the Enantiomers of HIV-1 Protease Show Reciprocal Chiral Substrate Specificity, Science, 256, 1445-1448. (Title correction, Science, 257, 147.)

Ne'eman,Y., 1972: Lectures Delivered at the Erice Summer School on High Energy Astrophysics and its Relation to Elementary Particle Physics, for reference see P.C.W.Davies 1977.

Norden,B., J.O.Liljenzin, and R.K.Tokay, 1985: Stereo Selective Decarboxylation of Amino Acids in the Solid State, with Special Reference to Chiral Discrimination in Prebiotic Evolution, J.Mol.Evol., 21, 364-370.

Rehr,J.J., 1991: Crystal Effect in ß Decay, Nature, 354, 436-437.

Soós,J., Cs. Bagyinka, 1976: On the Differential ß-decay in Optical Isomers, Radiobiochem.Radioanal.Letters, 27, 169.

Thiemann,W., and K.Wagener, 1970: Is there an Energy Difference between Enantiomeric Structures?, Angewandte Chemie Int.Ed. English, 9, 740-741.

Weyl,H., 1918: English Version 1952: Space, Time, Matter, Dover, New York.

Weyl,H., 1949: Philosophy of Mathematics and Natural Science, Princeton Univ. Press, Princeton p 116.

Wigner,E.P., 1957: Relativistic Invariance and Quantum Phenomena, in W.J.Moore, and M.Scriven (Eds.), Symmetries and Reflections. Ind., Univ.Press, Bloomington, Ind. USA, 51-81.

EXTRATERRESTRIAL INTELLIGENCES (*)

Jean Heidmann (§)
Observatoire de Paris
F-92195 Meudon, France

1. INTRODUCTION

1.1 Life in the cosmos...

Recent developments in the fields of radioastronomy, solar system space exploration and investigations of the origins of life suggest that life is likely to be the natural result of evolution in the cosmos. From this perspective it is reasonable to conclude that the appearance and evolution of life on the Earth may very well have occurred on other planets orbiting some of the 300 thousand million stars in our Milky Way galaxy. Confronted with this tremendous possibility, astronomers find themselves on the front line of research because their domain of study is the cosmos.

1.2 ...and its evolution

The existence of extraterrestrial life does not necessarily imply that any lifeform would have evolved to the level of intelligence and technology experienced by humans on Earth. The complex trail of evolution on the cosmic scale may be classified into five stages :

1 - A cosmic stage that begins with the appearance of matter soon after the Big Bang, proceeds to the formation of stars and planets several billion years later, and continues through the synthesis of chemical elements like carbon, the centerpiece for life as we know it.

2 - An organic chemistry stage, with the formation of the first organic molecules that are the stepping stones towards our own life ; these molecules have been discovered in interstellar space by radioatronomers and in comets by space probes.

3 - A prebiotic chemistry stage, in which complex "building blocks" are produced, such as nitric bases that form the rungs of the DNA double helix ladder or amino acids, the essentials elements of proteins ; amino acids have been collected in meteorites that have fallen on Earth ; some facts indicate that a prebiotic chemistry could be proceeding even now on Titan, the large moon of the planet Saturn.

4 - A primitive biological stage like the one of bacteria which dominated the first few billion years of the Earth ; scientists engaged in space exploration hope to discover this level of biological activity, perhaps in a different form, buried in the frozen permafrost of Mars.

5 - Finally, an "advanced" stage of life, perhaps more advanced than ours, is plausible ; nothing in our investigations of the universe indicates that Homo Sapiens is unique or that we occupy a special place in the cosmos ; an effective approach to answer the question, "are we alone?" is to conduct

ISBN 0-937194-31-X

comprehensive "SETI" (Search for ExtraTerrestrial Intelligence) observations that are designed to search for artifical radio signals eventually produced by extraterrestrial civilizations.

1.3 Bioastronomy and SETI

Building on the discoveries of modern ground-based and space astronomy, the international community of scientists formed a new branch of astronomy, bioastronomy, which is dedicated to the study of thepossible existence of life in the universe and the search for evidence of intelligent life. Echoing the noblest human aspirations, bioastronomy has become a very popular subject. Renowned institutions such as the International Academy of Astronautics and the National Academies of the United States and the Soviet Union were the first to support bioastronomical research. In 1982 the International Astronomical Union created a new commission devoted to biostronomy. In 1989 the US national Aeronautics and Space Administration initiated the NASA Microwave Observing Project, and in 1990 the U.S. Congress voted to provide the funds required to design and build the special purpose digital processing systems capable of searching tens of millions of radio frequency channels.

2. THE FIRST HYPOTHESIS, OR LIFE ON EARTH

"Life is a natural process of evolution in the cosmos"

Thirty years ago, Stanley Miller and Harol Urey carried out an experiment which sowed the seeds of much developments : electrical discharges in a simple gaseous mixture in the presence of liquid water rapidly produce what have been called the building blocks of life, organic molecules which form the basis of the macromolecules in living creatures.

Since then, astronomers and geologists have come up with valuable information on the primaeval atmosphere of the Earth and other bodies such as Mars, Titan and Triton, and on the physical conditions wich prevailed at early times, and meanwhile the experiments for producing the building blocks of life have progressed enormously. The physico-chemical conditions existing on the surface of grains of interstellar dust, or on the nuclei of comets, have also been reproduced in the laboratory.

So what are these simple organic molecules ? I shall briefly mention only the main ones which we astronomers encounter in our researches ; hydrogen cyanide, HCN, found in the atmosphere of Titan ; formadehyde, H_2CO, detected in interstellar space ; the C-H hydrocarbon bond observed in comets ; the amino acids, NH_2- RCH- COOH, occuring in certain meteorites. Other molecules, more complex but not observed: nitrogenous bases such as adenine, DNA, deoxyribonucleic acid, formed from a double chain of nucleotides (a deoxyribose sugar, phosphoric acid, H_3PO_4 , and a nitrogenous base) ; proteins, chains of amino acids stripped of an H and OH , ...NH-RCH-CO...

The earliest indications of life on Earth come from fossil organic molecules, 3800 million years old, in which the isotope ratios are different from those naturally occuring, suggesting an origin in some biological activity which was already flourishing but of which we know nothing. Life must therefore have appeared rapidly.

The oldest fossils are the stromatolites, stacks of alternate layers of sediment and single-celled organisms which had invented the cell membrane. 3500 million years ago they were very successful ; several forms of these single-cell creatures were already quite diversified.

A thousand million years later, oxygen was already being produced by photosynthesis. Then, 1400 million years ago, the eukaryotic cells appeared, a thousand times larger and extremely complex in comparison, probably resulting from a symbiosis of several types of cells to produce veritable cellular factories with different specialized workshops : a nucleus for DNA, mitochondria for breathing, chloroplasts for photosynthesis, Golgi bodies for excretion, ribosomes for synthesizing proteins and even flagella for movement.

The oxygen content of the oceans increased as a result of this biological activity, and yet more complex forms of life appeared : the first multicellular organisms, the Ediacarians, soft-bodied marine creatures resembling domed leaves of up to a metre in size. They were the rulers of the marine world from 670 to 550 million years ago.

Then the oxygen became plentiful enough for a layer of ozone to form which protected the land from ultra-violet radiation. The result was that the continents became colonized in the Cambrian period, 550 milion years ago. The explosive growth of the number of phyla exceeds even the best science fiction. To mention only the animal kingdom : sponges, worms, sea anemones, insects, starfishes, molluscs, and vertebrates. The vertebrates include fishes, reptiles, birds, and mammals. Mammals include carnivores, insectivores, cetaceans, and so on through to the primates. The primates include lemurs, monkeys, and anthropoid apes.

This astonishing burgeoning of life, which took only a tenth of the age of life on the Earth, has led in the course of a few million years - only a thousandth of the age of the planet - to Autralopithecus, who walked upright, and Homo Sapiens sapiens, who has walked on the Moon.

3. THE SECOND HYPOTHESIS, OR BIOASTRONOMY

"What happened on Earth may have happened elsewhere"

3.1 The comic stage

In the beginning was the Big Bang. Thanks to it, space is large enough not to contract straight away, and a minor asymmetry between quarks and antiquarks has left us the small residue of matter without which life could never have appeared.

Several hundred million years after the initial cataclysm, matter collectected to form the first stars. As a result of nuclear reactions inside them the chemical elements were produced, including carbon, of fundamental importance for organic molecules.

Then the planets came into existence. Recent very sensitive methods have shown that the stars HD 114762, beta Pictoris, epsilon Eridani, and gamma Cephei are serious candidates for a planetary system. But several more years of research will be needed so that we can be sure.

Space exploration has provided us with information in the physicochemical state of the surface of numerous bodies in the solar system : Mars with its once warm humid climate, Titan with an atmophere similar to that of the primaeval Earth, Europa, with a possible ocean beneath a layer of ice ; very different environments, and all of interest for the study of cosmic life.

3.2 The organic stage

Eighty different kinds of molecules have been dicovered by radio astronomers in interstellar space, the majority of them organic, containing up to 13 atoms. They probably have been synthesized on the surface of grains of silicate material covered with ice, frozen methane and ammonia, by bombardment with ultra- violet light from the stars. For example : $CH_4 \rightarrow CH_3$, $NH_3 \rightarrow NH_2$ and $CH_3 + NH_2 \rightarrow CH_3NH_2$ (methylamine).

The grains themselves could have been produced during the formation of a planetary system, in particular in its comets; the Giotto space probe detected abundant quantities of H, C, N and O atoms in comet Halley. Also in meteorites, where other processes of organic chemistry could take place : 90 amino acids have been identified in meteorite Murchison, 8 of which occur in terrestrial proteins. Polycyclic aromatic hydrocarbons have ben found , and possibly polyoxymethylene.

3.3 The prebiotic stage .

Two processes are envisaged that lead to greater complexity:

- the formation, in a primitive atmosphere, of reactive precursors : nitriles of the forms RCN (including hydrogen cyanide, HCN) and aldehydes of the form RCHO (including formaldehyde, HCHO, or very possibly their import by comet and micrometeorite bombardment;
- the formation, in oceans or lakes, of the buiding blocks of life from their precursors amino acids, sugars, and nitrogenous bases. Thus the polymerization of HCN leads to adenine, while the “formose” reaction converts formaldehyde into sugars, including deoxyribose.

Laboratory experiments and theoretical models confirm these possibilities, but according to François Raulin they require a primitive reducing atmosphere composed of methane, hydrogen, and water vapour. The thermal springs produced by plate tectonics, already active two thousand million years ago, could have been favourable sites for such developments.

One of the most exciting results of bioastronomy has been the discovery by the Voyager probes of a ‘prebiotic planet’, Saturn’s largest satellite Titan. With a diameter of 5000 km and an atmosphere of nitrogen at a pressure of 1.5 bar, it is a dream model of the primaeval Earth, deepfrozen at -180°C. There we find aerosol fogs of hydrocarbons and organonitrogen compounds created by prebiotic atmospheric chemistry.

French scientists are working on the Cassini/Huygens space mission, which is intended to parachute a probe onto Titan in 2005 in order to discover more about this major stage in the origin of life.

3.4 The primitive biological stage

The origin of proteins, nucleotides, and membranes, fundamental staging posts in the appearance of genuine life, is being investigated in the laboratory. Unfortunately the traces of early life on our planet have been obliterated by geological processes, but bioastronomers still have a hope in planet Mars.

Even if the two Viking crafts left behind on Martian soil in 1976 have provided no trace of a primitive biological process, the two orbiters have revealed signs of astonishing flowss of water, in particular a giant dried-up river bed, Kasei Vallis ; 3000 million years ago, it must have carried a thousand times as much water as the Amazon today.

During the first few thousand million years of its existence, Mars had an appreciable atmophere and surface water in liquid form. So when we consider that primitive life appeared on Earth in less than 100 milllion years, we cannot reject the hypothesis that a form of life could have developed there, too. A large part of the water is now locked up in permafrost, a mixture of ice and rock which could be up to 7 km thick.

Will the new expeditions to Mars, starting with the Phobos probes and now Mars Observer, reveal another type of biological life, independent of that on Earth, similar or different , fossilized or frozen in the Martian subsoil ?

3.5 All of Bioastronomy

Adding up the stage of evolution of advanced life, as set out above, we obtain the complete panorama of bioastronomy. On a par with astrophysics, cosmochemistry, and planetary geology and climatology, it is the results of applying, at a cosmic level, sciences initially developped on Earth : an exercice which often produces very fruitful results. The new paradigm of a biological universe is in full blossoming these last years.

At its general meeting in 1982, the International Astronomical Union set up a 51st commission with the title 'Bioastronomy : the search for extra-terrestrial life', which puts this new discipline on an official status.

Its objectives are :

(1) to search for planets in other stellar systems

(2) to investigate the evolution of planets and the possibility of life on them

(3) to search for molecules of biological interest in the cosmos

(4) to detect primitive biological activity

(5) to detect artificial extraterrestrial signals

(6) to search for signs of advanced civilizations

(7) interdisciplinary collaboration with other international organizations in biology, astronautics, space law, etc.

4. THE THIRD HYPOTHESIS, OR THE ADVANCED STAGE

"Human intelligence is not the last word from the cosmos"

4.1 Intelligence

For centuries the human male, steeped in his own superiority, refused to admit any intelligence in animals, or even in women. If, occasionally impressed by the forces of nature, he acknowledged the existence of superior intelligences, then they belonged to a world outside the real one, and therefore were not in competition with him. These attitudes could be at the root of the opposition in certain quarters to the possible existence of extraterrestrial intelligence : a real intelligence superior to our own.

Let us look at the facts. The observable universe contains a hundred thousand million galaxies, each with tens of thousand million stars, of which about ten per cent are similar to our Sun. Furthermore, the majority of them are older than the Sun, up to twelve billion years old.

So when we consider the seething evolution which took place here in less than 5000 million years, when we think only of the last few rapid strides in the cosmic marathon which have been taken by Australopithecus and the Apollo astronauts, we cannot reasonably reject the third hypothesis out of hand. In short, it is hard to believe that the 20th century on planet Earth represents the *nec plus ultra* in the long history of the vast cosmos.

We owe a debt to the radio astrophysicist Nicolai Kardashev who gave the problem a scope which is refreshingly free from a constraining anthropomorphism, with his concept of type I, II and III civilizations. These would be characterized, from the energy and spatial point of view, by the handling of their own planet, star, and galaxy, respectively. The last possibility, while physically possible, is in the realms of the fantastic.

4.2 SETI

If we embark on predicting the characteristics of the advanced stage in the evolution of life, even confining ourselves to the future of our own, we unfortunately launch into pure speculation. Who, given information about the neolithic age, could have deduced the present state of civilization ? However, that does not entitle scientists to give up in despair, because we now have a way of attacking the problem from the observational point of view. Only one a way, incidentally. It is called SETI, Search to Extraterrastrial Intelligence.

In its current, limited, phase, this observational tool consists essentially of the search for artificial extraterrestrial radio signals, which would betray the existence of stages of evolution more advanced than our own, whether the signals were specifically intended to make contact or are simply 'leakage' from astrotechnological equipment .

Why radio ? Because for the last thirty years, thanks to the theoretical work of Guiseppe Cocconi and Philip Morrison, and the first attempts at 'listening in' by Franck Drake, we know that progresses made in radio astronomy allows us to detect such signals (if they exist) at interstellar distances. If other civilizations have technical means at least equal to our own, then over a million stars, up to a thousand light years away, are within our reach. Our own generation has made a major technological leap.

Electromagnetic waves are the fastest known means of transmitting information across interstellar distances. But electromagnetic waves with frequencies lower than 1000 megahertz are drowned by all kinds of natural emissions from our galaxy. And those with frequencies greater than 10,000 MHz, by virtue of their quantum corpuscular nature, have a low efficiency - for the case we are concentrating on deliberate radiation.

So to be thoroughly pragmatic and reasonable, and wishing to approach the problem according to the means at our disposal, the search for signals is now carried out in the 'galactic radio window' from 1 to 10 gigahertz, the decimetre waveband.

4.3 The NASA super-receiver

Although it now possesses an adequate sensitivity, SETI is still faced with a major problem. If signals are intended to attract attention the frequency band used must be as narrow as possible to keep the signal/noise ratio a maximum. It is also a way to promise an a priori artificial characteristic for the signal. Now the stray electrons in interstellar space cause a frequency dispersion of the order of a tenth of a Hertz, making that the narrowest practical bandwith. So the galactic radio window will contain 100 thousand million possible communications channels.

That is the enormous challenge. The best present-day astronomy receivers possess only 1000 simultaneous observing channels, a mere nothing compared to the multitude of channels to be explored in the galactic window.

For that reason the SETI project office at NASA made the decision, about ten years ago, to commission an R&D for a receiver possessing ten million simultaneous channels.

Its main element is the MCSA or multi-channel spectrum analyzer, which has the job of chopping the radio band received into ten million slices. It is based on a sequence of mathematical operations called Fourier transforms, implemented on a computer using a modular architecture of hundreds of parallel processors mounted round specially designed VLSI (very large scale integration) chips.

The MCSA is followed by a real-time signal analyzer which recognizes any regularities which could reveal an intelligent origin. To give one an idea : an observation of 1000 seconds would produce a screen picture of 1000 lines, each of 10 million points, and each point having an intensity encoded from 1 to 100. Among these billion values, forming a snow storm on the screen, we try to locate a few dozen snowflakes which are aligned in a regular manner, before making the next observation - there is no question of storing the data, which are essentially 'noise'.

Complete prototypes with 100,000 channels were already functioning, as well as a MCSA with 2 million channels and a simplified analyzer. After much work Jill Tarter, the project manager, switched on the super-receiver on 12 october 1992, the 500th anniversary of the discovery of an unknown world by Christopher Colombus, and thus launched on new waves a new caravel destined for a new world, a cosmic one this time.

Which capes and islands will the cosmic *Santa Maria* set sail for ? Two kinds of programmes are planned, which will last about ten years : the target search and the scan search.

The target search will take aim on the 1000 stars most like the sun ,to a distance of up to 100 light years : it will use maximum sensitivity (10^{-28} W/m^2) and 1 -Hz channels, and is intended to locate signals coming from civilizations which could have evolved under planetary conditions comparable to our own.

The scan search will systematically sweep the entire sky, but with reduced sensitivity and channels of 32 to 1024 Hz, the object being to locate possible powerful radio beacons, intentional or otherwise, of an unknown nature that is not necessarily planetary.

4.4 SETI based on pulsars

The NASA receiver represents considerable progress, but the number of channels to be explored is so great that innumerable sucessive shifts of its active band will still have to be made to cover the width of the galactic radio window. That, incidentally, is why the NASA listening programme will extend over several years in spite of the compromises that have been made.

Imagine one channel represented by a line only 1 mm wide ; then the galactic window would be 100,000 km wide, while a receiver with ten million channels covers only 10 km. So it would be necessary to reposition the 10-km viewport 10,000 times to cover the entire window.

In the face of this challenge, I proposed a listening strategy based on pulsars. Pulsars are powerful natural interstellar radio 'lighthouses' distributed throughout the galaxy. The blips they send out as a result of their rotation have impressively regular periods, while their lives exceed a million years. So might they not have been used as common reference frequencies by civilizations engaged in intentional interstellar communications ?

If so, there are two problems to be resolved :

1 - Which reference pulsars do we select ?

2 - How do we convert their rotational frequencies, which are generally low (1 to 1000 Hz) into frequencies in the SETI window ?

The answer to the first question depends on the type of search ; I have distinguished three types:

(a) NASA target search of the 1000 stars most like the sun : use the nearest pulsar.

(b) Searches in cluster of stars : use the most powerful pulsar in the cluster.

(c) Scan searches of whole sky : use the pulsar nearest to the direction being searched at any given moment.

As to the second question : we multiply the rotational frequency of the chosen pulsar by a distinctive and universal mathematical number somewhat less than 10 (10 being the relative width of the SETI window) a sufficient number of times such that the result falls (necessarily) inside the window.

Let us take an example. For NASA'S search of 1000 stars, the nearest pulsar is PSR 1929+10 in Aquila, 260 light years away, with a rotational frequency 4.41468 Hz. Mutiply it by 2 pi (6.2831853...) eleven times in sucession. The result is the SETI frequency of 2.65998 GHz.

I propose that we start exploring the galactic window using such frequencies based on the pulsars, so as to increase the probability of early success.

4.5 The MegaSETI project

The NASA super-receiver will have to be connected to a large radio telescope. The largest, at Arecibo in Puerto Rico, could be used only part time, taking turns within the entire astronomical community.

I therefore made the following suggestion in early 1985 : the Paris Observatory possesses the second largest radio telescope in the world for decimetre waves, at Nançay near Bourges in Central France. So could NASA not make a copy of its super-receiver, when it is finished, for installation at Nançay in return for an agreed amount of telescope time, for joint work on SETI ?

Since then, cooperation with fellow astronomers in America has begun to set this project, called MegaSETI, in motion. They have assessed our radio telescope and its site favourably, and it is intended to modernize it with an improvement similar to that applied to the Arecibo telescope : installing a double-mirror focal system. This will benefit all radio astronomers by updating the radio telescope to the status of a new generation one, competitive for the next twenty years.

4.6 The SETI contex

SETI is now supported by several pretigious international oraganizations. Initiated thirty years ago by the two major powers, SETI has become an international undertaking. There are the American activities by NASA, by the Planetary Society, and at Harvard, Ohio and Berkeley Universities. On the ex-Soviet side, Kardashev, now director of the large radio telescope at Samarkand has put forward an ambitious programme. Argentina is equipped with two million channels. At Nançay, besides the MegaSETI project, François Biraud has headead cosmic searches will Jill Tarter since 1981.

International collaboration is desirable, indeed indispensable. Arecibo and Nançay share the coverage of the sky ; this is necessary in order to verify likely signals and to ensure continuous 'listening-in' in spite of the rotation of the Earth. We also feel that if a signal is discovered, then it should belong to the whole of humanity, as one of our more valuable new acquisitions.

We appreciate the help given by legislators and senior civil servants in working out a 'Declaration of principles on extraterrestrial intelligence', which has already been endorsed by the International Academy of Astronautics, the International Institute of Space Law and the International Astronomical Union.

So as to complement this cooperation from a practical point of view, I have proposed setting up a ' SETI global network' which would serve as a hot line between the scientists, engineers, and technicians engaged in the SETI work in the field. The development of this network is endorsed by the SETI committee of the International Academy of Astronautics and the Bioastronomy commission of the International Astronomical Union.

On the budget side, NASA has operated on two million dollars per year for its SETI work since 1982. Recently Congress requested 100 million dollars over ten years, and, for the fiscal years 1991-2-3, 12 to 14 million dollars were granted. In France, apart from the valuable and considerable asset represented by Nançay radiotelescope and its staff, SETI subsists only on a minor part of the budget of a few institutions. Greater commitment is not only desirable but highly recommended.

Bioastronomers held their first international symposium at Boston in 1984, a second one on Lake Balaton in Hungary in1987, and at my suggestion at the foot of the Mont Cenis pass in the Alps in 1990, when the theme was 'Bioastronomy : the exploration widens'. The next one will be in 1993 in California.

SETI also holds international meetings as part of the Planetary society congresses, COSPAR (Commitee on Space Research) congresses, and the congresses of the International Astronautical Federation and the International Academy of Astronautics. In June 1994 the SETI committee of the IAA will organize a symposium on 'SETI and society' in Chamonix Mont Blanc in order to ask the international learned communities from all fields what would be their reactions in case we discover we are not alone in the universe.

EPILOGUE

Are we going to find a terrestrial planet orbiting a star ? or amino acids in interstellar space, adenine on Titan, or frozen bacteria in the Martian subsoil ? Shall we pick up an artificial extraterrestrial signal, be it only:

II/ II/ II/ II///II/II/ II/II II II II II///II/II/ ?

Has the cosmic odyssey which took place (and is still taking place) on Earth has its equivalents at other times and other places ? Will others one day enlighten us with a more evolved and deeper understanding of the cosmos, a more precise knowledge of our future ?

Above all, if all these enterprises make progress, shall we become better instructed, wiser and more far-sighted to take intelligent hold of a destiny which could still go for thousands of millions of years of hapiness in store for us ? It is too late to turn back now : the technological, scientific, cultural, philosophical, and ethical spin-offs of bioastronomy and SETI will be immense.

For further reading

Cosmic Odyssey, J.Heidmann, Cambridge University Press, 1989.

Bioastronomy ; the search for extraterrestrial life, J. Heidmann & M.J. Klein, Springer-Verlag, 1991.

Intelligences Extraterrestres, J.Heidmann, Ed.Odile Jacob, 1992 (being translated in English by Cambridge University Press).

(*) This paper is adapted from a text in French by J.Heidmann published in N°44 of the periodical Espace Information (Feb.1990) which is produced by the Publications Department of CNES, Centre Spatial de Toulouse, 18 avenue Belin, 31055 Toulouse Cédex. Tel . (33)61 27 31 31, and has been translated into English by Hugh Casement, D-8011 Forsten. whom I thank.

(§) Jean Heidmann is a Member of the International Academy of Astronautics and of its SETI Committee and Secretary of the Bioastronomy Commission of the International Astronomical Union.

BIOCHEMICAL MARKERS IN PRECAMBRIAN SEDIMENTS - INDIAN SUBCONTINENT[1]

Sneha S. Rane, A.V. Patankar,, M.S. Chadha
Bhabha Atomic Research Centre
Bombay, Maharashtra, 400 085, INDIA

B. Udayraj and S.M. Naqvi
National Geophysical Research Institute
Hyderabad, Andhra Pradesh, 500 007, INDIA

ABSTRACT

Biochemical markers are present in sediments as extractable and insoluble organic compounds. They were isolated from Precambrian sediments - carbonaceous shales and carbon phyllites > 1.5 Ga. The characterization of extractable compounds was carried out using Gas Chromatography, whereas techniques like elemental analyses, pyrolysis combined with Gas Chromatography, and Gas Chromatography - Mass Spectrometry and X-ray diffraction were used for the kerogens. Extractable organic compounds showed no peaks on GC; however, the presence of graphite was noted in a few kerogen samples from elemental analysis and XRD patterns.

1. INTRODUCTION

Studies on some ancient rock samples from Southern India have revealed the presence of microbiota (Taylor et al., 1988; Meijerink et al., 1984; Venkatachala et al., 1989 and Naqvi et al., 1987). It was therefore of interest to explore the presence of biochemical remains in such rock samples.

It is well known that most of the palaeontological and geochemical information in Precambrian sediments has been destroyed. However, some information may still be retained as molecular fossils in the organic matter of sedimentary rocks of this era. Two approaches have been adopted in the search for such biological markers. One is to examine the solvent extracts of the rock and the other is to investigate the kerogen, an insoluble macromolecular material with no regular structure, obtained by treating the rock samples with acids.

[1]This work was supported by Department of Science and Technology, New Delhi.

ISBN 0-937194-31-X

Various problems such as low concentrations of indigenous organic matter, in situ contamination even prior to collection and extraction of samples, laboratory contamination and the possibility of migration of younger samples through ground water are associated with Precambrian samples which generally provide ambiguous evidence of early biochemical processes. Despite this warning, considerable research on extractable organic matter from Precambrian sediments have been reported (Hayes et al., 1983).

The potential contamination problem associated with soluble extracts has resulted in adopting the second method of investigating the kerogen. Moreover, it has also been concluded (Hoering, 1967; Mckirdy, 1974b) that kerogen is likely to be syngenetic and uncontaminated and hence may be a more reliable source of molecular fossils than extractable organic matter. Since kerogen cannot be extracted with solvents, alternative methods for converting kerogen into molecules that may be analyzed by modern instrumental methods have been reviewed by Hayes et al. (1983).

The molecular fossils bonded to kerogen are highly unstable and get transformed easily. This intrinsic instability of complex molecular structures in kerogen sets a limit to this degradative approach. Hunt (1979) indicated schematically the spontaneous alteration of kerogen involving hydrogen disproportionation reactions where the substrate lost hydrogen in the form of petroleum. The residual, hydrogen-depleted material condenses and aromatizes eventually to form graphite.

Leventhal et al. (1975) have studied the elemental composition and flash pyrolysis of kerogen in pre-Phanerozoic sediments. They suggested that carbonaceous materials even in least metamorphosed rocks contain amorphous carbon or graphite and do not yield molecules with extensive structure.

Wedeking et al. (1983a) have investigated elemental analysis and X-ray diffraction characteristics of kerogenous residues isolated from cherts, carbonates and shales belonging to the Proterozoic and Archaen sediments. They concluded that Precambrian materials have suffered severe alteration involving development of significant degrees of graphitic structure and extensive dehydrogenation and the chances of syngenetic chemical fossils in Archaen sediments analyzed appeared to be remote.

On the other hand, well preserved organic matter in Precambrian sediments have been reported. Hoering (1978) has identified normal, isopronoid, iso- and anteiso-hydrocarbons from the Nonesuch Shale ($\sim$1.0 Ga, northern Michigan, USA) although he was unable to detect steranes and triterpane hydrocarbons. Grantham (1986) has reported the origin of crude oils from South Oman from late Precambrian source rocks.

The research on Precambrian kerogen puts forth a possibility that oil generating reactions may not have proceeded to absolute completion, there may be a part of kerogen that is completely dehydrogenated while other parts may retain traces of intact structures. Thus, all evidence points to the fact that a further search requires selective methods combined with considerable sensitivity.

The present study was extended to explore both the methods of investigating the extractable organic compounds and kerogen to determine the presence of molecular fossils in Precambrian sediments. The rock samples were extracted with 3:1 benzene:methanol before and after treatment with acids and only with benzene during the acid treatment (scheme-1). The extractable organic compounds obtained as four different extracts (scheme-1) were analyzed on gas-liquid chromatography. On the other hand, various instrumental techniques viz. elemental analysis, pyrolysis-gc, pyrolysis-gcms and X-ray diffraction were applied for the characterization of kerogen.

2. EXPERIMENTAL

2.1 DESCRIPTION OF SAMPLES

The carbonaceous shale samples selected for analysis were obtained from the Cuddapah Basin, and black carbon phyllites from Sandur Basin, both of India. All the samples analyzed were $>$ 1.5 Ga of age (Taylor et al., 1988 and Meijerink et al., 1984).

All the six samples collected from the Cuddapah Basin were black shales (Table-1). All the nine samples analyzed were supplied by National Geophysical Research Institute (NGRI), Hyderabad, India.

2.2 CONTAMINATION CONTROLS

The degree of contamination of the samples in the geological environment prior to collection has not been assessed. However, recognizable laboratory contamination was reduced by carefully distilling all the solvents and by eliminating all organic materials (e.g., plastic, rubber, paper etc.) from the experimental procedures.

2.3 ANALYTICAL PROCEDURE

After brushing the surfaces of the specimens under running distilled water to remove loose or friable material, the samples were shattered by hammering and pieces of 2-3 cm. in size were collected. These pieces were sonicated with double distilled water and 0.1N HCl for 15-20 min. washed again with double distilled water until they were free of acid, dried overnight in an oven at 50°C and crushed to a powder of - 240 mesh size using steel mortar and

SCHEME-1

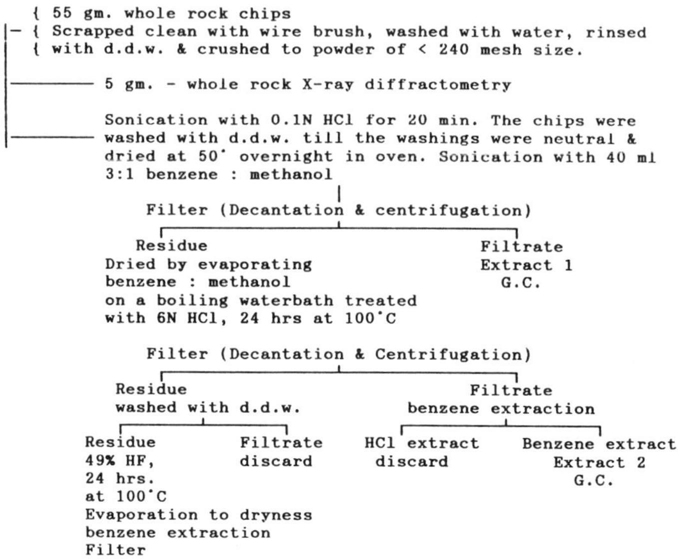

Extract 1 : To confirm the presence/absence of surface contamination

Extract 2 : To check if any extractable matter is liberated during HCl treatment.

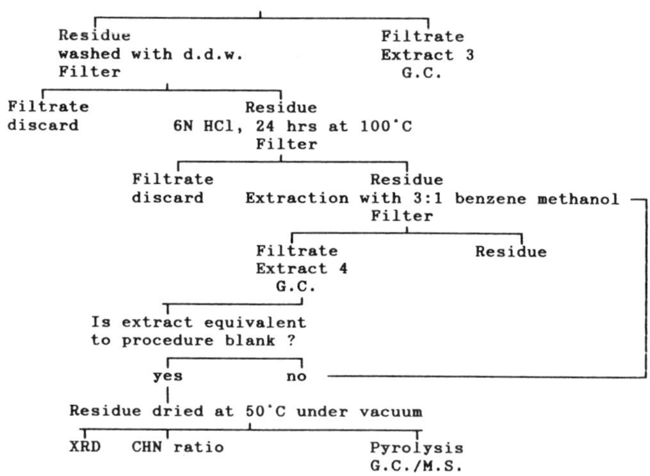

Extract 3 : To check if any extractable matter is liberated during HF treatment.

Extract 4 : To check if any extractable matter is present/absent on complete acid treatment.

pestle and agate mortar, and collected. Preparation of kerogen from rock powder by acid treatment (Walters, 1981; Wedeking et al., 1983b) is described in scheme-1.

Table 1. Description of samples[a]

S. No.	Geological code number	Colour	Type	Place
1.	TAD-1 (T-1)	Pale brown	Carbonaceous Shale	Cuddapah basin
2.	VEM-12 (V-12)	Cream	"	"
3.	VEM-43 (V-43)	Grey	"	"
4.	CUM-3 (C-3)	Grey	"	"
5.	CUM-6 (C-6)	Black	"	"
6.	CUM-7 (C-7)	Grey	"	"
7.	CM-151	Black	Carbon phyllites	Sandur basin
8.	CM-152	Black	"	"
9.	CM-153	Black	"	"

[a]Samples with serial numbers 1-6 were received in the form of rock powder with particle size less than 240 mesh, Samples with serial numbers 7-9 were received as rock pieces of about 5 to 10 centimeters in size.

Neoformed fluorides (Durand and Nicaise, 1980) were removed using a solution of $AlCl_3$ (Wedeking et al., 1983). It was presumed that neoformed fluorides were present in those samples which gave kerogen 2%. Thus, fluoride-containing samples, previously rinsed to neutrality, where shaken with approximately 200 ml of 1.2 M $AlCl_3$ (prepared by dissolving 250g of $AlCl_3.6H_2O$ in 750g water) for 24 hrs at 25°C, after which the aqueous phase was removed and discarded. The resulting residue was washed with distilled water and dried in vacuo and weighed.

2.3.1 Elemental analysis

Elemental analysis was carried out using elemental analyzer 1106 of "Carlo Erba Instruments" with data processor DP110 PRC-200. Replicate analyses of individual kerogen samples and of a standard compound, acetanilide, indicated that the accuracy of these measurements was ±0.3%.

2.3.2 Gas chromatography (GC)

Gas chromatographic analysis was carried out on a "Shimadzu" GC-16A with flame ionization detector (FID). The conditions for GC were : Analytical columns, 3% OV-17 and 2% OV-17 on gas chrom Q with a particle size of 80-100 mesh, N_2 gas, 40 ml/min. as a carrier, temp. programming from 80-260°C at 4 and 8°C/min. The sensitivity of GC determined in terms as minimum measurable quantity was 25 nanograms using tricosene as a standard.

2.3.3 Pyrolysis-GC and pyrolysis-GCMS

The pyrolyzer used was a "Shimadzu" PYR-2A with B-10 type transformer. The temperature range was from 0-1000°C. Pyrolysis was carried out at 800°C for 10-15 min. in the presence of pure N_2 and the pyrolyzates were collected (i) in 3 ml n-hexane at room temperature and further concentrated to 0.1 ml by evaporating the solvent by passing a slow stream of N_2 gas and (ii) under dry ice, diluted with 0.1 ml n-hexane. About 1-2 microlitres of this pyrolyzate solution was injected in GC. Wherever GC analysis showed some peaks, GC-MS analysis of that sample was also carried out under identical conditions of GC on "Shimadzu" GC-MS QP-1000A. The sensitivity of GC-MS was determined to be 250 nanograms using tricosene as a standard.

2.3.4 X-ray diffraction analysis

X-ray diffraction analysis was carried out using Philips PW 1050, Ni filtered Cukα radiation λ = 1.5418 A°. The patterns were recorded in the 2θ range, 7 - 66° with a chart speed of 20 mm/min. One cm. on the chart corresponds to 1° (2θ). Spectroscopic grade graphite was used as a standard. No interference was noted from blank glass slides.

3. RESULTS AND DISCUSSION

3.1 ELEMENTAL ANALYSIS

Except few sediments, reports of elemental analyses of Precambrian kerogens are relatively rare. Mckirdy and coworkers (Mckirdy and Hahn, 1982; Mckirdy and Powell, 1974b) have summarized these few reports and tabulated the data in terms of the observed atomic ratio of hydrogen to carbon. More than half of the reports indicated H/C ratios below 0.5 and thereby suggested that most materials must have been heavily altered. However, this picture underestimated the general extent of alterations of Precambrian organic materials for two reasons. Firstly, most investigators do not submit graphitic materials for elemental analysis; and secondly, systematic errors may arise in those samples that are rich in inorganic material as they carry excess hydrogen in the form of water of hydration, bound to the inorganic contaminants. This error will shift the H/C ratios to higher, rather than lower values.

Oehler (1977) demonstrated the irreversible absorption of hydrocarbons by kerogens. The pyrolytic release of benzene and water at 600°C from some of the Isua graphitic residues proves that some of the measured hydrogen is not part of the graphitic structure.

Table 2 summarizes the results of elemental analyses of all the nine samples studied. The measured hydrogen to carbon atomic ratio of all the samples studied lies in the range of 0.0009 to 0.1506. In four samples viz. C-6, CM-151, CM-152, and CM-153 out of the nine samples investigated (Table 2) in addition to low H/C ratio, the percentage of carbon was greater than 50. The significance of low H/C ratio combined with high values of percentage carbon may be explained on the basis of Fig. 1, which shows that these kerogens are more closely related to graphite than to the oil shales where chemical fossils are abundant (Mckirdy and Hahn, 1982).

The other five samples viz. T-1, C-3, C-7, V-43 and V-12 may not contain graphite as they have low percentage of carbon, although, their H/C ratios are similar to the four graphitic samples mentioned above (Table 2).

3.2 GAS CHROMATOGRAPHY (GC)

Smith et al. (1970) have investigated the presence of extractable organic matter in Precambrian cherts. They concluded that unless chemical criteria clearly indicative of a Precambrian origin can be established, the extractable constituents of very ancient sediments will provide very little or no interpretable evidence of early biological processes. They further suggested that early biochemical evolution may be studied in a better way by analyzing the insoluble kerogen-like fraction that comprises of organically preserved microfossils.

Gas chromatographic analyses of the extracts 1,2,3 and 4 (Scheme-1) for all the samples showed no peaks. This may be either due to the total absence of extractable organic matter especially hydrocarbons, or their presence in trace quantities which falls below the sensitivity of the instrument used for this work.

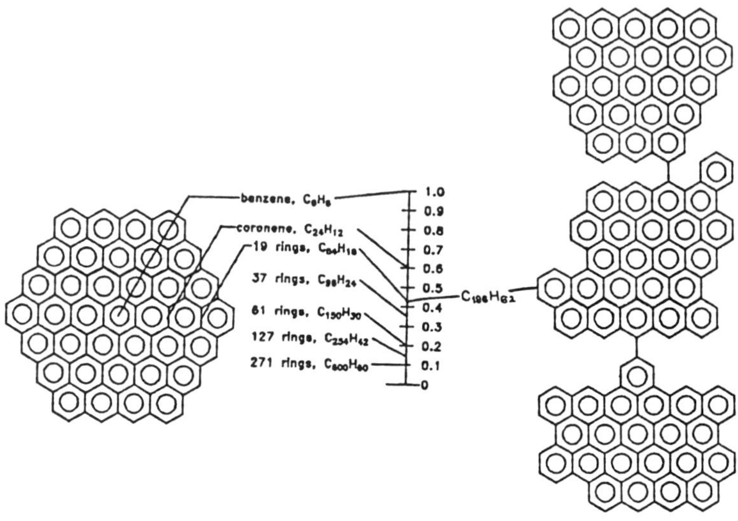

FIG. 1 Representative structures of aromatic hydrocarbons with low H/C ratios. Note that an H/C ratio of 1/n will be obtained from a fully aromatic hexagonal structure with concentric rings of carbon atoms (Hayes et al., 1983). Copyright by Princeton University Press.

Table 2. Elemental analysis of kerogen[3]

S. No.	Geological code number	Amount of kerogen recovered (g)	Elemental Analysis %N	%C	%H	H/C ratio
1.	TAD-1 (T-1)	1.0843	0.0000	0.4583	0.0457	0.1009
2.	VEM-12 (V-12)	0.2931	0.0000	0.9562	0.1514	0.1506
3.	VEM-43 (V-43)	1.0320	0.0150	6.2130	0.3300	0.0530
4.	CUM-3 (C-3)	1.0120	0.0000	1.2700	0.0300	0.0236
5.	CUM-6 (C-6)	1.0352	0.1840	68.1420	0.7720	0.0115
6.	CUM-7 (C-7)	0.6210	0.0000	4.5000	0.0350	0.0770
7.	CM-151	1.6580	0.0240	49.4300	0.6400	0.0130
8.	CM-152	0.3429	0.2150	89.8650	0.0850	0.0009
9.	CM-153	1.6020	0.0000	83.0100	0.0000	-

[3] Samples were treated once with acid and if necessary with 1.2M $AlCl_3$ solution.

3.3 PYROLYSIS-GC, PYROLYSIS-GCMS

The thermal dissociation of organic matter into simple compounds and the subsequent analysis of these compounds can be used to determine the component parts of large, insoluble structures which defy conventional methods of analysis. The location of the initial bond rupture in the thermally excited molecule is characteristic of that molecule. The initial dissociation into free radicles is often followed by further decompositions.

Pyrolysis of all the samples studied (Table-1) produced no organic molecules which could be considered chemical fossils. The virtual absence of organic molecules from the samples agrees well with their H/C ratios. Thus pyrolysis-GC supports the evidence obtained from elemental analyses that biological markers are absent in five samples (viz., T-1, C-3, C-7, V-43, V-12) or their concentration lies below the sensitivity level (25ng) of the GC used for this study; whereas the remaining four samples (viz., C-6, CM-151, CM-152 and CM-153) may contain amorphous carbon or graphite. It also proves that the sample preparation procedure has removed all extractable organic contaminants as none were observed by pyrolysis-gas chromatographic analysis. These results are in agreement with those reported by Hoering et al. (1987). According to them, small amounts of hydrocarbon were found in certain lower Cambrian and upper Precambrian sediments with normal alkanes predominating; however, the yields diminished with increasing age even for well preserved rocks where the minerals showed little effect of metamorphism. Sample with ages greater than 1.6 Ga yielded no detectable hydrocarbons.

3.4 X-RAY DIFFRACTION

The technique of measurement of graphite crystallinity has been applied in the kerogen analysis as only crystalline materials provide well defined X-ray diffraction patterns. The first report correlating graphite crystallinity to the increasing levels of metamorphic alterations of the Early Proterozoic Biwabik Iron Formation (of northern Minnesota, USA) came in 1964 (French, 1964). Most of the literature cited suggests correlation between various X-ray parameters and other indicators of maturation and metamorphism. Although fully ordered graphite displays many peaks in its diffraction pattern, interest has centered on the main (002) reflection because it is the first to appear in the diffraction pattern of kerogen as it develops from an amorphous to an ordered crystalline structure. The location of this peak is related to the spacing between the layers of aromatic rings in graphite. The sharpness of the peak is related to the regularity and the size of the crystallites. Highly disordered material displays a broad ill-defined peak, while fully ordered material in which all the ring planes are evenly separated by the interlayer distance characteristic of perfectly crystalline graphite displays a very sharp peak if the crystallites are at least a few thousand angstroms in diameters.

Besides this, the crystallinity of graphite materials may be correlated with the burial history of the host rock. Some development of graphite structure apparently begins under conditions of very low grade metamorphism, just beyond the stage of organic geochemical catagenesis. The width of the 002 peak is generally less than 5° 2θ in the chlorite zone of the greenschist facies, with the width dropping to and below 1° 2θ by the close of greenschist facies metamorphism. Nearly complete dehydrogenation takes place in this interval, with H/C values dropping from barely post metagenetic levels to values near zero as the greenschist facies is traversed. Higher levels of metamorphism lead to the development of fully ordered graphite.

Wedeking et al. (1983a) have investigated carbonization of Precambrian kerogens. They have reported that there is marked discontinuity in the graphitization of natural kerogens in the range of H/C ratios below 0.1. They have also reported that when a certain degree of crystallinity is attained, corresponding to a range of W1/2 from approximately 5° to approximately 3° (2θ) regardless of sample origin, the development of well ordered graphitic structure is sudden and virtually complete (in agreement with the transition noted by McKirdy et al., 1975). Numerous workers (Oberlin et al., 1980 and McKirdy et al., 1975) have also noted that graphitization does not proceed smoothly and continuously, but can involve sudden changes above a certain degree of crystallinity close to that of true graphite. This behaviour probably occurs after breakage of the covalent cross-linked structure is complete, resulting in rapid development of a parallel structure associated with increased mobility of crystallites. Evidently, when an H/C ratio of ~ 0.1 is reached, most (if not all) crossed-linked structure is lost and graphitization proceeds to completion.

In the present study, four samples (viz., C-6, CM-151, CM-152 and CM-153) out of the nine samples studied gave sharp (002) reflection with W1/2 $<$ 2° (2θ) very similar to that of spectroscopic graphite (Fig. 2, Table 3). The lower values of W1/2 combined with low H/C ratio ($<$ 0.1) (Table 3) suggests that in these samples, graphitization has proceeded to completion and hence all the information about biochemical markers has been lost.

4. CONCLUSION

Precambrian samples from the Cuddapah and Sandur Basin of India with an age of $>$ 1.5 Ga were studied for the presence of

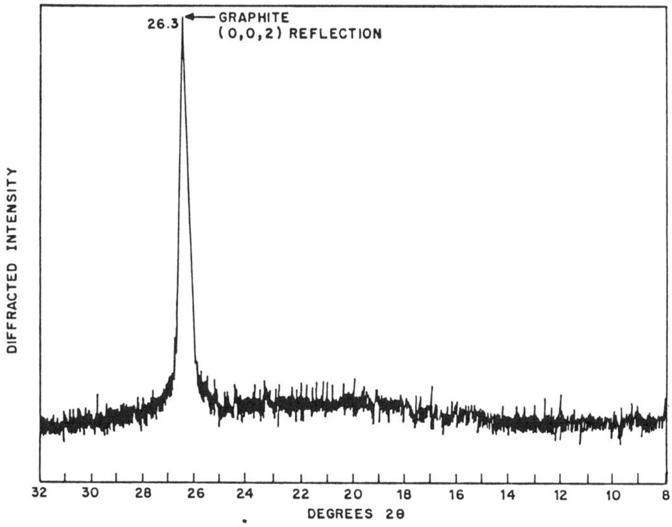

FIG.2B X-RAY DIFFRACTION OF CM-152 USING CuK$_\alpha$, λ = 1.5418 Å

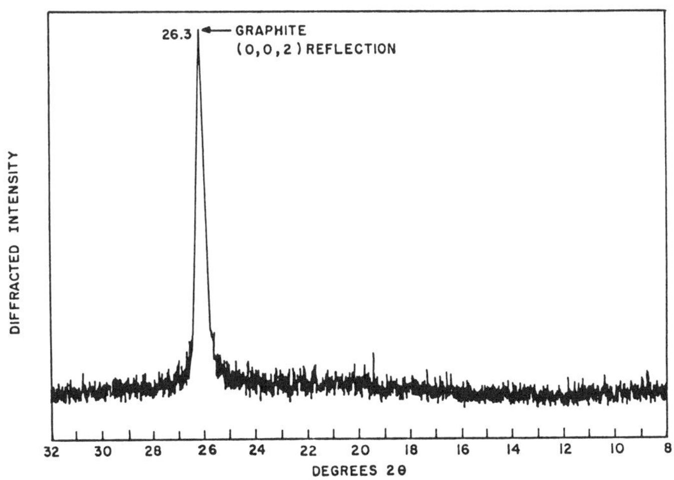

FIG.2A X-RAY DIFFRACTION OF SPECTROSCOPIC GRADE GRAPHITE (STANDARD SAMPLE) USING CuK$_\alpha$, λ = 1.5418 Å

biochemical markers which are usually present as extractable organic matter and the insoluble matter, kerogen. Extractable organic compounds were absent from the samples studied. Four kerogen residues (viz., C-6, CM-151, CM-152 and CM-153) out of the nine samples studied showed low H/C ratio ($\sim$0.1) and low W1/2 ($<$ 2°,2θ) and thereby indicated the virtually complete graphitization of these samples. The results are in agreement with those reported by Wedeking et al. (1983a).

ACKNOWLEDGEMENTS

Our grateful thanks are extended to Department of Science and Technology for financial support and associateship given to S.S.R. We also thank Shri D.R. Suryavanshi, Bio-Organic Division, BARC for carrying out elemental analysis and Dr. P.N. Moorthy, Head, Applied Chemistry Division, BARC for his help on XRD analysis.

Table 3. Structural characteristics of kerogen[+]

S.No.	Code No.	X-ray diffraction parameters		H/C ratio
		$W_{1/2}$ 2θ	d_{002},A	
1.	Spectroscopic grade graphite	1.40	3.4269	-
2.	C-6	0.50	3.3022	0.0115
3.	CM-151	1.15	3.3510	0.0130
4.	CM-152	1.80	3.4269	0.0009
5.	CM-153	1.00	3.3634	-

[+] Correlations between X-ray diffraction parameters and H/C ratio of graphitic samples.

REFERENCES

Durand, B. & Nicaise, G. (Eds) 1980 : Procedures for kerogen isolation, kerogen, Technip, Paris.

French, B.M.; 1964: Graphitization of organic material in a progressively metamorphosed Precambrian Iron Formations, Science, 146, 917-918.

Grantham, P.; 1986: The occurrence of unusual C_{27} and C_{29} Sterane prodominences in two types of Oman crude oil, Organic Geochemistry, 9, 1-10.

Hayes, J.M.; Kaplan, I.R. & Wedeking, K.W.; 1983: Precambrian Organic geochemistry, Preservation of the record. Schopf J.W. (Ed), Origin and evolution of the Earths Earliest Biosphere, Princeton University Press, Princeton, NJ.

Hoering, T.C.; 1978: Molecular fossils from the Precambrian Nonesuch shale. C. Ponnamperuma (Ed), Comparative Planetology, Academic press, New York.

Hoering, T.C. and Navale, V.; 1987: A search for molecular fossils in the kerogen of Precambrian sedimentary rocks, Precambrian Research, 34, 247-267.

Hoering, T.C.; 1967: The Organic geochemistry of Precambrian rocks. P.H. Abelson (Ed), Researches in Geochemistry, Wiley, NY.

Hunt, J.M.; 1979: Petroleum Geochemistry and Geology, Freeman and Co., San Francisco, CA.

Leventhal, J.; Suess, S.E. and Cloud, P.; 1975: Nonprevalence of biochemical fossils in kerogen from the pre-Phanerozoic sediments, Proc. Nat. Acad. Sci. U.S., 72, 4706-4710.

Mckirdy, D.M. and Hahn, J.H.; 1982: Composition of kerogen and hydrocarbons in Precambrian rocks. H.D. Holland and M. Schidlowski (Eds), Mineral Deposits and the evolution of Biosphere, Dahlem Konferen Zen Dahlem workshop Rept. 16, Berlin, Springer.

Mckirdy, D.M. and Powell, T.G.; 1974a: Metamorphic alteration of carbon isotopic composition in ancient sedimentary organic matter : New evidence from Australia and South Africa, Geology, 2, 591-595.

Mckirdy, D.M.; 1974b: Organic geochemistry in Precambrian research, Precambrian Research, 1, 75-137.

Mckirdy, D.M.; Sumartojo, J.; Tucker, D.H. & Gostin, V.; 1975: Organic mineralogic and magnetic indications of metamorphism in the Tapley Hill Formation, Adelaide, Geosyndine, Precambrian Research, 2, 345-373.

Meijerink, A.M.J.; Rao, D.P. and Rupke, J.; 1984: Statigraphic and structural development of the Precambrian Cuddapah Basin, S.E. India, Precambrian Research, 26, 57-104.

Naqvi, S.M.; Venkatachala, B.S.; Shukla, M.; Kumar, B.; Natarajan, R. and Sharma M.; 1987: Silicified Cyanobacteria from the cherts of Archaean Sandur Schist belt, Karnataka, India, Jour. Geol. Soc. India, 29, 535-539.

Oberlin, A.; Boulmier, J.L. and Villey, M.; 1980: Electron microscopic study of kerogen microtexture. Selected criteria for determining the evolution stage of kerogen, Durand, B. (Ed), Kerogen Techniq, Paris.

Oehler, J.H.; 1977: Irreversible contamination of Precambrian kerogen by ^{14}C-labelled organic compounds, Precambrian Research, 4, 221-227.

Smith, J.W.; Schopf, J.M. and Kaplan, I.R.; 1970: Extractable organic matter in Precambrian cherts, Geochim. Cosmochim. Acta, 34, 659-675.

Taylor, P.N.; Chadwick, B.; Friend, C.R.L.; Ramakrishnan, M.; Moorbath, S. and Viswanathan, M.N.; 1988: New age data on the geological evolution of Southern India, Workshop on the Deep Continental Crust of Southern India. Jour. Geol. Soc. India, 31, 155-157.

Venkatachala, B.S.; Shukla, M.; Sharma, M.; Naqvi, S.M.; Srinivasan, R. and Uday Raj, B.; 1990: Archaean microbiota from the Donimalai formation, Dharwar supergroup, India, Precambrian Research, 47, 27-34.

Walters, C.C.; 1981: Laboratory of Chemical Evolution, Organic Geochemistry of the 3,000 million year old metasediments from Isua, Greenland, Ph.D. dissertation, Univ. of Maryland, 115.

Wedeking, K.W. and Hayes, J.M.; 1983a: Carbonization of Precambrain kerogens, Advances in Organic Geochemistry, 546-553.

Wedeking, K.W.; Jayes, J.M. and Matzigkeit, U.; 1983b: Procedures of Organic Geochemical Analysis, J. William Schopf (Ed), Earth's Earliest Biosphere, Its Origin and Evolution, Princeton University Press, Princeton, New Jersey.

PRACTICABILITIES AND LIMITS OF STEREOSPECIFIC AUTOCATALYSIS: AN EXPERIMENTAL APPROACH

T. Buhse and W. Thiemann
Universität Bremen
Fachbereich Chemie, W-2800 Bremen 33 (Germany)

D. Lavabre and J.-C. Micheau
Université Paul Sabatier
Laboratoire des IMRCP, UA au CNRS N° 470, F-31062 Toulouse (France)

ABSTRACT

Spontaneous amplification of enantiomeric excess generated by a stereospecific and autocatalytic chemical reaction network is predicted by several theoretical models for the process of molecular chiral symmetry breaking. A practical approach to these models is presented here by the examination of a non-linear chemical system provided with the basic kinetic characteristics of these models. Although laboratory experiments give no evidence for amplification capacity so far, it is shown by kinetic analysis and computer experiments that the dynamics of the reacting system allows powerful amplification of enantiomeric excess in principle, but it also marks out the existence of a fundamental and strong limit for such kind of processes in general: the high sensitivity towards even a small loss of absolute configuration of the involved chiral reacting species. It is also demonstrated that the strength of the kinetic feedback mechanism affects the amplification capacity in a minor degree, and that the system possesses several bifurcation points regarding stationary values of the enantiomeric excess under flow conditions although the reacting system shows only weak autocatalysis and no kinetic bistability.

1. INTRODUCTION

Two main aspects within the scope of the pronounced molecular homochirality of living matter have attracted scientific interest for more than one century:

1. The primary source of chiral-symmetry-breaking states in chemical systems caused by physical forces, and

2. The dynamics of amplification of initial enantiomeric excess (*ee*) caused by non-linear chemical kinetics, where *ee* is defined as a function of the concentrations of the R- and S-enantiomers:

$$ee = \frac{([R] - [S])}{([R] + [S])} \tag{1}$$

ISBN 0-937194-31-X

Until now both aspects are far from being completely understood in their full range, and in contrast to the development of theoretical models there is - especially concerning the concepts of spontaneous amplification of *ee* by non-linear chemical dynamics - a striking deficiency of available experimental data (Bonner, 1991; Mason, 1989; and De Min et al. , 1988).

In the past we have performed quite a number of experiments and theoretical models for proving an energy difference between enantiomeric species, resulting from a parity violation of weak interaction (Thiemann, 1974a; 1974b; 1975; 1981; and Thiemann and Darge, 1974). In this paper we focus on the possibility of amplifying such a difference in an homogeneous solution by selecting a suitable chemical system with autocatalytic characteristics, - independent on the assumption whether we start from an inherent disparity or from a spontaneous inhomogeneity due to stochastic fluctuations of the molecular moieties involved. This could serve as a model for shaping the chiral biosphere of today in the course of prebiological evolution on the early Earth. It is in our opinion irrelevant whether the system described below has anything to do with the biochemical molecules which we find today as the building blocks of the biosphere.

In general most of the theoretical models proposed so far to explain spontaneous amplification of *ee* in chemical systems (Frank, 1953; Seelig, 1971; Decker, 1979; Iwamoto and Seno, 1979; 1982; and Kondepudi and Nelson, 1984; 1985) invoke non-linear phenomena which are closely related to the concepts of selforganization and dissipative structures developed for systems far from equilibrium by the pioneering work of Glansdorff and Prigogine (1977). But already 40 years ago it was Frank (1953) who presented as one of the first a model of an open autocatalytic chemical system in which a small *ee* (either induced by physical forces or by stochastic fluctuations) is strongly amplified up to 100% while the almost initial racemic state of the system becomes instable.

$$A + R \longrightarrow 2\,R \tag{2a}$$

$$A + S \longrightarrow 2\,S \tag{2b}$$

$$R + S \longrightarrow \text{inactive products} \tag{3}$$

Although the above kinetic scheme is entirely symmetric, two autocatalytic reaction sequences (in which an achiral substrate molecule (A) reacts either with an R- or S-enantiomer that serves as a catalyst of its own production) and a third step of mutual inhibition (eliminating the two enantiomers in respect to one another irreversibly as an inactice product) are sufficient to give rise to a strictly homochiral final state starting with a not strictly racemic mixture ($ee \neq 0$, but no matter what quantity) and performing the calculations under conditions of a flow system in which a critical substrate input concentration is exceeded.

These concepts - as demonstrated by Frank's model - which are in general cited as stereospecific autocatalysis run counter to the common laboratory experiences of chemists, and only partial success has been made by attempts to illustrate them practically (Micheau et al. , 1987; Wynberg, 1989; and Alberts and Wynberg, 1990), and solely one system was recently reported by Kondepudi et al. (1990) which demonstrates a distinct effect of spontaneous chiral symmetry breaking by sodium chlorate crystallization.

Our tasks to design a chemical system which should demonstrate the practicabilities and limits of the concepts of stereospecific autocatalysis were made by the following attempts:

1. Search for literature data on a reacting system in which both autocatalysis and chiral effects are simultaneously present - there are only a few examples for such a system and most of the contributions deal either with stereochemical aspects related to organic chemistry or with autocatalytic kinetics related to the subject of physical chemistry.

2. Looking for a simple and homogeneous reacting system to clarify the role of the "pure" kinetics in this concept: There have been made several experiments on heterogeneous reacting systems (e.g. crystallization) which are more difficult to be analyzed because of the considerable influence of involved physical transport processes and thermodynamical effects.

3. Regarding the minimal criteria for stereospecific autocatalysis of such a reacting system: starting with an achiral substrate molecule which reacts to give a chiral intermediate which catalyses its own formation by reaction (directly or indirectly) with the substrate molecule in a feedback manner - and which is on the other hand removed continuously from the system to give kinetical stable products.

4. Performing a computer analysis of experimental data on basis of a kinetic scheme followed by simulations using this scheme and the kinetic parameters which were obtained by fitting of experimental data.

2. THE MODEL SYSTEM

It is well established that a wide range of organic compounds react with molecular oxygen *via* a radical chain mechanism at low temperatures in a relatively slow oxidation process. This type of reaction - cited as autoxidation reaction - which occupies a central role in biochemical mechanisms as well as in technological heat energy processes is often reported to show autocatalytic behaviour (Emanuel et al. , 1967; Howard, 1973; and Mill and Hendry, 1980), and in specific cases critical phenomena like bistability (Trainin and Zaitsev, 1989; 1990), oscillations (Colussi et al. , 1990), and propagating chemical waves (Boga et al. , 1990) were predicted or observed.

For our model system we have chosen an autoxidation reaction using a simple hydrocarbon - tetralin (RH) - as an achiral substrate which is well-known to be oxidized under mild conditions in liquid phase (Woodward and Mesrobian, 1953). RH and molecular oxygen react to give tetralin hydroperoxide (ROOH) which itself is chiral and which is partially decomposed by producing in the main part peroxyl radicals (ROO·) which consequently are chiral as well and which react with RH leading again to formation of further ROOH as well as to yield final products like tetralol (ROH) and tetralone (RO).

TABLE 1. AUTOXIDATION OF TETRALIN

Schematized chemical pathway for the autoxidation of tetralin (RH) leading to chiral tetralin hydroperoxide (ROOH) as a reactive intermediate and to α-tetralol (ROH) and α-tetralone (RO) in the final stages of oxidation. Autocatalysis is provided by the decomposition of ROOH to give chiral peroxyl radicals (ROO•) which catalyse the formation of further ROOH.

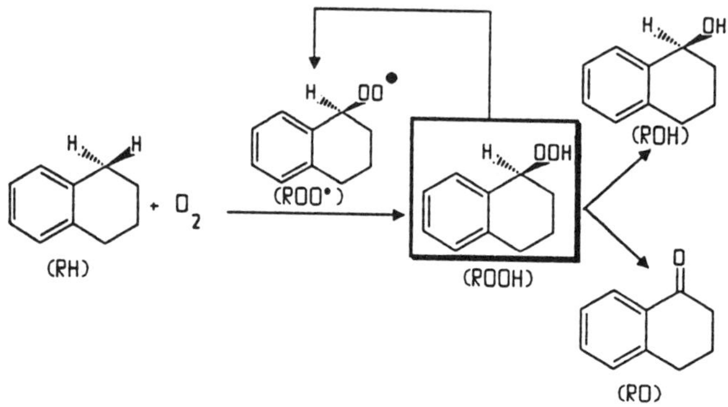

2. EXPERIMENTS AND COMPUTATIONS

Autoxidation experiments using a solution of RH as the substrate and $Cr(acac)_3$ as a catalyst which were dissolved in n-heptane and provided with a constant inlet-stream of O_2 were performed in a well-stirred double wall batch reactor at 60 °C. A plot of the time-evolution of the ROOH concentration, which was detected by a method described by Banerjee and Budke (1964), shows the characteristic S-shaped curvature of self-acceleration (Fig. 1) which indicates an autocatalytic formation of the ROOH species. Since herewith the basic criterion of stereospecific autocatalysis is realized (a chiral intermediate is produced from an achiral substrate by an autocatalytic pathway), two different experimental set-ups were performed:

a) Looking for an occurrence of optical activity regarding the R- or S-ROOH enantiomers while the autoxidation reaction was performed. These experiments were made under the hypothesis that the ROOH species which are present in the initial stage of the experiment are not in a completely racemic state because of statistical fluctuations. An amplification of this very small initial *ee* as a result of the inherent feedback mechanism should have been detected by using the very sensitive method of measuring the circular dichroism indicated by an increase of ellipticity at the wavelength of ROOH absorption. Taking aliquots of the reaction mixture after several time periods up to 500 min reaction time in none of the samples any optical activity was detected, which means that the system remains at the (almost) racemic state.

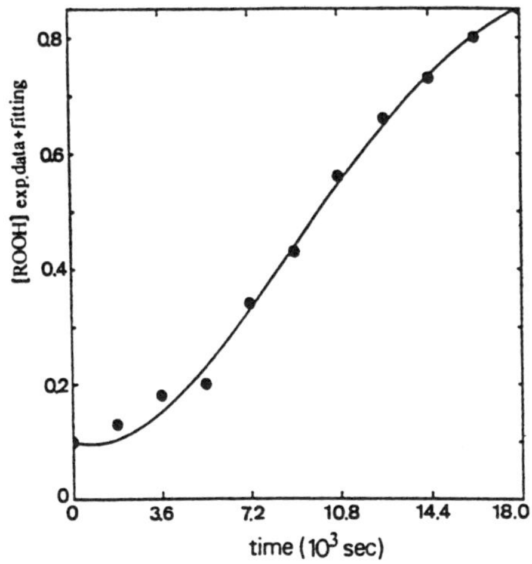

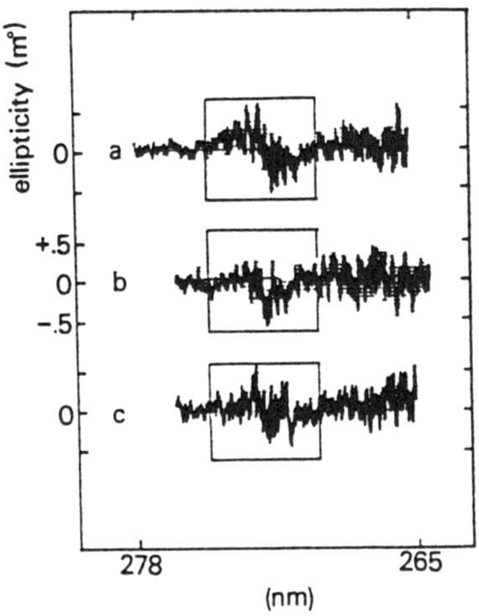

FIGURE 1.
Evolution of [ROOH] against time of a typical autoxidation experiment. $[RH]_0$ = 5 M, $[ROOH]_0$ = 10^{-2} M, $[Cr(acac)_3]$ = 10^{-3} M, O_2-inlet = 4 l h^{-1}. The solid line shows the fitting of experimental data by the 3-step model, using the following numerical values for the kinetic parameters :
$k_1 = 1.3 \cdot 10^{-4}$, $k_2 = 2.8 \cdot 10^{-4}$, $k_3 = 3.8 \cdot 10^{-3}$, and $k_4 = 3.5 \cdot 10^{-3}$.

FIGURE 2.
CD spectra of aliquots of reacting mixture (100 μl/100 ml n-heptane) taken after a) t = 0 min, b) t = 180 min, and c) t = 300 min running time of a system containing (R)-ROH as an asymmetric inducer. $[RH]_0$ = 5 M, $[ROOH]_0$ = $2 \cdot 10^{-2}$ M, $[Cr(acac)_3]_0$ = 10^{-3} M, $[R\text{-}ROH]_0$ = $6 \cdot 10^{-3}$ M, O_2-inlet = 4 l h^{-1}.

b) Testing the effect of an initial *ee* by adding an enantiomerically pure inducer to the reaction mixture. Since it was impossible to obtain R- or S-ROOH directly by synthesis R-ROH was used for that purpose. Although ROH is often reported to be a kinetical inactive compound of autoxidation processes there is some evidence for the occurrence of ROH(OOH) hydroperoxide species and related peroxyl radicals (Mizukami and Imamura, 1978) that might conserve the optical activity of R-ROH and could be able to interact with the key compounds in our experiment. Nevertheless as shown in Fig. 2 also in this case there is no effect of amplification of the initial optical activity detectable while the reaction is proceeding.

In the following a computer analysis was performed in order to find out the reason for the collapse of initial *ee* resp. arguments for the non-appearance of spontaneous chiral symmetry breaking in our model system, as well as to establish the kinetic and stereochemical parameters which are responsible for the above experimental results, and to carry out calculations by varying these parameters and by simulating an open system. Starting from the generally accepted kinetic scheme of an liquid-phase autoxidation as shown at Table 2 we designed a simplified 3-step model (Table 3) which fits the experimentally evaluated time evolution of the ROOH concentration (Fig. 1) and which takes into account the key-role of ROO· radicals which becomes determinant at comparable high O_2 concentrations like those which were present under our experimental conditions.

TABLE 2. KINETIC SCHEME OF AUTOXIDATION

RH	$\xrightarrow{O_2}$	free radicals	[I1]
M^{n+} + ROOH	$\longrightarrow$	$M^{(n+1)}$ + RO· + OH^-	[I2]
$M^{(n+1)}$ + ROOH	$\longrightarrow$	M^{n+} + ROO·⁻ + H^+	[I3]
RO· + RH	$\longrightarrow$	ROH + R·	[P4]
R· + O_2	$\longrightarrow$	ROO·	[P5]
ROO· + RH	$\longrightarrow$	ROOH + R·	[P6]
2 ROO·	$\longrightarrow$		
2 R·	$\longrightarrow$	inactive species	[T7]
2 RO·	$\longrightarrow$		

TABLE 3. 3-STEP MODEL

Simplification of the generally accepted kinetic scheme of autoxidation as 3-step model.

ROOH	$\longrightarrow$	ROO$^•$	[1]
ROO$^•$ + RH	$\longrightarrow$	ROOH + ROO$^•$	[2]
2 ROO$^•$	$\longrightarrow$	inactive species	[3]

Since the 3-step kinetic scheme - which seems to be a sufficient approach to the chemical kinetics of an liquid-phase autoxidation - does not reflect the stereochemical interactions of the model system, a mirror image of each kinetic step was constructed and stereospecific parameters (α, β, and γ) were introduced as a degree of the stereoselectivity of these respective elementary reactions (see Table 4).

In the case where all the three stereospecific parameters are set to the value of equal to 0.5 - which means there is no stereoselectivity in each step - the above extended scheme can be reduced to the simple 3-step model. But two additional assumptions must made to reflect the chiral interactions:

1. Considering the possibility of kinetic resolution based on a difference in reactivity between the symmetric recombination of two R- or two S-ROO· radicals and the asymmetric recombination of them (R-ROO· + S-ROO·) we construct two pairs of peroxyl radical recombination steps and assign them two different rate constants k_3 and k_4.

TABLE 4. 3-STEP MODEL WITH CHIRAL INTERACTIONS

Parameters α, β, γ ($0 < \alpha, \beta, \gamma < 1$) are introduced as a measurement of the stereoselectivity of these respective elementary reactions. ex:

$\alpha = 1$: inversion of configuration
$\alpha = 0.5$: no stereoselectivity
$\alpha = 0$: retention of configuration.

$$S\text{-ROOH} \xrightarrow{k1} \alpha R\text{-ROO}^{\bullet} + (1-\alpha) S\text{-ROO}^{\bullet} \quad [S1]$$

$$R\text{-ROOH} \xrightarrow{k1} \alpha S\text{-ROO}^{\bullet} + (1-\alpha) R\text{-ROO}^{\bullet} \quad [R1]$$

$$S\text{-ROO}^{\bullet} + \text{RH} \xrightarrow{k2} \beta R\text{-ROOH} + (1-\beta) S\text{-ROOH} + \gamma R\text{-ROO}^{\bullet} + (1-\gamma) S\text{-ROO}^{\bullet} \quad [S2]$$

$$R\text{-ROO}^{\bullet} + \text{RH} \xrightarrow{k2} \beta S\text{-ROOH} + (1-\beta) R\text{-ROOH} + \gamma S\text{-ROO}^{\bullet} + (1-\gamma) R\text{-ROO}^{\bullet} \quad [R2]$$

$$R\text{-ROO}^{\bullet} + R\text{-ROO}^{\bullet} \xrightarrow{k3} \quad [R3]$$

$$S\text{-ROO}^{\bullet} + S\text{-ROO}^{\bullet} \xrightarrow{k3} \quad \text{inactive species} \quad [S3]$$

$$R\text{-ROO}^{\bullet} + S\text{-ROO}^{\bullet} \xrightarrow{k4} \quad [RS4]$$

$$S\text{-ROO}^{\bullet} + R\text{-ROO}^{\bullet} \xrightarrow{k4} \quad [SR4]$$

$$R\text{-ROO}^{\bullet} \underset{k\text{-}5}{\overset{k5}{\rightleftharpoons}} S\text{-ROO}^{\bullet} \quad [5]$$

2. Taking into account the loss of optical activity by racemization an additional step 5 was introduced into the kinetic scheme (note that the model reaction proceeds *via* a free radical chain process !). Although the stability of peroxyl radicals against racemization is expected to be comparatively high, obviously the rate of racemization for the tetralyl radicals (R$^\cdot$) (which are for simplicity not explicitly included in the above scheme) is not negligible and might be the main source for loss of optical activity in the system.

On basis of the 3-step model including chiral interactions the following set of differential equations were used for our computations:

$$\frac{d[S\text{-ROOH}]}{dt} = -k_1[S\text{-ROOH}] + k_2(1-\beta)[S\text{-ROO}^\cdot][\text{RH}] + k_2\beta[R\text{-ROO}^\cdot][\text{RH}] \quad (4.1)$$

$$\frac{d[R\text{-ROOH}]}{dt} = -k_1[R\text{-ROOH}] + k_2(1-\beta)[R\text{-ROO}^\cdot][\text{RH}] + k_2\beta[S\text{-ROO}^\cdot][\text{RH}] \quad (4.2)$$

$$\frac{d[\text{RH}]}{dt} = -k_2[\text{RH}][R\text{-ROO}^\cdot] - k_2[\text{RH}][S\text{-ROO}^\cdot] \quad (4.3)$$

$$\frac{d[S\text{-ROO}^\cdot]}{dt} = k_2\alpha[R\text{-ROOH}] + k_1(1-\alpha)[S\text{-ROOH}] + k_2\gamma[R\text{-ROO}^\cdot][\text{RH}] + k_2(1-\gamma)[S\text{-ROO}^\cdot][\text{RH}] - k_2[S\text{-ROO}^\cdot][\text{RH}] - k_3[S\text{-ROO}^\cdot]^2 - k_4[S\text{-ROO}^\cdot][R\text{-ROO}^\cdot] - k_5[S\text{-ROO}^\cdot] + k_{-5}[R\text{-ROO}^\cdot] \quad (4.4)$$

$$\frac{d[R\text{-ROO}^\cdot]}{dt} = k_1\alpha[S\text{-ROOH}] + k_1(1-\alpha)[R\text{-ROOH}] + k_2\gamma[S\text{-ROO}^\cdot][RH] + k_2(1-\gamma)[R\text{-ROO}^\cdot][RH] - k_2[R\text{-ROO}^\cdot][RH] - k_3[R\text{-ROO}^\cdot]^2 - k_4[R\text{-ROO}^\cdot][S\text{-ROO}^\cdot] - k_5[R\text{-ROO}^\cdot] + k_{-5}[S\text{-ROO}^\cdot] \quad (4.5)$$

and

$$ee = \frac{([R\text{-ROOH}] - [S\text{-ROOH}])}{([R\text{-ROOH}] + [S\text{-ROOH}])} \quad (5)$$

Calculating the above rate equations introducing numerical values for the rate constants $k_{1,2,3,4}$ which were obtained by numerical fitting of experimental data, and considering the same initial concentrations of RH and ROOH like those in our laboratory experiment, it can be seen very drastically that only the cases where

- the stereospecifity of each elementary reaction is very close to 100% (α, β, and γ are almost equal to 0)
- the ratio of k_3/k_4 is kept in a small domain and must not be equal to 1
- and practically no racemization occurs in respect of the peroxyl radicals (k_5 resp. k_{-5} are very close to 0)

an increase of an initial enantiomeric excess of [*R*-ROOH], which was set as $ee_0 = 0.2$, is observable under batch conditions.

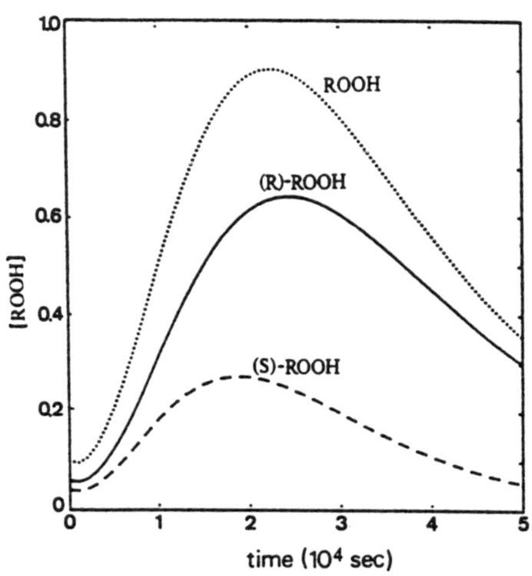

FIGURE 3. Computation of [ROOH] (dotted line), [*S*-ROOH] (dashed line), and [*R*-ROOH] (solid line) against time under batch conditions using parameters which lead to an increase of enantiomeric excess: $[RH]_0$ = 5 M, $[R\text{-ROOH}]_0 = 6 \cdot 10^{-2}$ M, $[S\text{-ROOH}]_0 = 4 \cdot 10^{-3}$ M ($ee_0 = 2 \cdot 10^{-1}$), $k_1,..,k_4$ are the same as in Fig. 1, $k_5 = 0$, and $\alpha = \beta = \gamma = 0$.

Since it is predicted by theoretical models - e.g. by means of Frank's model (1953) - that an open system is one of the boundary conditions for stereospecific autocatalysis, we introduced flow terms into the above rate equations in order simulate a continuous flow reactor system (CSTR). For this purpose we designed a CSTR containing an ee_0 = 0.2 (*R*-ROOH) at the initial time and then providing a constant flux of substrate (RH) and of racemic ROOH into the system and the corresponding quantity of reaction products out of it. By varying the flowrate (k_f) - resulting in different stationary concentrations of the reacting species - two scenarios were investigated: 1. the behaviour of a system using conditions which would lead to an increase of *ee* under batch conditions (eg. those as specified in Fig. 3), and 2. the time-evolution of *ee*, if a system is transfered into a CSTR which shows a decrease of *ee* under batch conditions (eg. α, β. or $\gamma > 0.05$).

As shown in Figs. 4a and 4b, in the 1. scenario (system which shows increase of *ee* under batch conditions) k_f triggers the system either into a racemic or chiral-symmetry-breaking state. But concerning the 2. scenario the system shows the same behaviour as in a closed system - no value of k_f was found that could be able to compensate the strong attraction caused by the racemic state.

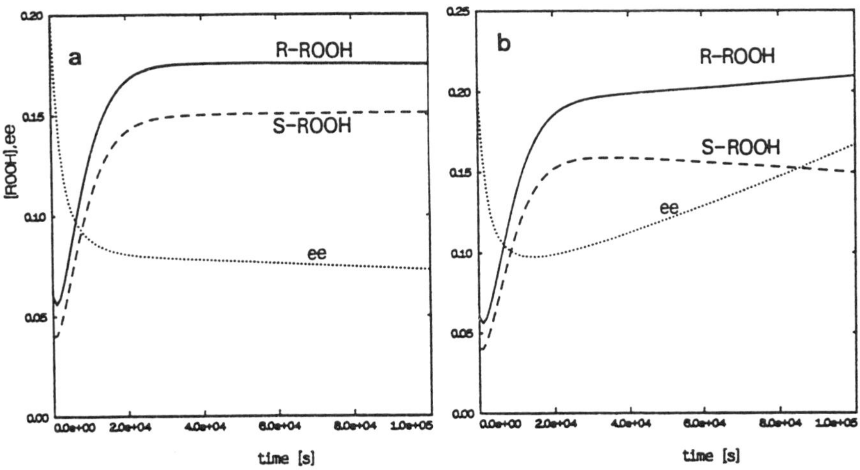

FIGURE 4. Simulations of the evolution of [*R*-ROOH] (solid line), [*S*-ROOH] (dashed line) in M, and of *ee* (dotted line) in a flow system by variation of the flow rate k_f and by maintaining the initial concentrations and rate constants (see Fig. 3) which are leading to an increase of *ee* under batch conditions.

Input concentrations of RH = 5 M and of *rac*-ROOH = 0.1 M.

a) $k_f = 2.15 \cdot 10^{-4}\ s^{-1}$: decrease of *ee* ;

b) $k_f = 1.95 \cdot 10^{-4}\ s^{-1}$: increase of *ee*.

(Bifurcation point is between $1.95 \cdot 10^{-4}$ and $2.15 \cdot 10^{-4}\ s^{-1}$).

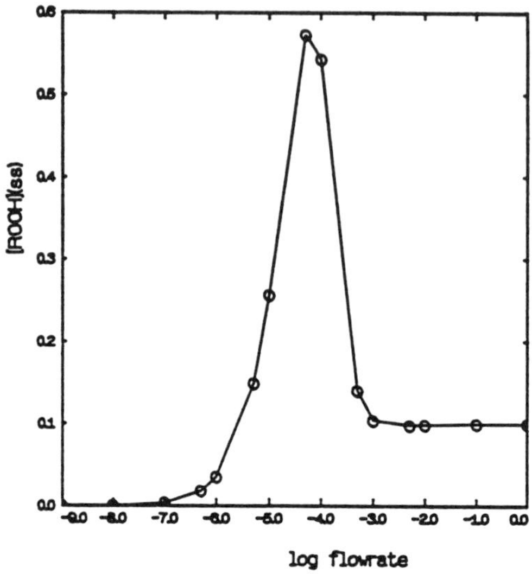

FIGURE 5. Simulation of a flow-diagram (k_f in s^{-1}) regarding [ROOH] in M by the 3-step model. ROOH accumulate for value of flow rate around $6.3 \cdot 10^{-5}$ s^{-1} ($10^{-4.2}$). Note that autocatalysis is too weak to give rise to unstable steady state (bistability).

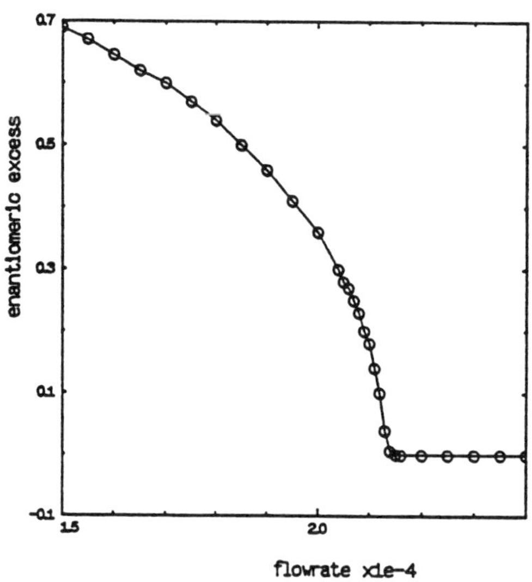

FIGURE 6. Bifurcation diagram which shows stationary values of *ee* against the flowrate obtained by a simulation varying k_f (in s^{-1}) by using the same parameters as mentioned in Fig. 4.

As indicated by the almost symmetrical shape of a simulated flow-diagram (Fig. 5) the model system shows only weak (possibly quadratic) autocatalysis under the fixed boundary conditions, which generally means that such a system cannot exhibit kinetic bistability or multiple steady states and - as a resulting of that - not any related critical phenomena.

Nevertheless in spite of predicted kinetical monostability the 3-step model exhibits several bifurcation points regarding the stereochemical aspects of the system: the flow-rate (see Fig. 6) as well as the stereospecific parameters and k_5 resp. k_{-5} act as an bifurcation parameter for the stationary values of *ee* in a CSTR. Near the bifurcation points the characteristic anomalous behaviour of the relaxation time - "critical slowing down" - (Dewel et al. , 1984) can be observed expressed as significant extended time-periods until the system reaches a stationary value of *ee*.

4. CONCLUSIONS

In a similar manner as recently reported by Gutman et al. (1992) and by Gutman and Todorović (1992), who performed a reexamination of the mechanism of Frank (1953) with racemization, our computations demonstrate as well the impressive basic characteristics of non-linear dynamics: a simple kinetic system only exhibiting weak chemical feedback terms pronounced as quadratic autocatalysis has the potential to generate very unexpected and complex behaviour expressed in the striking fact that - in our case - the simplified model of an autoxidation reaction possesses several bifurcation points regarding the stationary values of *ee* against the flow-rate, the stereospecific parameters, and the rate of racemization. This phenomenon is exceptional by the fact that the overall dynamics of the system show no bistability and only weak autocatalysis.

This kind of critical behaviour, corresponding to a transition of the model system between two steady states - between a racemic and a chiral-symmetry-breaking state and *vice versa* -, is only observable under flow conditions. But concerning several theoretical models of stereospecific autocatalysis, in which an open system is set as one of the boundary conditions, it is an interesting aspect that our calculations show an increase of *ee* in a CSTR only by using those parameters which lead to an increase of *ee* under batch conditions as well. In the other case an amplification effect cannot be achieved under flow conditions where the corresponding batch system shows no amplification capacity, because the chiral species is a reactive intermediate. That means if one tries to verify chiral symmetry breaking by experiment and applies a chemical system which obeys kinetics comparable to those investigated in this paper, it is sufficient to rule out the amplification capacity by a batch experiment.

The are two main factors defining the potential of amplification of enantiomeric excess by stereospecific autocatalysis:

1. Stereospecifity. One of the basic features resulting from our above analysis is the unexpected - and by several theoretical models even unrecognized - strong influence of stereospecifity resp. racemization on a non-linear system with the capacity to give rise to chiral- symmetry-breaking states (in our model

system both factors are expressed by the stereospecific parameters α, β. and γ and by the rate constants k_5 resp. k_{-5} of the racemization rate of peroxyl radicals). This very high sensitivity is illustrated by the fact that if only one stereospecific parameter is slightly increased from its value of 0 (which means the stereospecifity is slightly decreased from its maximum value of 100%) and all the others are kept respectively at 100%, there will be a pronounced decrease of intial *ee* - no matter whether the simulations are performed under flow or under batch conditions. The value of e.g. α near the bifurcation point is about 0.05 which means that the stereospecifity of chiral interactions in which α is involved must exceed 95% to achieve an amplification effect (note that in this case β and γ are kept at the value of 0). The influence of k_5 and k_{-5} on steady state values of *ee* is as drastic as well: if it exceeds the value of about $3 \cdot 10^{-5}$ (near the bifurcation point) the system will evolve into a racemic state.

2. Autocatalysis. Since it is well-known from the analysis of non-linear systems the strength of the feedback mechanism (expressed in terms ranging from quadratic = weak to cubic = strong autocatalysis) gives rise either to monostability or to multiple steady states (Gray and Scott, 1990), obviously these features were also regarded by constructing theoretical models of chiral symmetry breaking systems. As already mentioned, our model system presented in this paper exhibits only the characteristics of weak autocatalysis, and one might think that stronger autocatalysis could be able to compensate the high sensitivity towards stereospecifity. But as shown by calculations of Iwamoto and Seno (1982), who investigated the stereospecific processes of an open model system exhibiting cubic autocatalytic steps, they resulted in a similar value of about 95% stereospecifity as a minimum value for the occurrence of an optically active steady state.

Since 95% stereospecifity takes more and more the shape of a critical value for stereospecific autocatalysis (no matter if we look for a system exhibiting weak or strong autocatalysis) the limits of this concept are evidently indicated, and because of this further work should concentrate more on reacting systems which show - of course - non-linear kinetics as a basic criterion, but also template reactions or related effects (Park et al. , 1992), which possess higher stereospecifity than any common liquid-phase system. In the described reaction of autoxidation of tetralin in an homogeneous liquid-phase system obviously stereospecifity is well below 95%, due to partial reaction steps that involve radical intermediates. So therefore the experimental results showing no amplification of *ee* in our system is in agreement with our computational modelling; this conclusion can be generalized for all typical homogeneous systems.

REFERENCES

Alberts, A. H., and H. Wynberg, 1989: The role of the product in asymmetric C-C bond formation: stoichiometric and catalytic enantioselective autoinduction, J. Am. Chem. Soc., 111, 7265-7266.

Banerjee, D. K., and C. C. Budke, 1964: Spectrophotometric determination of traces of peroxides in organic solvents, Anal. Chem., 36, 792-796.

Boga, E., G. Peintler, and G. Nagypál, 1990: Propagating reaction front in the cobalt(II)-catalyzed autoxidation of benzaldehyde, J. Am. Chem. Soc., 112, 151-153.

Bonner, W. A., 1991: The origin and amplification of biomolecular chirality, Origins of Life Evol. Biosphere, 21, 59-112.

Colussi, A. J., E. Ghibaudi, Z. Yuan, and R. M. Noyes, 1990: Oscillatory oxidation of benzaldehyde by air. 1. Experimental observations, J. Am. Chem. Soc., 112, 8660-8670.

Decker, P., 1979: Spontaneous generation and amplification of molecular asymmetry through kinetical bistability in open systems. In D. C. Walker (Ed.), Origins of optical activity in nature, Elsevier Scientific Publ., Amsterdam, 109-124.

De Min, M., G. Levy, and J.-C. Micheau, 1988: Chiral resolutions, asymmetric synthesis and amplification of enantiomeric excess, J. Chim. Phys., 85, 603-619.

Dewel, G., P. Borckmans, and D. Walgraef, 1984: Relaxation kinetics in a bistable chemical system, J. Phys. Chem., 88, 5442-5445.

Emanuel, N. M., E. T. Denisov, and Z. K. Maizus, 1967: Liquid-phase oxidations of hydrocarbons, Plenum Press, New York.

Frank, F. C., 1953: On spontaneous asymmetric synthesis, Biochim. Biophys. Acta, 11, 459-463.

Glansdorff, P., and I. Prigogine, 1977: Thermodynamics of structure, stability, and fluctuations, Wiley-Interscience, New York.

Gray, P., and S. K. Scott, 1990: Chemical oscillations and instabilities. Non-linear chemical kinetics, Clarendon Press, Oxford.

Gutman, I., D. Todorović, M. Vucković, and P. Jungwirth, 1992: Modelling spontaneous chiral stereoselection: the Frank mechanism with racemization, J. Chem. Soc. Faraday Trans., 88, 1123-1127.

Gutman, I., and D. Todorović, 1992: The critical time in Frank's chiral amplification model, Chem. Phys. Letters, 195, 62-66.

Howard, J. A., 1973: Homogeneous liquid-phase autoxidations. In J. K. Kochi (Ed.), Free radicals. Vol 2, Wiley Interscience, New York, 3-62.

Iwamoto, K., and M. Seno, 1979: Effects of fluctuation on dissipative structures. II. Appearance of optically active states as dissipative structures, J. Chem. Phys., 70, 5858-5864.

Iwamoto, K., and M. Seno, 1982: On a chemical system related to absolutely asymmetric synthesis, J. Chem. Phys., 76, 2347-2351.

Kondepudi, D. K., and G. W. Nelson, 1984: Chiral-symmetry-breaking states and their sensitivity in nonequilibrium chemical systems, Physica, 125 A, 465-496.

Kondepudi, D. K., and G. W. Nelson, 1985: Weak neutral currents and the origin of biomolecular chirality, Nature (London), 314, 438-441.

Kondepudi, D. K., R. Kaufmann, and N. Singh, 1990: Chiral symmetry breaking in sodium chlorate crystallization, Science, 250, 975-976.

Mason, S. F., 1989: The development of concepts of chiral discrimination, Chirality, 1, 183-191.

Micheau, J.-C., M. de Min, and M. Gimenez, 1987: Dissipative structures and amplification of enantiomeric excess (an experimental point of view), Biosystems, 20, 85-93.

Mill, T., and D. G. Hendry, 1980: Kinetics and mechanisms of free radical oxidation of alkanes and olefins in the liquid phase. In C. H. Bamford and C. F. H. Tipper (Eds.), Comprehensive chemical kinetics. Vol. 16, Elsevier Scientific Publ., Amsterdam, 1-87.

Mizukami, F., and J. Imamura, 1978: Liquid-phase oxidations of hydrocarbons with molecular oxygen. I. The effect of metal ions and their ligands on product distribution in the oxidation of tetralin in acetic acid, Bull. Chem. Soc. Jpn., 51, 1404-1412.

Park, T. K., Q. Feng, and J. Rebek, Jr., 1992: Synthetic replicators and extrabiotic chemistry, J. Am. Chem. Soc., 114, 4529-4532.

Seelig, F. F., 1971: Systems-theoretic model for the formation of optical antipodes in strongly asymmetric yield, J. Theor. Biol., 31, 355-361.

Thiemann, W., 1974a: The origin of optical activity, Naturwissenschaften, 61, 476-483.

Thiemann, W. (Ed.), 1974b: International symposium on generation and amplification of asymmetry in chemical sytems, Berichte der KFA, Jül-Conf-13, Kernforschungsanlage Jülich, Germany.

Thiemann, W., 1975: Life and chirality beyond the earth, Origins of Life, 6, 475-481.

Thiemann, W. (Ed.), 1981: Generation and amplification of chirality in chemical systems. Proceedings of the 2nd international symposium, University of Bremen, July 16-18, 1980, Origins of Life, 11, 5-194.

Thiemann, W., and W. Darge, 1974: Experimental attempts for the study of the origin of optical activity on earth, Origins of Life, 5, 263-283.

Trainin, V. M., and A. S. Zaitsev, 1989: Critical phenomena in the catalyzed liquid-phase oxidation of hydrocarbons in an open system. 1. Analysis of bistability in a kinetic scheme, Kinet. Catal. Engl. Transl., 30, 624-632.

Trainin, V. M., and A. S. Zaitsev, 1990: Critical phenomena in the liquid-phase catalytic oxidation of hydrocarbons in an open system. 2. Critical effects observed upon alteration of the initial conditions, Kinet. Catal. Engl. Transl., 31, 40-46.

Woodward, A. E., and R. B. Mesrobian, 1953: Low temperature autoxidation of hydrocarbons. The kinetics of tetralin oxidation, J. Am. Chem. Soc., 75, 6189-6195.

Wynberg, H., 1989: Autocatalysis - the next generation of asymmetric syntheses?, Chimia, 43, 150-152.

IONIZING RADIATION AND CHEMICAL PROCESSING OF WATERS ON EARLY EARTH

Ivan G. Draganić and Sofija I.Vujošević
The Institute of Nuclear Sciences "Vinča"
P.O.B. 522, 11000 Belgrade, Yugoslavia

ABSTRACT

We consider chemical action of radiations emitted by radioactive decay of ^{40}K in early ocean, and the role of radiation of natural nuclear reactors in chemical processing of underground waters of early Earth. The sources of ionizing radiation and doses delivered are surveyed, as well as some basic facts on the action of ionizing radiation on water. We review the results of computer simulations of chemical processes induced by ionizing radiation in early ocean and underground waters. Significant, radiation-induced oxidizing capacity of primitive waters and the implications to chemistry of early Earth are considered.

CONTENTS

In memory of Dr. Zorica D. Draganić. The death has prevented her participation on this work. Missing greatly her gently and stimulating presence we tried to complete the task in the spirit in which she would approach it.

ISBN 0-937194-31-X

1. INTRODUCTION

Only modest amount of radioactivity dispersed in rocks is usually taken into account when considering the Earth's inventory of energy sources (Miller and Orgel, 1974). Almost unknown is the Oklo phenomenon (Neuilly et al.,1972), geological arrangements of uranium and water which were real natural nuclear reactors and have acted as efficient localized sources of energy over a large geological time-scale (IAEA, 1975; IAEA, 1978).

The ionizing radiation is generally taken as a minor partner in the group of energy sources when considering the chemistry on the early Earth. This may be justified if comparing only its amount with those of other available sources. And neglecting its advantage: efficiency in producing chemically reactive species (ions, free-radicals, radical-ions) in conditions hostile for a chemistry that we know in laboratory, such as those in ocean depths and underground waters of early Earth. In this work, after presenting general facts on the sources of ionizing radiation, and water radiolysis, we review the results of computer simulation of radiation-induced chemistry in model aqueous systems of interest to chemical evolution processes on the early Earth.

Beyond the scope of present study was radiolytic behavior of some of compounds which are produced by laboratory experiments simulaiting primitive atmosphere or hydrosphere, e.g. aldehydes, cyanides, nitriles, carboxylic and amino acids. Their Radiation chemistry is well established, provides a wealth of data of interest to chemical evolution studies, and merits a separate treatment. A kinetic analysis shows that the results of computer modelling examined in this work, in particular the oxidizing capacity of primitive waters, will not be essentially affected by their potential presence in expected concentrations (millimolar or less).

2. SOURCES AND DOSES OF IONIZING RADIATION.

2.1 RADIOACTIVE POTASSIUM-40 IN EARLY OCEAN

This radioactive isotope (half-life 1250 Myr) constitutes 0.012 per cent of total potassium on the contemporary Earth, but represented 0.1 per cent of potassum in nature 3800 Myr ago. The presence of potassium salts in primitive ocean waters, and the radioactive decay of ^{40}K, has provided an internal source of radiation in Archaean ocean (Draganić el al., 1991).

The amount of energy delivered by radioactive potassium depended on the concentration of potassium salts in primitive ocean waters. The latter were recycled many times during geological history and may not have retained the information of their past potassium content. A reliable estimate is, nevertheless, possible. It is based on inferences drawn from the nature of ancient sediments and the products of their metamorphism. Various observations suggest that the main components of the lithosphere and the hydrosphere were extant by the time of Isua water-laid sediments of West Greenland (Ernst, 1983). A detailed analysis of Isua sediments shows that the average major element composition is in the normal range of averages for younger Archaean and Proterozoic formations

(Dymek and Klein, 1988). It has been noted that if conditions at Isua were representative of the state of Earth some 3800 Myr ago, than the nature of weathering, solution transport, and sedimentation was already similar to that at present time (Holland,1984). Therefore, it has been assumed that 3800 Myr ago the potassium content was 0.38 g per kilogram of ocean water and, consequently, 0.38 mg kg^{-1} of potassium - 40. To calculate the energy deposited by ^{40}K radiation, gamma (11%) and beta rays (89%), the average energy liberated per one decay (0.455 MeV) was taken into account (Weast, 1987). The calculated absorbed dose rate is 1.84×10^{15} eV kg^{-1} yr^{-1} (2.95×10^{-4}J $kg^{-1}yr^{-1}$). Because of slow radioactive decay of ^{40}K, this value varies only $\pm$ 5 per cent if the considered age (3800 Myr) shifts $\pm$ 100 Myr.

2.2 NATURAL NUCLEAR REACTORS AND UNDERGROUND WATERS

Natural nuclear reactors were geological arrangements of uranium and water which have led to a fission chain process in the uranium deposits on Precambrian Earth, as was the case at the fossil reactor site discovered in Oklo (Gabon, Africa). The fissile ^{235}U has a half life of 710 Myr. Its present concentration in natural uranium is 0.7 per cent and a chain fission process cannot start even in the richest uranium deposits. On the early Earth it was up to 40 times higher and, as suggested long before the discovery of the Oklo phenomenon (Kuroda, 1956), could have led to the chain fission process in the nature.

From the point of view of nuclear reactor physics and chemistry, the fossil natural nuclear reactors site at Oklo is well understood. The operational time of individual reactors was between 0.6 and 0.8 Myr, while the total duration of the Oklo phenomenon (six reactors cores) was up to several million years. It has been also concluded that none of the geological and geochemical characteristics of the Oklo deposit appear unique and the nuclear reactor site at Oklo could not have been a singular occurrence on the Precambrian Earth (IAEA, 1975; IAEA 1978).

A rough estimate of the amount of terrestrial uranium that has taken part in natural nuclear reactors, based on slight variations of ^{238}U / ^{235}U in presently accessible uranium ores with 1000 ppm of UO_2 or more, suggests that thousands of nautral nuclear reactors were operating between 1800 and 4100 Myr ago. If 1 km thick layer of Earth's crust is taken into account, and its average uranium abundance 3 ppm, than 100 million reactors sites of the Oklo type could have been active in the past of our planet (Draganić et al., 1983).

Individual natural nuclear reactors were operating for a million years or more. Comparing to man-made reactors it occurred at very low power: in average about 1 kW only. Routine procedures for in-pile dose evaluations show (Draganić et al., 1983) that they have delivered dose rates of the order of 10^{17}eV g^{-1}per hour or less. These were considerably lower than in the man-made nuclear reactors, but are similar to those available in irradiation units in laboratories, and could have been very convenient for radiation chemistry.

Table 1 gives dose-rates calculated for a natural nuclear reactor operating 3200 Myr ago at 1 kW power.

Table 1.DOSE-RATES IN A NATURAL NUCLEAR REACTOR 3200 Myr AGO

ENERGY SOURCE	DOSE-RATE, $x10^{16}eV\ g^{-1}h^{-1}$
Fission fragments	22.78
Prompt neutrons	0.69
Prompt gamma rays	0.87
Fission products gamma rays	0.87
Fission products beta rays	1.06
Gamma rays due to neutron captures	0.81
TOTAL:	27.08

In the calculation the diameter of reactor core was assumed to be 2 meters. If the considered criticality time was taken to be before or after 3200 Myr, it could have been smaller or larger depending on the abundance of fissile ^{235}U in natural uranium; this would not affect the dose rates calculated for 1 kW power of the reactor.

Water had an important role in making of uranium deposit a natural nuclear reactor. It served to moderate neutrons and help the propagation of the chain fission process. Also, it removed the heat and controlled the fission chain process. At the same time, water was radiation processed and the substances present were chemically altered. Underground water drain was of help to circulate efficiently the material subjected to irradiation.

3. RADIOLYSIS OF LIQUID WATER

3.1 REACTIONS IN PURE WATER

A convenient expression for the extent of chemical change induced by ionizing radiation is the radiation chemical yield, usually called the G - value. Traditional definition of G(P) is the number of chemical species P which are formed or decomposed per 100 eV energy absorbed.

Radiation chemical behaviour of water and aqueous systems belongs to the best understood domains of radiation chemistry (Draganić and Draganić, 1971; Farhatazis and Rodgers, 1987; Spinks and Woods, 1990). When a water molecule is ionized by an energetic photon or charged particle, the electron liberated has sufficient energy to ionize further water molecules. This leads to the formation of clusters of ions along the track of the ionizing radiation. The size of clusters (spurs, tracks), their distribution and structure, depend on the radiation type, energy and intensity, i.e. on the density of primary events along the pathway of incident radiation. For low density radiation, such as gammas and fast electrons, the free radical yields are higher and those of molecular yields subsequently lower. The reason is an easier escape of free radicals (H, OH, e_{aq}^-) from the spurs, and a lesser chance for their recombinations to form molecular products (H_2, H_2O_2).

Some nanoseconds after passage of the radiation pure water contains, in addition to molecules of H_2 and H_2O_2, also chemically reactive species, the free radicals: oxidizing (OH) and reducing (H, e_{aq}^-). Diffusing away, these react among themselves:

Reaction	Rate constant
$OH + OH \rightarrow H_2O_2$	$k_1=6.0x10^9$
$OH + e_{aq}^- \rightarrow OH^- + H_2O$	$k_2=2.5x10^{10}$
$OH + H \rightarrow H_2O$	$k_3=2.5x10^{10}$
$OH + HO_2 \rightarrow H_2O + O_2$	$k_4=7.9x10^9$
$OH + O_2^- \rightarrow OH^- + O_2$	$k_5=1.0x10^{10}$
$OH + H_2O_2 \rightarrow H_2O + O_2^- + H^+$	$k_6=2.7x10^7$
$OH + HO_2^- \rightarrow H_2O + O_2^-$	$k_7=7.5x10^9$
$OH + H_2 \rightarrow H_2O + H$	$k_8=4.0x10^7$
$e_{aq}^- + e_{aq}^- \rightarrow 2OH^- + H_2$	$k_9=3.0x10^9$
$e_{aq}^- + H \rightarrow OH^- + H_2$	$k_{10}=2.0x10^{10}$
$e_{aq}^- + O_2^- \rightarrow HO_2^- + OH^-$	$k_{11}=1.2x10^{10}$
$e_{aq}^- + H_2O_2 \rightarrow OH + OH^- + H_2O$	$k_{12}=1.6x10^{10}$
$e_{aq}^- + H^+ \rightarrow H + H_2O$	$k_{13}=2.2x10^{10}$
$e_{aq}^- + O_2 \rightarrow O_2^- + H_2O$	$k_{14}=2.0x10^{10}$
$H + H \rightarrow H_2$	$k_{15}=1.0x10^{10}$
$H + HO_2 \rightarrow H_2O_2$	$k_{16}=2.0x10^{10}$
$H + O_2^- \rightarrow HO_2^-$	$k_{17}=2.0x10^{10}$
$H + H_2O_2 \rightarrow OH + H_2O$	$k_{18}=6.0x10^7$
$H + OH^- \rightarrow e_{aq}^-$	$k_{19}=1.5x10^7$
$H + O_2 \rightarrow O_2^- + H^+$	$k_{20}=2.0x10^{10}$
$HO_2 \rightarrow O_2^- + H^+$	$k_{21}=8.0x10^5$
$HO_2 + HO_2 \rightarrow O_2 + H_2O_2$	$k_{22}=7.5x10^5$
$HO_2 + O_2^- \rightarrow O_2 + HO_2^-$	$k_{23}=1.0x10^8$
$O_2^- + H^+ \rightarrow HO_2$	$k_{24}=5.0x10^{10}$
$H_2O_2 + OH^- \rightarrow HO_2^- + H_2O$	$k_{25}=5.0x10^8$
$HO_2^- + H_2O \rightarrow H_2O_2 + OH$	$k_{26}=5.735x10^4$
$H_2O \rightarrow H^+ + OH^-$	$k_{27}=2.599x10^{-5}$
$H^+ + OH^- \rightarrow H_2O$	$k_{28}=1.4385x10^{11}$

3.2 PRESENCE OF SOLUTES

In addition to reactions among themselves (eqs. 1-28), the difusing radicals react also with solutes present in the bulk of water causing a large variety of chemical changes. For example, oxidations ($OH + Fe^{2+} \rightarrow Fe^{3+} + OH^-$), reductions ($H + Ce^{4+} \rightarrow Ce^{3+} + H^+$), abstraction reactions ($OH + CH_3OH \rightarrow H_2O + CH_2OH$), chain reactions etc.

Quantitative kinetic presentation of a radiation induced chemical process in water containing several solutes requires usually dozens of reactions in addition to those given for pure water. It is worth noticing that reliable computer simulations of radiolyses of liquid aqueous systems are a daily routine in laboratory. They are based on experimental evidence on thousands of short-living intermediates and measured rate constans.

4. COMPUTER SIMULATION OF RADIOLYSIS OF OCEAN WATER 3800 Myr AGO

4.1 RADIATION-INDUCED PROCESS IN SEA WATER

The simulation was based on the energy deposited 1.84 x 10^{15} eV kg^{-1}yr^{-1}, the reactions in pure water (eqs, 1-28), the constituents of sea water (Stumm and Morgan, 1981), and their reactions with primary products of water decomposed (Draganic et al., 1991):

$OH + Cl^- \rightarrow Cl + OH^-$	$k_1=4.3x10^9$
$e_{aq}^- + Cl \rightarrow Cl^- + H_2O$	$k_2=1.0x10^{10}$
$e_{aq}^- + Cl_2^- \rightarrow 2Cl^- + H_2O$	$k_3=1.0x10^{10}$
$e_{aq}^- + ClOH^- \rightarrow Cl^- + OH^- + H_2O$	$k_4=1.0x10^{10}$
$e_{aq}^- + Cl_2 \rightarrow Cl_2^- + H_2O$	$k_5=1.0x10^{10}$
$H + Cl \rightarrow Cl^- + H^+$	$k_6=1.0x10^{10}$
$H + Cl_2^- \rightarrow 2Cl^- + H^+$	$k_7=8.0x10^9$
$H + ClOH^- \rightarrow Cl^- + H_2O$	$k_8=1.0x10^{10}$
$H + Cl_2 \rightarrow Cl_2^- + H^+$	$k_9=7.0x10^9$
$HO_2 + Cl_2^- \rightarrow 2Cl^- + O_2 + H^+$	$k_{10}=4.0x10^9$
$HO_2 + Cl_2 \rightarrow Cl_2^- + O_2 + H^+$	$k_{11}=1.0x10^9$
$O_2^- + Cl_2^- \rightarrow 2Cl^- + O_2$	$k_{12}=1.2x10^{10}$
$H_2O_2 + Cl_2^- \rightarrow 2Cl^- + O_2^- + 2H^+$	$k_{13}=1.4x10^5$
$H_2O_2 + Cl_2 \rightarrow HO_2 + Cl_2^- + H^+$	$k_{14}=1.9x10^2$
$OH^- + Cl_2^- \rightarrow ClOH^- + Cl^-$	$k_{15}=7.3x10^6$
$H^+ + ClOH^- \rightarrow Cl + H_2O$	$k_{16}=2.1x10^{10}$
$Cl^- + Cl \rightarrow Cl_2^-$	$k_{17}=2.1x10^{10}$
$Cl^- + ClOH^- \rightarrow Cl_2^- + OH^-$	$k_{18}=9.0x10^4$
$ClOH^- \rightarrow OH + Cl^-$	$k_{19}=6.1x10^9$
$Cl_2^- \rightarrow Cl + Cl^-$	$k_{20}=1.1x10^5$
$Cl_2^- + Cl_2^- \rightarrow Cl_2 + 2Cl^-$	$k_{21}=7.0x10^9$
$OH + Br^- \rightarrow BrOH^-$	$k_{22}=5.0x10^9$
$e_{aq}^- + Br \rightarrow Br^- + H_2O$	$k_{23}=1.0x10^{10}$
$e_{aq}^- + Br_2^- \rightarrow 2Br^- + H_2O$	$k_{24}=1.3x10^{10}$
$H + Br_2 \rightarrow Br_2^- + H^+$	$k_{25}=1.0x10^{10}$
$HO_2 + Br_2^- \rightarrow Br_2 + HO_2^-$	$k_{26}=3.8x10^9$
$O_2^- + Br_2 \rightarrow H^+ + Br^- + Br + O_2$	$k_{27}=1.5x10^8$
$H_2O_2 + Br \rightarrow Br^- + O_2^- + 2H^+$	$k_{28}=2.5x10^9$
$H_2O_2 + Br_2^- \rightarrow 2Br^- + O_2^- + 2H^+$	$k_{29}=1.9x10^6$
$BrOH^- \rightarrow Br^- + OH$	$k_{30}=3.0x10^7$
$BrOH^- \rightarrow Br + OH^-$	$k_{31}=4.2x10^6$
$BrOH^- + H^+ \rightarrow Br + H_2O$	$k_{32}=4.4x10^9$

Reaction	Rate constant
$BrOH^- + Br^- \rightarrow Br_2^- + OH^-$	$k_{33}=1.9\times10^8$
$Br + Br^- \rightarrow Br_2^-$	$k_{34}=5.4\times10^9$
$Br_2^- \rightarrow Br + Br^-$	$k_{35}=2.1\times10^4$
$Br_2^- + Br_2^- \rightarrow Br_2 + 2Br^-$	$k_{36}=1.8\times10^9$
$OH + SO_4^- \rightarrow SO_4^{2-} + OH^-$	$k_{37}=1.6\times10^6$
$SO_4^- + HCO_3^- \rightarrow SO_4^{2-} + CO_3^- + H^+$	$k_{38}=9.1\times10^6$
$OH + CO_3^{2-} \rightarrow CO_3^- + OH^-$	$k_{39}=4.0\times10^8$
$OH + HCO_3^- \rightarrow CO_3^- + H_2O$	$k_{40}=8.5\times10^6$
$e_{aq}^- + CO_2 \rightarrow CO_2^- + H_2O$	$k_{41}=7.7\times10^9$
$e_{aq}^- + HCO_3^- \rightarrow CO_3^- + H$	$k_{42}=6.0\times10^5$
$H + CO_2 \rightarrow CO_2^- + H^+$	$k_{43}=1.0\times10^6$
$H + HCO_3^- \rightarrow CO_2^- + H_2O$	$k_{44}=3.4\times10^4$
$O_2^- + HCO_3^- \rightarrow HO_2^- + CO_3^-$	$k_{45}=1.0\times10^6$
$O_2^- + CO_3^- \rightarrow CO_3^{2-} + O_2$	$k_{46}=4.0\times10^8$
$H_2O_2 + CO_2^- \rightarrow CO_2 + OH + OH^-$	$k_{47}=7.0\times10^5$
$H_2O_2 + CO_3^- \rightarrow CO_3^- + O_2^- + 2H^+$	$k_{48}=4.3\times10^5$
$H^+ + HCO_3^- \rightarrow CO_2 + H_2O$	$k_{49}=1.0\times10^{10}$
$OH^- + HCO_3^- \rightarrow CO_3^{2-} + H_2O$	$k_{50}=1.0\times10^9$
$H_2O + CO_2 \rightarrow HCO_3^- + H^+$	$k_{51}=70.0$
$H_2O + CO_3^{2-} \rightarrow HCO_3^- + OH^-$	$k_{52}=3.6\times10^3$
$H_2O + CO_4^{2-} \rightarrow HO_2^- + OH^- + CO_2$	$k_{53}=0.2$
$O_2 + CO^- \rightarrow CO + O_2^-$	$k_{54}=2.0\times10^9$
$CO_2^- + HCO_3^- \rightarrow HCO_2^- + CO_3^-$	$k_{55}=2.0\times10^3$
$CO_2^- + CO_2^- \rightarrow C_2O_4^{2-}$	$k_{56}=4.0\times10^8$
$CO_2^- + CO_3^- \rightarrow CO_2 + CO_3^{2-}$	$k_{57}=5.0\times10^7$
$CO_3^- + HCO_2^- \rightarrow HCO_3^- + CO_2^-$	$k_{58}=1.5\times10^5$
$CO_3^- + CO_3^- \rightarrow CO_4^{2-} + CO_2$	$k_{59}=7.0\times10^6$

In the computation the radiation-induced process was followed over a geological time scale of one hundred million years (100 Myr). The beginning was deliberately taken at 3800 Myr ago. This is the age of oldest known sedimentary rocks and most likely of ocean waters whose salt content and major chemical elements were not essentialy different from the contemporary ones.

Action of radiation leads to water decomposition with one H_2 formed per H_2O decomposed. Assuming the mass of water $M_{ocean}= 1.4 \times 10^{24}$g, the total amount of molecular hydrogen generated over 100 Myr was 3.8×10^{18}g H_2. This amount is similar to hydrogen released by volcanic activity on early Earth over the same period of time (Holland, 1984).

In addition to molecular hydrogen, the oxygen (Figure 1) and hydrogen peroxide (Figure 2), are produced in significant amounts. Because of uncertainty concerning the amount of early atmospheric oxygen, in the calculation two extreme cases were taken into account: 10^{-2} and 10^{-6} PAL (Present Atmospheric Level).

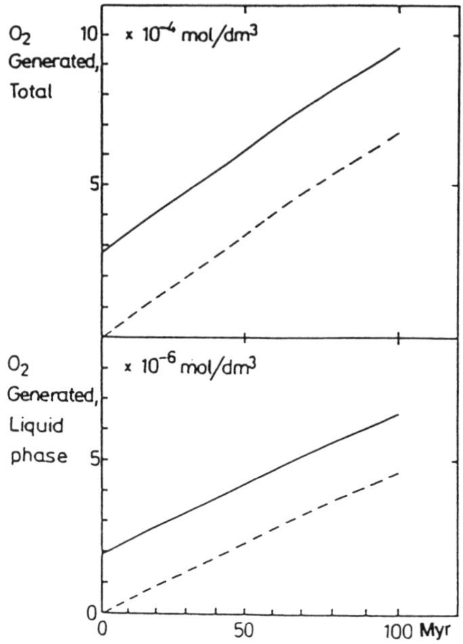

Figure 1. Oxygen generated by ^{40}K radiation in ocean waters 3800 Myr ago. $M_{ocean}=1.4x10^{24}g$. Concentration of initially present oxygen due to the liquid-gas phase equilibrium:
(-) 10^{-2} PAL and (- -) 10^{-6} PAL.

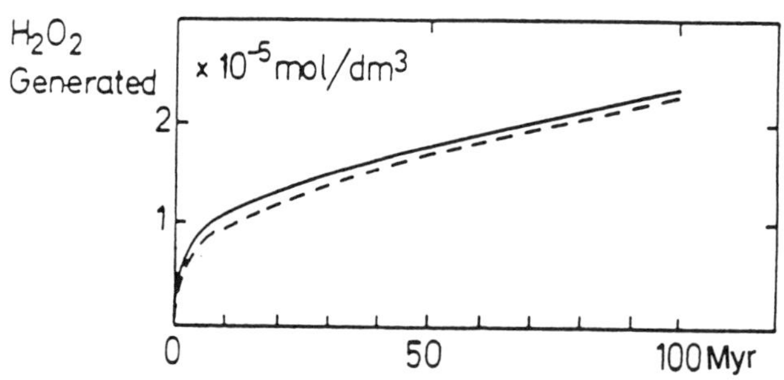

FIGURE 2. Hydrogen peroxide generated by ^{40}K radiation in ocean waters 3800 Myr ago. $M_{ocean}=1.4x10^{24}g$. Concentration of initilly present oxygen due to the liquid-gas phase equilibrium:
(-) 10^{-2} PAL and (- -) 10^{-6} PAL.

Critical examination of chemical kinetics of the simulated process shows: the mechanism is not sensitive to concentration changes if the concentration ratio of the main constituents is not affected. For example, some local dilution or a local concentration (due to frezing or evaporation). Also, because of dominance of radiation chemical reactivity of chloride ion, the absence of minor constituents, or a presence of new compounds (millimolar concentration or less), would not significantly affect the reaction mechanism.

4.2 INTRINSIC OXIDIZING CAPACITY

Figure 1 shows a significant radiolytic generation of total molecular oxygen over 100 Myr: 3.0×10^{19} in the ocean which water mass is assumed to be 1.4×10^{24}g. For comparison, 1.2×10^{21}g is present in the contemporary atmosphere. At the same time 1.1×10^{18} H_2O_2 g were also generated (Fig. 2).

A survey of reaction mechanism shows that irradiated system contains both the strongly reducing (H, e_{aq}-) and oxidizing (OH) species produced by direct water decomposition. In addition, many other oxidizing radicals appear as intermediates, e.g.: HO_2-, O_2-, ClOH-, BrOH-, Cl_2-, Br_2-, Cl, Br, SO_4-. Their reactions in the model system yield eventually O_2 and H_2O_2. Nevertheless, they would also efficiently react with oxidizible substances if present in the system. These reactions could affect the generation of molecular oxygen and hydrogen peroxide but water decomposition would not be affected and, because of their oxidizing character, disappearance of intermediates will be followed by oxidations, hence also contributing to the total oxidizing capacity of early ocean water.

The oldest known terrestrial sediments, 3800 Myr-old rocks in the Isua Supracrustal Group of West Greenland, contain primary oxidized iron minerals (Appel, 1980; Dymek and Klein, 1988). Two sources of oxygen are considered plausible for the oxidation processes at the time of their precipitation and deposition; photolysis of water vapor by solar UV radiation (Canuto et al., 1983), and a biotic source, photosynthetic generation of oxygen by some microorganisms (Chapman and Schopf, 1983). The data on Figs. 1 and 2 show that the contribution of radiolysis of ocean water, the generation of O_2 and H_2O_2 and the abundance of oxidizing short-living intermediates, can not bee neglected when considering the origin of primary oxidized iron minerals.

Particular interest is that the formation of radiolytically produced oxidizing agents has taken place throughout the ocean volume, at variance with the action of products of water photolysis which was limited to the ocean surface layers only. We believe that the radiation induced oxidizing capacity of early ocean waters may have played a significant role in early chemistry of Earth before the emergence of oxygen due to the photosynthetic microorganisms.

5. COMPUTER SIMULATION OF RADIOLYSIS OF UNDERGROUND WATERS ON EARLY EARTH

5.1 RADIATION-INDUCED PROCESS IN SANDSTONE WATERS

Action of mixed nuclear radiations on a sandstone underground water as a model system was examined in the conditions which very likely were existing in natural nuclear reactors. Chemical composition of sandstone water was taken according to Stumm and Morgan (1981). Only radiation-chemically active constituents were included in the reaction scheme. The rate constants of reactions take into account enviromental conditions (high temperature and pressure); primary radical yields are relevant to mixed nuclear reactor radiation. The reactions considered (Bjergbakke et al.,1989) are the following:

$OH + OH \rightarrow H_2O_2$ $k_1=6.7\times10^{10}$
$OH + e_{aq}^- \rightarrow OH^- + H_2O$ $k_2=2.8\times10^{11}$
$OH + H \rightarrow H_2O$ $k_3=2.8\times10^{11}$
$OH + O_2^- \rightarrow OH^- + O_2$ $k_4=1.1\times10^{11}$
$OH + H_2O_2 \rightarrow H_2O + O_2^- + H^+$ $K_5=4.2\times10^{8}$
$OH + H_2 \rightarrow H_2O + H$ $k_6=1.4\times10^{9}$
$e_{aq}^- + e_{aq}^- \rightarrow 2OH^- + H_2$ $k_7=3.9\times10^{11}$
$e_{aq}^- + H \rightarrow OH^- + H_2$ $k_8=2.8\times10^{11}$
$e_{aq}^- + O_2^- \rightarrow HO_2^- + OH^-$ $k_9=1.3\times10^{11}$
$e_{aq}^- + H_2O_2 \rightarrow OH + OH^- + H_2O$ $k_{10}=1.6\times10^{11}$
$e_{aq}^- + H^+ \rightarrow H + H_2O$ $k_{11}=2.5\times10^{11}$
$e_{aq}^- + O_2 \rightarrow O_2^- + H_2O$ $k_{12}=2.2\times10^{11}$
$H + H \rightarrow H_2$ $k_{13}=1.1\times10^{11}$
$H + O_2^- \rightarrow HO_2^-$ $k_{14}=2.2\times10^{11}$
$H + H_2O_2 \rightarrow OH + H_2O$ $k_{15}=2.3\times10^{9}$
$H + O_2 \rightarrow O_2^- + H^+$ $k_{16}=2.2\times10^{11}$
$HO_2 + O_2^- \rightarrow O_2 + HO_2^-$ $k_{17}=4.13\times10^{8}$
$O_2^- + O_2^- \rightarrow HO_2^- + H^+ + O_2$ $k_{18}=4.1\times10^{5}$
$H_2O_2 \rightarrow H_2O + O$ $k_{19}=2.0\times10^{-2}$
$O + O \rightarrow O_2$ $k_{20}=2.8\times10^{11}$
$OH + CO_3^{2-} \rightarrow CO_3^- + OH^-$ $k_{21}=4.0\times10^{9}$
$OH + HCO_3^- \rightarrow CO_3^- + H_2O$ $k_{22}=1.5\times10^{8}$
$O_2^- + CO_3^- \rightarrow CO_3^{2-} + O_2$ $k_{23}=4.0\times10^{9}$
$H_2O_2 + CO_3^- \rightarrow CO_3^{2-} + O_2^- + 2H^+$ $k_{24}=4.3\times10^{6}$
$H_2O + CO_4^{2-} \rightarrow HO_2^- + OH^- + CO_2$ $k_{25}=2.0$
$CO_3^- + CO_3^- \rightarrow CO_4^{2-} + CO_2$ $k_{26}=7.0\times10^{7}$
$CO^2 + e_{aq}^- \rightarrow CO_2^- + H_2O$ $k_{27}=7.7\times10^{10}$
$CO_2 + H \rightarrow CO_2^- + H^+$ $k_{28}=1.0\times10^{7}$
$HCO_3^- + H \rightarrow CO_2^- + H_2O$ $k_{29}=3.4\times10^{5}$
$CO_2^- + HCO_3^- \rightarrow HCO_2^- + CO_3^-$ $k_{30}=2.0\times10^{4}$
$CO_2^- + CO_2^- \rightarrow C_2O_4^{2-}$ $k_{31}=8.0\times10^{9}$
$CO_3^- + HCO_2^- \rightarrow HCO_3^- + CO_2^-$ $k_{32}=1.0\times10^{6}$
$HCO_2^- + OH \rightarrow CO_2^- + H_2O$ $k_{33}=3.0\times10^{10}$
$CO_2^- + O_2 \rightarrow CO_2 + O_2^-$ $k_{34}=2.0\times10^{10}$
$CO_2^- + OH \rightarrow CO_2 + OH^-$ $k_{35}=1.0\times10^{11}$
$CO_2^- + CO_3^- \rightarrow CO_2 + CO_3^{2-}$ $k_{36}=1.0\times10^{11}$
$C_2O_4^{2-} + OH \rightarrow CO_2^- + CO_2 + OH^-$ $k_{37}=1.0\times10^{8}$
$C_2O_4^{2-} + CO_3^- \rightarrow CO_2^- + CO_2 + CO_3^{2-}$ $k_{38}=1.0\times10^{8}$
$CO_2^- + H_2O_2 \rightarrow CO_2 + OH^- + OH$ $k_{39}=1.0\times10^{11}$
$CO_2^- + O_2^- \rightarrow CO_2 + O_2^{2-}$ $k_{40}=1.0\times10^{11}$

The simulation has assumed that the irradiation took place only within one month, that the reactor power was 3 kW and the dose-rate 8.1 x 10^{17} eV g^{-1} $hour^{-1}$. It can be seen that the main products are hydrogen peroxide, oxygen and hydrogen. The efficiency of their formation depends on the possibility of gases to escape out of radiation zone, which was also considered in the computer modelling (FIGURE 3).

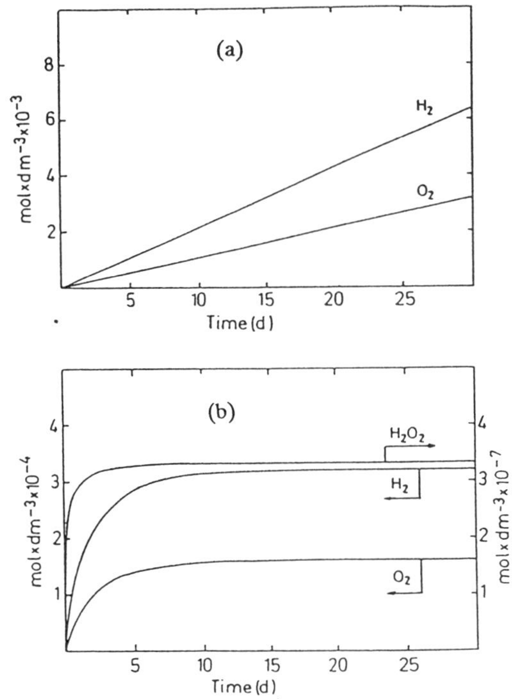

FIGURE 3. Radiolytic products in the irradiated air-free sandstone water 3200 Myr ago:
(a) Production of hydrogen and oxygen when gases are allowed to escape the radiation zone;
(b) equilibrium concentration of products in a closed radiation zone.

5.2 AMOUNTS AND TYPES OF RADIATION-PROCESSED MATERIAL IN UNDERGROUND WATERS

The calculation for one nuclear reactor, operating only one month, shows (Fig.3) that significant amounts of oxigen, hydrogen and hydrogen perohide could have been produced on geological time scale. Many of chemically reactive free radical intermediates could have been also involved in reactions with various inorganic and organic compounds, dissolved in waters which were flowing through or in the vicinty of natural nuclear reactors. Computer modelling of such a complex chemistry is feasible with the available kinetic data, but a critical choice of types and amounts of potential participants in the process is still missing.

In principle, one should be able to estimate the amount of underground waters that were radiation processed in the Earth's past. Reasonably well known are the dose-rate in reactor core (about 10^{17}eV $g^{-1}h^{-1}$), its operational time (about 1 Myr), and the amount of water needed for the maintenance of uranium chain fission: several tons (depending on uranium concentration, reactor diameter and its age). Yet, the rough estimates of the number of natural nuclear reactors operating in the Earth's past vary considerably, from a conservative value of about 1000 to more realistic 100 millions. A reliable estimate needs a better insight into the amount of terrestrial uranium that may have taken part in natural nuclear reactors and more information on the variations of $^{235}U/^{238}U$ in it.

6. SOME IMPLICATIONS AND FUTURE WORK

6.1 IONIZING RADIATION AND EARLY OCEAN

Compared to amounts of energy provided by other sources on early Earth (Miller and Orgel, 1974; Chyba and Sagan, 1991), that due to radioactive potassium was extremly modest. However, in many respects ^{40}K was an ideal in situ source. Because of the ubiquitous presence of potassium in the ocean,the energy was deposited homogeneously throughout the water volume, including deep ocean. In addition, because of the very nature of radioactive decay, the process of energy deposit occurred undisturbed and uninterrupted so that the amount of energy available for chemistry over a geological time-scale was significant.

It is certain that on the geological time scale the role of oxidizing intermediates could have been significant not only in local chemistry but also in shaping the environment where oxygen tolerant life forms evolved.

The annual generation of oxidizing species in primitive oceans was low, about ten picomoles per dm^3 (about 10^{-11}mol dm^{-3} yr^{-1}). Depending on the local supply of oxidizable substrates, they could have been more or less efficiently consumed leading to the synthesis of new compounds. Assuming a low radiation chemical yield, G(P) about 0.15 and an average molecular mass M_p=75 a.m.u., about 10^{14} g of organics could have been synthesized over 100 Myr in the ocean.

Archean oceans have very likely contained some other compounds, inorganic and organic, in addition to those taken in our computer modelling. These could have been supplied by processes of different origin, duration and intensity: intensive weathering (Veizer, 1983), lightening and coronal discharge in the atmosphere (Chyba and Sagan, 1991) or volcanic eruptions (Miller and Orgel, 1974). They were delivered locally, and have been involved in chemistry before efficient mixing occurred i.e. providing some kind of solution more complex and concentrated then the bulk of ocean water considered in the computer simulation.

Of a particular interest are recent considerations of a large delivery of water and organic matter during the period of heavy bombardment 4500 Myr to 3900 Myr ago. It has been proposed: comets have brought vast quantities of water to the primitive Earth (Fernandez and Ip, 1983), the Earth may have acquired an exogenous ocean of water (Chyba, 1987; 1990) and cometary delivery of organics was significant (Chyba et al., 1992).

It has been suggested that the supply of organic molecules by exogenous delivery, impact-shock syntheses and endogenous production of organic molecules was impressive (Chyba and Sagan, 1992). However, the suggested stedy-state concentrations in the ocean 4000 Myr ago would have been about 10^{-3}g per g for a reducing atmosphere, and 10^{-6}g per g for an intermediate oxidation state atmosphere. These concentrations are to low to affect the final outcome of chloride-ion dominated radiolysis of ocean water. The concentrations should be even lower by 3800 Myr ago, if the bombardment declined and 10^7 yr is assumed to be a mean life-time of organics in ocean water. Therefore, it seems very likely that the findings of computer simulation (Figs. 1-2) would remain essentially the same.

Nevertheless, if not affecting the main feature of radiation-induced process in the sea water, potential presence of compounds of endogenous and exogenous origin may have lead to the formation of biogenic molecules, minor in amounts but major in the importance for chemical evolution processes. This is why a future work should concern the computer modelling of extended present model system. It is feasible with existing computer program (Rasmussen et al., 1984) and available data on radiolytic behaviours of potential new constituents e.g. NH_3, CH_4, SO_2, H_2S, HCN, nitriles, amines, aldehydes, carboxylic and amino acids. However, a critical choice of compounds and their amounts to be used in the reaction mechanism is missing. It requires a better understanding of chemical composition of primitive hydrosphere.

6.2 IONIZING RADIATION AND EARLY UNDERGROUND WATERS

Radiation-induced oxidizing capacity of underground waters has to be taken into account when approaching the oxidizing processes in primitive waters, or looking for origins of primary oxidized iron minerals on early Earth. Future work should provide a reliable evalution of the number of natural nuclear reactors operating on the early Earth and, consequently, the amounts of water and materials which were radiation processed.

It is worth noticing that presence of long-lived fission products ensured that the core of a natural nuclear reactor was a place of localized radiations for long periods of time after the chain fission process in the reactor was over (hundreds of years or more). Also, some of radionucleides would migrate and continue to irradiate after fixation in the environment, or after uptake by Precambrian life.

The dose rates were low, but combined with the possibility of irradiation during large time periods could have been effective in producing evolutionary changes in primitive life. An examination of major events in early biological evolution shows that the main evolutionary changes of procariotic life resulted from biochemical rather than morphological inovations (Schopf, 1978). By initiating free radical processes in the early biochemistry, the ionizing radiation could have been a very effective mutagenic agent for Precambrian biota.

REFERENCES

Appel, P.W.U., 1980: On the early Archaean Isua iron-formation, West Greenland, Precambrian res., 11, 73-87.

Bjergbakke, E., Z.D.Draganić, K.Sehested and I.G.Draganić, 1989: Radiolytic products in waters - Part II. Computer simulation of some radiolytic processes in nature. Radiochim. Acta, 48, 73-77.

Canuto, V. M., J.S. Levine, T.R. Augustsson, C.L.Imhoff and M.S. Giampapa, 1983: The young Sun and the photochemistry of the early Earth. Nature, 305, 281-286.

Chapman,D.J. and W.J.Schopf,1983. Biological and biochemical effects of the development of an aerobic environment. In Earth's Earliest Biosphere. Its Origin and Evolution. J.W.Schopf(Ed.), Princeton University Press.

Chyba, C.F., 1987: The cometary contribution to the oceans of primitive Earth. Nature, 330, 632-635.

Chyba, C.F., 1990: Impact delivery and erosion of planetary oceans in the early inner Solar System. Nature, 343, 129-133

Chyba, C.F., P.J. Thomas, L. Brookshaw and C.Sagan, 1990: Cometary delivery of organic molecules to the early Earth. Science, 249, 366-373.

Chyba, C.F. and C.Sagan, 1991: Electrical energy sources for organic synthesis on the early Earth. Orig.Life and Evol.Biosphere, 21, 3-17.

Chyba, C.F. and C.Sagan, 1992: Endogenous production, exogenous delivery and impact-shock synthesis of organic molecules: an inventory for the origin of life. Nature, 355, 125-132.

Draganić, I.G. and Z.D.Draganić, 1971: The Radiation Chemistry of Water, Academic Press, New York.

Draganić, I.G., Z.D.Draganić and D.Altiparmakov, 1983: Natural nuclear reactors and ionizing radiation in the Precambrian. Precambrian Res., 20, 283-298.

Draganić, I.G., E.Bjergbakke, Z.D.Draganić and K.Sehested, 1991: Decomposition of ocean waters by potassium-40 radiation 3800 Ma ago as a source of oxygen and oxidizing species. Precambrian res., 52, 337-345.

Dymek, F.R. and C.Klein, 1988: Chemistry, petrology and origin of banded iron-formation lithologies from the 3800 Ma Isua supracrustal belt, West Greenland. Precambrian Res., 39, 247-302.

Ernst, W.G., 1983: The early earth and the Archean rock record. In Earth's Earliest Biosphere. Its Origin and Evolution. J.W.Schopf (Ed.), Princeton Univer. Press.

Farhatazis and M.A.J. Rodgers, 1987: Radiation Chemistry. Principles and Applications. Verlag Chemie Publishers.

Fernandez, J.A., and W.H. Ip, 1983: On the time evolution of cometary influx in the regions of the terrestrial planets. Icarus, 54, 377-387.

Holland, H.D., 1984: The chemical evolution of the atmosphere and oceans, Princeton University Press.

IAEA, 1975: The Oklo Phenomenon. International Atomic Energy Agency, Vienna, 647 pp.

IAEA, 1978: Natural Fission Reactors. International Atomic Energy Agency, Vienna, 754 pp.

Kuroda, P.K., 1956: On the nuclear stability of the uranium minerals. J.Chem.Phys., 25, 781-782.

Miller, S.L. and L.E.Orgel, 1974: The Origins of Life on the Earth. Prentice-Hall.

Neuilly, M., J.Bussac, C.Frejacques, G.Nief, G.Vendryes and J.Yvon. 1972: Sur l'existence dans un passé reculé d'uranium d'Oklo. C.R.Acad.Sci. 275 D, 1847-1849.

Rasmussen, L.O. and E.Bjergbakke, 1984: CHEMSIMUL-A program package for numerical simulation of chemical reaction systems, Riso-R-393.

Schopf, J.W., 1978: The evolution of the earliest cells. Sci.Am., 239, 110-138.

Spinks, J.W.T. and R.J.Woods, 1990: An Introduction to Radiation Chemistry, Third Edition, John Willey & Sons, Inc.

Stumm, W. and J.J.Morgan, 1981: Aquatic Chemistry, Wiley, New York.

Weast, R.G. (Ed.), 1987: CRC HANDBOOK OF CHEMISTRY AND PHYSICS, CRC Press.

Veizer J., 1983: Geologiç evolution of the archean early Proterozoic Earth. In Earth's Earliest Biosphere. Its Origin and Evolution. J.W.Schopf (Ed.), Princeton Univer. Press.

CHEMICAL EFFECTS OF IONIZING RADIATION AND SONIC ENERGY IN THE CONTEXT OF CHEMICAL EVOLUTION [1].

A. Negrón-Mendoza and G. Albarrán
Instituto de Ciencias Nucleares, UNAM
A.P. 70-543 México, D.F. 04510 México.

ABSTRACT

Ionizing radiation and sonic energy are considered as sources for chemical evolution processes. These sources have still a modest place in the interdisciplinary approach to the prebiological synthesis of organic compounds.

Studies in Radiation Chemistry and Sonochemistry can provide a deeper insight into the chemical processes that may have importance for prebiotic chemistry. The present work concerns with the analysis of some chemical reactions induced by ionizing radiation or cavitation in aqueous media that may be relevant to chemical evolution studies.

TABLE OF CONTENTS

---

[1] We dedicate this article to the memory of Zorica Draganic.

ISBN 0-937194-31-X

1. INTRODUCTION.

The physico-chemical basis of the origin and evolution of life have been studied since the idea of chemical evolution was originated by Oparin (1924) and Haldane (1928). In this context, chemical evolution is the chemistry of the formation and the organization of bio-organic compounds under assumed primitive geological and astronomical conditions. Starting from Miller's work, experimental data under widely variety of conditions, ranging from extremely reducing to slightly oxidizing, and with various energy sources have successfully produced organic monomers such as amino acids, fatty acids and nucleic acids that are essential for life (Miller and Orgel, 1974). Many questions are raised in this endeavor and various avenues toward answering the stated questions have been proposed.

For the synthesis of complex molecules a supply of energy was needed. Several sources of energy are effective in generating organic compounds from simulated experiments, and they have been proposed as being present in primitive environments. Among them are ionizing radiation and sonic energy.

The estimates of the abundance of these two sources on the Primitive Earth depend on the model used for their calculation. The comparative importance for their contribution to the synthesis was made by Miller and Urey (1959) and their values have been taken since then. There is a common agreement that they were minor partners, and they have still a modest place in the interdisciplinary approach for the prebiological synthesis of organic compounds. Still, there are many reasons to believe that radioactivity and sonic energy were more abundant that is admitted.

The purpose of this review is to act as a critical introduction for those interested in the chemical effects of ionizing radiation, and ultrasonic energy related to chemical evolution processes. For these kinds of studies, we have to choose the conditions that may have occurred on the Primitive Earth, and they are inevitably related to aqueous media, due to the fact that water is among the most abundant compounds in the universe, so the emphasis will only be on aqueous systems.

1.1 WHY THESE SOURCES?

There are great theoretical achievements in Radiation Chemistry, and Sonochemistry that could be helpful in gaining a deeper insight into the chemical processes that may have importance for prebiotic chemistry. In this way, the reactions induced by these sources can be followed by fast kinetic methods and give us a precisely imagine of the system. Since in dilute aqueous solutions the dominant paths are through the water, model systems for computer simulation also can be used. The energy deposition can be estimated rather easyily.

The synthesis of biomolecules induced by these sources does not preclude the formation of organic compounds by other means.

A brief overview of the physical and chemical events induced by acoustic cavitation or ionizing radiation is presented to explain the origin of their chemical reactivity.

2. CAVITATION- THE ORIGIN OF SONOCHEMICAL PROCESSES.

The chemical effects of high intensity ultrasound have their origin in acoustic cavitation: the creation, growth, and implosive collapse of gas vacuoles in solution by the sound field. When sound passes through a liquid, it consist of expansion (negative-pressure) waves and compression (positive-pressure) waves. These cause vapor and gas-filled bubbles to grow and recompress. Acoustic cavitation can lead, to an implosive collapse of cavities. This collapse generates transient hot-spots with local temperatures and pressures of several thousand K and hundreds of atmospheres, in these conditions, high energy chemistry can take place (Suslick, 1988).

2.1 SOURCES

Sonochemical processes have been suggested as widespread natural phenomena of any streaming or wavy body of water, as in waterfalls, fast streaming rivers, and particularly in stormy seas. Cavitation takes place when water impinges on water at somewhat low linear velocities (Anbar, 1968).

2.2 THE CHEMICAL EFFECTS OF ULTRASOUND IN AQUEOUS SOLUTIONS.

The effect of ultrasound in water has been exhaustively studied (Weissler et al.,1950; Weissler, 1958; and Margulis, 1976a). Much of the early work was concerned with the fact that a variety of inorganic substances underwent not only oxidation, but also reduction reactions when exposed to ultrasound (Margulis, 1976b; 1976c; and 1976d).

The very high temperature and pressure induced by ultrasonic cavitation in collapsing gas bubbles result in the thermal dissociation of water into OH radicals and H atoms. These radicals are the primary products of the sonolysis of water.

Spin trap experiments (Makino et al., 1982; and 1983) have conclusively provided the evidence for the formation of OH and H radicals, but data have also indicated the formation of other high energy species, such as hydrated electrons and hydroxyperoxy radicals, depending on conditions (Suslick, 1988). These radicals recombine and given rise to hydrogen peroxide and molecular hydrogen. When a solute is present, other free radicals are formed, that follow multitude of kinetic paths for their recombination and collisions with the water molecules. Consideration of the high vapor pressure of water relative to inorganic or dilute organic solutes would predict that these paths must always be dominated by secondary reactions.

The main reactions in the sonolysis of pure water are:

$$H_2O \xrightarrow{)))} H + OH \quad (1)$$

$$H + OH \longrightarrow H_2O \quad (2)$$

$$2H \longrightarrow H_2 \quad (3)$$

$$2OH \longrightarrow H_2O_2 \quad (4)$$

$$2OH \longrightarrow O + H_2O \quad (5)$$

$$2O \longrightarrow O_2 \quad (6)$$

$$1/2\ O_2 + 2H \longrightarrow H_2O \quad (7)$$

2.2.1 Cavitation in Chemical Evolution

Some pioneering experiments on the applications of cavitation to chemical evolution studies were conducted by Anbar (1966) and Anbar and Ponnamperuma (1968). In spite of diversity of the chemical effects of ultrasonic waves, research since then has been sparse and uneven.

The data obtained then is not so easy to compare due to the fact that sonicators used by the various research groups were often different in configuration and delivered different frequencies, intensities and wattages. To complicate the matter further, little was known of the role of these variables in affecting reaction rates. As a result, there are few generalizations upon which new investigations can rely. The uses of sonochemistry in prebiotic experiments must be considered an area that is in its early stages and needs much further study.

A summary of prebiotic experiments performed with ultrasound are listed in Table 1.

TABLE I. SIMULATION PREBIOTIC EXPERIMENTS USING CAVITATION.

SUSTRATES	PRODUCTS	REFERENCES
H_2O, NH_3, CH_4	amino acids (traces)	Anbar, and Ponnamperuma 1968
NH_3, CH_4	amino acids	Anbar, 1968
NaCN, NH_4, H_2O, CH_4	amino acids	Burke et al., 1974
CH_3CN, H_2O	carboxylic acids	Negrón and Albarrán, 1992
KCN, H_2O	carboxylic acids	Negrón and Albarrán, 1992

An additional source of hypersonic energy has been postulated to induce chemical reactions. This is related to the impact of solid bodies on the atmosphere and the hydrosphere of the young Earth. The reactions are a consequence of the formation of a shock around a large meteorite and its collision with water at hypersonic velocities that can yield a variety of organic compounds (Hochstin, 1963; Bar-Nun, et al., 1970; Shaviv and Bar-Nun, 1974; Park et al., 1975; Martell, 1984).

2.2.2 Heterogeneous Sonochemistry. A Solid Supported Reaction

In recent years, there has been a revival of attention focused on the application of ultrasound to accelerating synthetic methods. Several experiments were carried out in heterogeneous systems (Ley and Low, 1989; Mason and Lorimer, 1988; Suslich, 1988).

The effects of ultrasound in homogeneous systems are dominated by the enormous changes in pressure and temperature created at hot spots on implosion of the cavitation bubbles. In two phase systems many other factors must be taken into consideration, although many of them have not been conclusively established. At solid/liquid interfaces the effects of cavitation are seen in the well known cleaning effect of ultrasound. These effects are responsible for the huge increase in rates of reactions. The overall rate is hence a combination of the rates at which species diffuse to and away from the surface, and also is a reflection of the number of reaction sites available (Ley and Low, 1989). Therefore sonication produces highly active, unpassivated surfaces. Although the relative contributions of the effect outlined have not been fully established, the beneficial effects of this energy in such systems are reflected in the large number of examples published.

For prebiotic processes heterogeneous reactions are of particular importance. Here the presence of clay minerals or other active surfaces may have an important role. Enhanced reactions induced by ultrasound in the presence of a heterogeneous catalysis can be expected. The potential advantages of such type of catalysis involve the ability to generate high energy species that can be trap in the solid catalyst and increased selectivity.

As an example of an induced reaction by ultrasound is the synthesis of amino acids. There are many reports on prebiotic synthesis of these compounds (Miller and Orgel, 1974). A classical method for the synthesis of amino acids is the Strecker synthesis. A new approach using heterogeneous sonochemistry has been proposed. Aminonitriles have been prepared in high yields by the direct reaction of KCN and NH_4Cl in the presence of alumina. The optimal conditions of the synthesis were when the mixture was sonicated. The yield of the reaction was poor under stirred conditions (Hanafusa, 1987). The increases in yield may be due to the formation of active ion aggregates on the alumina that block those surface sites associated with its Lewis acid nature. This is an interesting reaction that represents a whole range of supported reactions that are enhanced with ultrasound. Other reactions are presented in Table 2.

Some advantages of the use of ultrasound as a source in prebiotic experiments are the following: The use of ultrasound waves gives rise faster reaction rates, lower temperature or pressure requirements, higher yield and narrower products distribution. Also there is an enhancement of radicals reactions and an increase in the catalysis efficiency.

TABLE 2. SONOCHEMICAL SYNTHESIS THAT MAY HAVE IMPORTANCE IN PREBIOTIC CHEMISTRY.

SUSTRATES	PRODUCTS	REFERENCE
$H_2 + N_2$	NH_3	Sokol'skaya and El'piner, 1957
H_2 + CO	HCO	Sokol'skaya, 1978
N_2 +(CO, CH_4 or HCHO)	amino acids	Sokol'skaya, 1975 1978
Acetate	succinic, glyoxylic and glycollic acids, O_2, CO, CO_2	Gutierrez, et al., 1986
Thymine	hydroxylated products	Yu et al., 1980
Amino acids	H_2, CO, CO_2, NH_3, carboxylic acids	Staas and Spurlock, 1975
Aliphatic aldehydes, carboxylic acids	CO, CO_2, CH_4, ethylene carboxylic acids	Reifsneider and Spurlock, 1973

3. RADIOCHEMISTRY AND IONIZING RADIATION

3.1 SOME DEFINITIONS

Radioactivity is the property of some nuclei to undergo a spontaneous disintegration accompanied by the emission of radiation. The term radiation is used to denote the emission of any type of ray or particle from a source. If the radiation has sufficient energy to produce ions by either direct or indirect means, it is called ionizing radiation.

The study of the process by which high energy radiation is absorbed in matter and the chemical changes that are produced forms the subject of Radiation Chemistry. The net result is excitation and ionization, producing short lived chemical intermediates. During the process free radicals are formed and these can lead to the formation of new products. The decomposition due to radiation is called radiolysis.

The primary excitations and ionizations produced by radiation, and the subsequent reactions of the species formed, can be significantly different in condensed phases and in the gas phase.

Radioactivity was discovered at the end of the last century. Since then it has continued to make an impact over various fields of research due to the omnipresence of radioactivity.

3.2 DISTRIBUTION AND ABUNDANCE

Natural radioactivity has its origin in primordial radionuclides and their daughter products. These are distributed in soils, rocks and the hydrosphere. Primordial radionuclides were made by violent nuclear processes in the stars. Because of subsequent circulation of matter, they became incorporated into the Primitive Earth about 4500 millon years ago (Draganic et al., 1990).

Since its formation, our planet has been permanently exposed to ionizing radiation of both terrestrial and extraterrestrial origin. The terrestrial sources of ionizing radiation consist mainly of radiations emitted during the decay of several radionuclides. In this aspect, the inventory of the present Earth does not differ from the young Earth (Draganic et al., 1990).

High energy in the astronomical environments of the Earth came from: a) Cosmic rays of non-solar origin (galactic cosmic rays); b) Geomagnetically trapped corpuscular radiation; c) Particles arising from solar flares (that is, cosmic rays of solar origin) Haffner, 1967; Duley, 1984.

The use of ionizing radiation as an energy source has been proposed before (Ponnamperuma, 1966; Negrón-Mendoza and Ponnamperuma, 1976; Negrón-Mendoza, 1983; Draganic and Draganic, 1978). It stems from the calculations of the decay of radioactive elements such as potassium-40, uranium-235, uranium-238 and thorium-232. All them have half-lives of nearly 10^9 years or more. Several other naturally radioactive isotopes of long half-life have been discovered. For example rubidium-87, indium-115, lanthanum-138, etc., but these elements appear to be too rare to be significant sources of energy (Birch, 1954).

Under the conditions existing on the Earth's surface at the present time, the radiation effects of the above radionuclides are quite negligible. However, they might have been important sources of energy in the prebiotic epoch in inducing chemical changes. Indeed it has been calculated that radiation energy per unit time was three to four times greater on the primitive Earth (3.5 x 10^9 years ago) than it is now (Bullard, 1954). The amounts of energy liberated only from the decay of potassium-40 as beta and gamma radiation over of 3.5 x 10^9 years were about 2.05 x 10^{20} calories per year (Birch, 1954). It is quite important to note also that ^{40}K is evenly distributed in the crust, especially in persilisic rocks, clays and oceans (Eisenmund, 1963; Rankama, 1956). Furthermore, 3 x 10^9 years ago, or some 350 millon years after the solidification of the Earth crust, the amount of ^{40}K was 4.78 times higher than the present value, with a value calculated to be in sea water of nearly 300 pCi/l. In the other hand, it has been shown that clays are quite effective fixing potassium selectively (Rankama, 1956). Because of this it is possible to infer that high levels of radioactivity existed in the past.

Before radiation energy is finally degradated to heat it must produce physical and chemical changes. Some of these may be significant to the history of the Earth. It is possible that the effect of radiation on hydrocarbons could have been responsible for generation or modification of petroleum in certain deposits (Swallow, 1973).

Table 3 shows different sources of ionizing radiation on the Primitive Earth (Draganic and Draganic, 1978).

TABLE 3. THE SOURCES OF IONIZING RADIATION ON THE PRIMITIVE EARTH*

SOURCE	REMARK
Certain	
Potassium-40	half-life 1.25 x 10^9 years
Thorium-232	half-life 1.39 x 10^{10} year
Uranium-235	half-life 0.71 x 10^9 years
Uranium-238	half-life 4.5 x 10^9 years
Plutonium-244	half-life 7.6 x 10^7 years
Probable	
Fossil nuclear reactors	The same radioelements as in man-made nuclear reactors.
Possible	
Superheavy elements in the island of stability with atomic numbers around 114.	Half-lives estimated on theoretical grounds to be between 10^8 and 10^9 years.

* From Draganic and Draganic, 1978

3.3 RADIOLYSIS OF WATER AND AQUEOUS SOLUTIONS AND ITS RELEVANCE IN PRIMORDIAL OCEANS

Water is one of the simplest molecules and its radiolytic behavior is well understood for each of its physical stages (Draganic and Draganic, 1971). The predominate presence of water conferred a privileged status to aqueous environment as a setting for chemical evolution processes.

The absorption of ionizing radiation by water molecules leads to the formation of short-lived reactive species: H , OH, e_{aq}^-, H_2 and H_2O_2. These species initiate the chemical process by reacting among themselves and/or with solutes present in the solution. These reactions lead to the formation of stable products. The overall process is summarized in the following reaction:

$$H_2O \rightsquigarrow\text{---->} H + OH + e_{aq}^- + H_2 \quad H_2O_2 \qquad (8)$$

Radical and molecular yields may be determined from the amount of chemical reaction produced in an aqueous system by a known dose.
In dilute solutions indirect effects predominate. In solutions above 1 M, direct effects-the action on the solute itself-start to appear.

3.3.1 The Possible Role of Ionizing Radiation as an Energy Source in Chemical Evolution

Ionizing radiation has been considered as a minor partner of the various energy sources that were available on the primitive Earth. It is very efficient due to its omnipresence and its effectiveness in inducing free radical reactions.

In this context, the application of ionizing radiation to prebiotic synthesis is based on its unique qualities: its specific way of energy deposition, its distribution. The reactions are through free radicals and independent over a wide range of temperature, concentration and pH. The radiation furnishes pathways in the water in which chemical reactions occur. Contrary to any other form of energy, ionizing radiation can cause polymerization without needing catalysis.

Many reactions induced by radiation have been examined in the context of primordial synthesis. They have been focused on different aspects:
a) Synthesis of monomers and polymers of biologically important molecules.
b) Heterogeneous catalysis (mainly with clay minerals) in the presence of ionizing radiation.
c) Reactions relevant to cometary chemistry and interstellar space.
d) Spontaneous nuclear recoil effects produce hot atoms. The reactions of hot atoms can be produced by collisions of interstellar gas and dust clouds, shock waves (Rossler et al., 1984).

A list of some relevant experimental data focused on ionizing radiation can be found in Draganic and Draganic, 1978; Negrón-Mendoza, 1990; and Draganic and Vujosevic, 1992 and the references therein.

3.3.2 Applications of Radiation Chemistry Results to Other Fields

The knowledge obtained from Radiation Chemistry has been used for the chemistry of reactions in electric discharges or plasma jets. If a sufficiently high potential is applied across a gas it becomes electrically conducting because of ionization. The ionized and excited gas constitutes the plasma. Reaction of these ions and excited species proceed as in Radiation Chemistry. Subsequent reactions of the fragmentary chemical species are analogous to those induced by radiation (Swallow, 1973).

3.4 SOME FEATURES OF IONIZING RADIATION AND SONIC ENERGY

The sonolysis and radiolysis of many aqueous systems have common features such as the attack of free radicals generated from water, these cause the formation of similar products. Still, there are some differences: In sonolysis the chemistry that occurs is dominated by the fact that oxygen is also produced and acts as a scavenger for part of the intermediate free radicals. However, recently Draganic and Draganic (1991) pointed out that in oceans, due to the action of radionuclides, oxygen is also produced. This may lead to some more oxidizing products. The process of radical formation in sonolysis is thermal. In radiation the dominating process is ionization and excitation (Weissler et al. 1965; Gutierrez et al., 1986; Kendo et al., 1988).

4. REMARKS.

Our aim has been to stress the relevance of ionizing radiation and cavitation as tools for the study of the formation of compounds of potential interest to chemical evolution studies. Although the range of reactions that have been carried out by these means is extremely diverse, and much progress has been made, more work is required.

The study of reactions induced by sonic waves is still in its early stages. Acoustic cavitation represents a combination of extraordinary high temperature with exceptionally short residence times that gives an activation unobtainable for other sources in condensed media.

Studies in Radiation Chemistry can be of particular important for the the reactions in interstellar space and in cometary nuclei.

REFERENCES

Anbar, M., 1966: Chemical Reactions Induced by Sound, New Scientist, 365-377.

Anbar, M., and I. Pecht, 1967: The Sonolytic Decomposition of Organic Solutes in Diluted Aqueous Solutions III, J. Phys. Chem., 71, 1246-1249.

Anbar, M., and C. Ponnamperuma, 1968a: On the Possible Role of Sonochemical Processes in Chemical Evolution, Abstract of paper BIOL-100, 156th National Meeting of Am. Chem. Soc. Sep. 5-13.

Anbar, M., 1968b: Cavitation during Impact of Liquid Water on Water: Geochemical Implications, Science, 161, 1343-1344.

Bar-Nun, A., N. Bar-Nun, S. H. Bauer, and C. Sagan, 1970: Shock Synthesis of Amino Acids in Simulated Primitive Environments, Science, 168, 470-473.

Birch, F., 1954: Nuclear Geology, Faul (Ed.) Wiley Son, New York.

Bullard, E., 1954: The Earth as a Planet, G. P. Kuiper (Ed.) University Chicago Press, Chicago, 57-137.

Burke, S., W. K. Park, and C. Ponnamperuma, 1974: Cavitation Induced Organic Synthesis in the Primordial Ocean, In Abstract, The 9th Middle Atlantic Regional Meeting, Am. Chem. Soc. Pennsylvania, April 23-26.

Draganic, I. G., and Z. D. Draganic, 1971: The Radiation Chemistry of Water, Academic Press, New York, N. Y.

Draganic, I. G., and Z. D. Draganic, 1978: Radioactivity and Radiation - Chemistry on the Early Earth, In Plenary Lectures II Symposium on Nuclear Chemistry, Radiochemistry and Radiation Chemistry, C. Archundia (Ed.) Mexico, Dic. 13-15., 50-72.

Draganic, I. G., Z. D. Draganic, and J. P. Adloff, 1990: Radiation and Radioactivity on Earth and Beyond, CRC Press, Inc. Boca Raton, Florida.

Draganic, I. G., Z. D. Draganic, 1991: Decomposition of Ocean Waters by Potassium-40 Radiation 3800 Ma Ago as a Source of Oxygen and Oxidizing Species, Precambriam Res., 52,337-345.

Draganic, I.G. and S. Vujosevic, in press: Ionizing Radiation in Water in Chemical processings in the space and on the Early Earth. Available in the Institute of Nuclear Sciences Vinca, P.O.B. 522, 11000 Belgrade, Yugoslavia.

Duley, W. W., and D. A. Williams, 1984: Interstellar Chemistry, Academic Press, London.

Eisenbud, M., 1963: Environmental Radioactivity, Mc Graw Hill Book, New York.

Gutiérrez, M., A. Henglein, and Ch. H. Fischer, 1986: Hot Spot Kinetics of the Sonolysis of Aqueous Acetate solutions, Int. J. Radiat. Biol., 50, 313-321.

Haffner, J. W., 1967: Radiation and Shielding in Space, Academic Press, New York, N.Y.

Haldane, J. B. S., 1929: The Origin of Life, Ration. Ann., 148, 3-10.

Hanafusa, T., J. Ichihara, and T. Ashida, 1987: Useful Synthesis of Aminonitriles by Means of Alumina and Ultrasound, Chemistry Letters, 687-690.

Hochstim, A. R., 1963: Hypersonic Chemosynthesis and Possible Formation of Organic Compounds from Impact of Meteorites on Water, Proc. Nat. Acad. Sci., 50, 200-210.

Kendo, T., C. M. Krishna, and P. Riesz, 1988: Sonolysis, Radiolysis, and Hydrogen Peroxide Photolysis of Pyrimidine Derivatives in Aqueous Solutions: A Spin Trapping Study, Rad. Res., 116, 56-73.

Ley, S. V. and C. M. R. Low, 1989: Ultrasound in Synthesis, Springer-Verlag, Germany.

Makino, K., M. M. Mossoba, and P. Riesz, 1982: Chemical Effects of Ultrasound on Aqueous Solutions. Evidence for OH and H by Spin Trapping, J. Am. Chem. Soc., 104, 3537-3539.

Makino, K., M. M. Mossoba, and P. Riesz, 1982: Chemical Effects of Ultrasound on Aqueous Solutions. Formation of Hydroxyl Radicals and Hydrogen Atoms, J.Phys. Chem., 87, 1369-1377.

Margulis, M. A., 1976a: Energetics and Mechanism of Phonochemical Reactions. III. Oxidation of Iron(II) Sulphate in an Ultrasonic Field. Russian J. Phys. Chem., 50, 1358-1360.

Margulis, M. A., 1976b: Energetics and Mechanism of Phonochemical Reactions. IV Reduction of Cerium(IV) Sulphate in an Ultrasonic Field, Russian J. Phys. Chem., 50, 1361-1363.

Margulis, M. A., 1976c: The Energetics and Mechanism of Sonochemical Reactions. V. The Formation of Hydrogen Peroxide in the Sonolysis of Water and Aqueous Solutions of Formates, Russian J. Phys. Chem., 50, 1515-1518.

Margulis, M. A., 1976d: The Energetics and Mechanism of Sonochemical Reactions. VI. Determination of the Initial Sonochemical Yields, Russian J. Phys. Chem., 50, 1518-1521.

Martelli, G., 1984: Role of Hypervelocity Impacts in the Formation of Organic Molecules, Abstract, XXV COSPAR., Graz, 25 June-7 July, 314.

Mason T. J., and J. P. Lorimer., 1988: Sonochemistry: Theory, Applications and Uses of Ultrasound in Chemistry, Ellis Horwood Limited, England.

Miller, S. L., and L. E. Orgel, 1974: The Origins of Life on Earth, Englewood Cliffs, NJ: Prentice-Hall.

Miller, S. L., and H. C. Urey, 1959: Organic Compound Synthesis on the Primitive Earth, Science, 130, 245.

Negrón-Mendoza, A., and C. Ponnamperuma, 1976: Formation of Biologically Relevant Carboxylic Acids Acids During the Gamma Irradiation of Acetic Acid,Origins of Life, 7, 191-196.

Negrón-Mendoza, A., 1983: Tierra Primitiva y Síntesis Abiótica, in Homenaje a Oparin, M. Artís, M. Casanueva and N. Chávez, (Ed.) Correspondencia, México, D.F. 63-72.

Negrón-Mendoza, A., 1990: La Radiación: Fuente de Vida, ICYT., 12, 29-34.

Negrón-Mendoza, A., and G. Albarrán, 1992: Formación de Compuestos de Importancia Biológica por la Irradiación Ultrasónica de Soluciones Acuosas de Nitrilos y Cianuros Simples, Abstract del XXVIII Congreso Mexicano de Química Pura y Aplicada, Rev. Soc. Quim. Mex. 36, 31.

Oparin, A. I., 1924: Transl., 1967, as appendix in J. D. Bernal, The Origin of Life. Cleveland: World.

Park, W. K., A. R. Hochstim, and C. Ponnamperuma, 1975: Organic Synthesis by Quench Reactions, Origins of Life., 6, 99-107.

Ponnamperuma, C., 1966: Proceedings of the Third International Congress of Radiation Research, G. Silini (Ed.) Cortina d'Ampezzo, North Holand, Amsterdam, June-July, 700-713.

Rankama, K., 1956: Isotopic Geology, Pergamon Press, London.

Reifsneider, S. B., and L. A. Spurlock, 1973: Chemistry of Ultrasound. II. Irradiative Behavior of Aliphatic Aldehydes and Carboxylic Acids in an Aqueous Medium, J. Am. Chem. Soc., 95, 299-305.

Rossler, K., 1984: Synthesis of Organic Molecules by Atomic Carbon: Relevance to Cosmic Chemistry, Abstracts, 25th Plenary Meeting Committee on Space Research, Austria 25 June- 7 July.

Shaviv, A, and A. Bar-Nun, 1974: The Role of Shock Waves in the Chemical Evolution of Earth's Primitive Atmosphere, The Origin of Life and Evolutionary Biochemistry, K. Dose, S.W. Fox, G.A. Deborin, and T.E. Pavlovskaya (eds.) Plenum Press, New York, 445-458.

Sokol'skaya, A. V., I. E. El'piner, 1957: Synthesis of Ammonia and Cyanide by Ultrasonic Waves, Akust. Zhur., 3 293-294.

Sokol'skaya, A. V., 1975: Synthesis of organic compounds in an Ultrasonic Wave Field, Zh. Evol. Biokhim. Fiziol., 11 509-514.

Sokol'skaya, A.V., 1978: Synthesis of Glycine from Formaldehyde and Molecular Nitrogen in Aqueous Solutions in an Ultrasonic Field, J.Gen. Chem. USSR., 48, 1289-1292.

Staas, W. H., and L. A. Spurlock, 1975: Chemistry of Ultrasound. Part IV. Effects of Ultrasound on Some Amino-acids, J. C. S. Perkin I., 1675.

Suslick K. S., 1988: Ultrasound: Its Chemical, Physical, and Biological Effects, VCH Publishers, Inc., Federal Republic of Germany.

Swallow, A. J. 1973: Radiation Chemistry, an Introduction, Longman Group Limited, London.

Weissler, A., H. W. Cooper, and S. Snyder, 1950: Chemical Effect of Ultrasonic Waves: Oxidation of Potassium Iodide Solutions by Carbon Tetrachoride, J. Am. Chem. Soc., 72, 1769-1781.

Weissler, A., 1959: Formation of Hydrogen Peroxide by Ultrasonic Waves: Free Radicals, J. Am. Chem. Soc., 81, 1077-1081.

Weissler, A., I. Pecht, and M. Anbar, 1965: Ultrasound Chemical Effects on Pure Organic Liquids, Science., 150, 1288-1289.

Yu, T. J., R. G. Suttherland, and R. V. Verrall, 1980: The Sonolysis of Uracil, Can. J. Chem., 58, 1909-1915.

DIFFERENCES IN RADIOLYSIS BEHAVIOR OF D,L-AMINO ACID IN PRIMARY STAGE AND THERMODYNAMIC EQUILIBRIUM STATE

W. Q.Wang , J. L.Wu and J. Jiang
Peking University
Beijing 100871 , P. R. of China

ABSTRACT

A series of experiments were carried out to investigate the interactions of elementary particles with D- and L-amino acids.The annihilation of positrons β^+ from ^{22}Na in sublimed D-,L-,DL-leucine formed less triplet state positronium with D-isomer respect to L-isomer by a factor of 0.879, but this was not observed in crystallized samples. The ESR measured free radical formation in ^{90}Sr-^{90}Y β^--irradiated D- and L-alanine at 77K demonstrated 5.17% more of the relative radical concentration in D-isomer than in L-isomer. It should be noticed that in both β^+ and β^--irradiation, no stereoselective decompostion was found by the chirasil-val capillary column GC analysis. The γ-ray irradiation on sublimed D-, L- and DL-leucine again showed no asymmetric decomposition, either measured by GC analysis of D-, L-isomer, or the radiolytic products (CO_2, NH_3 and H_2). These results together with a careful reference survey lead to the conclusion that elementary particles do differentially interact with D- and L-enantiomers of amino acids. The subtle differences could be detected only when the system is far from equilibrium(the primary stage), but not in the thermodynamic equilibrium state. However the difference could not form a macroscopic excess of any particular isomer if an proper amplification mechanism is not invloved in.

1. INTRODUCTION

The origin of optically active molecules in nature, as part of the general asymmetries of nature is a question which has challenged scientists since the time of Pasteur. Following Wu's experimental confirmation of Lee and Yang's prediction that the parity principle might be violated for certain weak interactions such as those involved in the β-decay of radioactive nuclei, Vester and Ulbricht proposed that the chirality of molecules of biosphere may be originated through the differential interactions of enantiomers with spin-polarised electrons arising from β-decay, or circularly polarised bremsstrahlung produced during their

ISBN 0-937194-31-X

deceleration. Later, Yamagata claimed an intrinsic energy difference in the enantiomer molecules due to the parity non-conservation, which would manifest itself in the chirality of biomolecules. The so-called V-U hypothesis and the Yamagata hypothesis laid part of the foundation of the determinate abiotic mechanisms for the origin of chirality and have received much theoretical and experimental attention.

As a central concept, the parity nonconservative weak interaction should be separated into two types which correspond to the two hypotheses. The V-U hypothesis involves the charged-current parity nonconservation (CC-PNC) which leads to the polarized β-rays, while the Yamagata hypothesis involves the neutral-current parity nonconservation (NC-PNC) which results in the parity-violating energy difference (PVED) between two enantiomers.

Substantial experimental effort has been made to test the V-U and the Yamagata hypothesis and could be classed into two groups. One focused on the interaction itself of elementary particles with enantiomers, and the other investigated the interaction-induced asymmetric decomposition. Unfortunately the published data are inconclusive and controversial. A lot of work has been done to verify the existence of stereoselectivity of β^+, β^--particles and γ-ray radiolysis. However few positive results were obtained and more these could not be independently corroborated. Upon the differential interaction examination, less effort was presented. Several theoretical models were constructed to evaluate the possible interaction difference and tended to predict that the difference must be existed, though very small in magnitude. The experimental verification gave no consistent results again. The controversial experimental results, the poor reproducibility of some positive evidence and the theoretical prediction of the effects which are well below the threshold of experimental observation have spreaded the pessimistic view over the function of elementary particles in the origin of chirality of biosphere.

A fact received no sufficient attention in the previous studies is that a system dominated by homochiral molecules is highly macroscopically ordered, the formation of which demands the circumstance far from equilibrium and a nonlinear kinetic mechanism. Without external chiral factors and intrinsic energy difference between enantiomers, a racemic mixture will remain its racemic state even if the system is far from equilibrium because of the equal probability determination. With the presence of external chiral particles and/or the intrinsic energy differnce, the fate of the racemate is not doomed by equal chance of enantiomers, but rather favors the growth of one particular enantiomer in the sense of probability. The theoretical deduction proves that only very small influence is needed to produce the selection. On the other hand, a nonlinear kinetic mechanism such as autocatalyst is crucial to amplify a favored fluctuation to a state of homochirality.

From the view of dissipative structure, we, the authors, consider that there exists differential interaction of elementary particles with D,L amino acids. The differences are truly trivial in magnitude, but it could be detected when the system is far from equilibrium (the primary stage), by instant measurement as the interaction is still proceeding. When the system reaches its thermodynamic equilibrium state, the difference would not be detectable. If an effective amplification is involved in, the subtle difference would be enlarged to the extent of formation of macroscopically detectable chirality . Therefore the possibility that the asymmetry of elementary particles causes the formation of terrestrial excess of one particular isomer could not be excluded. This view may explain to some extent the long disputation and the controversy among reports. To test this postulation, a series of experiments were carried out, the results of which were described in detail below.

2. EXPERIMENTAL

2.1 Positron Annihilation

Positrons(e^+) , emitted with parity violation during the decay of neutron deficient radionuclide ^{22}Na, have a predominant "right handed" (spin forward) longitudinal polarization. Their initial high energy (~10^3 KeV) is rapidly degraded to near thermal energy (~10 eV) by collisions with matter.The positrons retain their longitudinal polarization while slowing down to form Ps. Garay et al. (1973, 1974) studied the annihilation of positrons with the D- and L-isomers of tryptophan, phenylalanine, tyrosine and dihydroxyphenylalanine, and Conte and Pieralice (1987) added alanine to the list. They reported that more triplet state positroniums (o-Ps) were formed in D-isomers of amino acids with respect to L-isomers. Garay's results were questioned by the inability of other investigators who, again using ^{22}Na positron sources, were unable to find any significant differences.

We investigated the annihilation of positrons from 5μC ^{22}Na in D- and L-leucine ,both sublimed and crystallized ,by using gamma spectroscopy. Much attention was paid to ensure the purity of the samples and optimal condition available. Since the difference in triplet state positronium (o-Ps) intensity was reported to be a basic effect, we choosed it to be the parameter of evalution. Samples were pressed to be a cylinder of 13.0 mm diameter and 2.0 mm thickness and placed in close touch with the ^{22}Na source. For each sample, the lifetime spectra were recorded two times after 0, 6, 12 and 240 hrs irradiation. Obtained data were presented in Table 1.

TABLE 1. INTENSITY OF O-Ps IN D- and L-LEUCINE

Sample No.	Irradiated Time(h)	Sublimed Samples D-Leu	L-Leu	D/L
1	0	17.15	20.14	0.852
2	0	14.62	17.47	0.837
3	0	23.76	26.46	0.898
1	6	17.02	19.50	0.872
2	6	14.07	15.60	0.902
3	6	22.58	25.40	0.889
1	12	15.39	17.66	0.871
2	12	13.69	15.14	0.904
3	12	22.66	25.40	0.892
1	240	14.42	16.34	0.882
2	240	13.94	16.20	0.860
3	240	22.90	26.20	0.874
			mean value =	0.879 ± 0.011

Sample No.	Irradiated Time(h)	Crystallized Sample D-Leu	L-Leu	D/L
1	0	24.44	25.28	0.967
2	0	19.91	21.64	0.920
1	12	23.04	22.88	1.007
2	12	20.62	20.21	1.020
1	12	21.66	22.57	0.960
2	12	18.80	20.40	0.922
			mean value =	0.965 ± 0.039

It is clearly seen in Table 1 that for sublimed samples the mean value of o-Ps intensity of L-leucine was higher than that of D-leucine, meaning more o-Ps positronium was formed in L-isomer. Interestingly this difference vanished in crystallized samples.

The study of the annihilation of polarized positrons in optical isomers has the advantage that it gives information about the interaction process itself rather than about the products of the interaction. The data obtained here supports that the differential interactions of positrons with isomers of amino acids were detectable in an open and far from equilibrium system. In addition, the results emphasized the distinction between crystallized and sublimed samples.The possible crystal effect to the stereoselectivity of elementary particles may explain some of the null results about positron annihilation in which crystallized samples of a variety of organic compounds were used.

With the presence of the selective interaction in mind, we turned to examine if asymmetric decomposition was induced. Sublimed D-, L- and DL-leucine were independently irradiated by a 20μC ^{22}Na source for 240 hrs. The chirasil-val capillary column gas chromatography was introduced to analyze the irradiated samples. No excess of any particular isomer was found.

2.2 ^{90}Sr-^{90}Y β^- -irradiation

The well-known V-U hypothesis based on the violation of parity during β-dacay is the firstly and most thoroughly studied deterministic mechanism. With the exception of the early reports of Garay (1968) and Bonner (1977) that small preferential degradation of D-isomer was observed, most other experiments gave no selectivity, including those of the exact duplication of Garay and Bonner's designs. On the other hand, Akaboshi and Conte (1986) used ESR measurement to study the interaction process and proved the asymmetrical interaction of β-particles from ^{90}Sr-^{90}Y-beta-decay with D- and L-alanines.

We reinvestigated the free radical formation in D- and L- alanine after ^{90}Sr-^{90}Y beta irradiation. The ESR spectra of radicals formed from γ and β-irradiated D- and L-alanine show the same line width, peak position and peak ratio, so it is reasonable to select the height of the central peak, H, as the evaluation standard for the relative concentration of radicals. To calibrate the possible influence of unknown factors, samples were firstly irradiated by ^{60}Co γ-rays with the absorbed dose of 6.0 x 10^2 Gy. Those samples with similar central peak height in ESR spectra were selected to perform the following bata irradiation at 77K with total dose of 2.1 x 10^3 Gy.

Table 2 shows the relative concentration of free radicals by gamma and beta irradiations expressed by H_β, H_γ, and $H_{\beta+\gamma}/H_\gamma$. As shown in the table, The average ratios $H_{\beta+\gamma}/H_\gamma$ were 1.495 for D- alanine and 1.348 for L-alanine. The asymmetric difference is 5.17% as defined in Eq. (1). It was similar with the results of Akaboshi and Conte. But it should be indicated that the difference in the relative free radical concentration does not definitely mean an asymmetrical decomposition as stated by Conte. The following description about beta radiolysis of D,L-leucine is concerning the problem.

TABLE 2. FREE RADICAL FORMATION IN D- AND L-ALANINE BY ^{90}Sr-^{90}Y BETA IRRADIATION

No.	Sample	H_β	$H_{\beta+\gamma}$	$H_{\beta+\gamma}/H_\gamma$	$(H_{\beta+\gamma}/H_\gamma)_D - (H_{\beta+\gamma}/H_\gamma)_L$
1	D-Ala	9.45	13.18	1.395	
	L-Ala	9.30	11.37	1.222	0.173
2	D-Ala	9.82	15.14	1.542	
	L-Ala	9.61	13.31	1.385	0.157
3	D-Ala	7.73	11.97	1.549	
	L-Ala	8.29	11.91	1.437	0.112
					mean value = 0.147 ± 0.032

$$\frac{(H_{\beta+\gamma}/H_\gamma)_D - (H_{\beta+\gamma}/H_\gamma)_L}{(H_{\beta+\gamma}/H_\gamma)_D + (H_{\beta+\gamma}/H_\gamma)_L} = 0.0517 \qquad (1)$$

Each 300 mg of sublimed D,L-leucine was exactly balanced and pressed into a cylinder of 13.0 mm diameter and 2.0 mm thickness. Samples were irradiated by ^{90}Sr-^{90}Y beta source to a total dose of 7.56 x 10^4 Gy, which caused about 13.2% degradation, as determined by the enantiomeric marker technique. The GS analysis was performed with the chirasil-val capillary column. Table 3 listed the obtained data.

TABLE 3. GC ANALYSIS OF DL-LEUCINE AFTER ^{90}Sr-^{90}Y β-IRRADIATION

Sample No.	Absorption Dose(x10^4Gy)	Decomposition Percentage	G Value	D/L
1	0	----	----	1.01
2	0	----	----	1.01
3	0	----	----	1.00
4	7.56	12.8%	125	0.99
5	7.56	14.2%	139	1.00
6	7.56	12.5%	122	0.99

The decomposition percentages for three DL-leucine samples are 12.8%, 14.2% and 12.5% respectively, and the corresponding G values are 125, 139 and 122. The large G values indicate a chain reaction. As shown in Table 3, the undecomposed leucine residues for all the three samples proved to be totally racemic within the error limit, suggesting no stereoselectivity.

The ESR measurement on the free radical formation in D- and L-alanine after ^{90}Sr-^{90}Y β-irradiation shows differential interaction, while the GC analysis on the beta decomposition of DL-leucine gives no asymmetricity. This demonstrates the necessity of differentiating the interaction itself and the interaction -induced decomposition.

2.3 ^{60}Co γ-irradiation

Yamagata first postulated that "parity-violating energy difference" (PVED) between two enantiomers has potential connection with the origin of biomolecular homochirality. The theoretically calculated PVED is extremely small (10^{-21} - 10^{-18} eV). Some investigators tried to amplify PVED by process such as crystallization, polymerization, reporting very slight polarimetric consequence. Differently, Merwitz (1976) manifested PVED by γ-irradiating D,L-amino acids. He reported that the rate of the γ-ray induced decarboxylation of D- phenylalanine was 2.7 times greater than that of L-phenylalanine. Such observation was subsequently extended by Norden (1985) to leucine enantiomers, with similar results. These enantiomeric rate differences were ascribed to PVED effects which were then amplified by a chain reaction. No mechanism for the chain reaction was suggested.

That the extremely small energy difference of enantiomers due to the presence of parity violating weak neutral current in electromagnetic interaction could cause so manifested chiral selectivity is doubtful. We carried out an irradiation experiment to clarify Merwitz and Norden's observation. Sublimed DL-, D- and L-leucine under vacumm were exactly balanced and put into sample tubes. Any air or volatile impurities were removed by flushing the tubes with N_2 of high purity. The samples were then sealed and subjected to ^{60}Co γ-irradiation with different doses. A possible asymmetric decomposition was determined by the chirasil- val capillary column gas chromatography. The relative yields of the radiolysis products (CO_2, NH_3 and H_2) were measured by the packed column gas chromatography.

As shown in Table 4, the peak area ratios D/L of DL-leucine after different absorbed dose of irradiation are all nearly 1.00, the value of samples undertaken no irradiation, within the limit of error. So are the ratios for the separately irradiated D- and L-leuicne (Table 5).

This suggests that no asymmetric decomposition was found. The relative yields D/L of radiolytic products (CO_2, NH_3 and H_2), listed in Table 6, are also very close to 1.00, indicating that the D- leucine cleavage rate was nearly the same with that of the L- leucine.

TABLE 4. DECOMPOSTIION OF DL-LEUCINE AFTER ^{60}Co γ-IRRADIATION MEASURED BY CHIRASIL-VAL CAPILLARY COLUMN GC.

Absorbed Dose(x10^{2}Gy)	Peak Area D-leu	Peak Area L-leu	Radiolysis Percentage	G Value (x10^3)	D/L
0	8410	8439	---	---	0.996
1.06	8544	8591	1.4%	9.77	0.995
2.02	7379	7441	6.2%	22.7	0.992
3.18	9140	9194	6.0%	13.9	0.994
5.91	6889	6935	7.1%	8.89	0.993
8.58	6815	6855	8.1%	6.99	0.994
9.24	9252	9352	9.4%	7.53	0.989

TABLE 5. DECOMPOSITION OF D- AND L-LEUCINE AFTER ^{60}Co γ-IRRADIATION MEASURED BY CHIRASIL-VAL CAPILLARY COLUMN GC.

Absorbed Dose(x10^{2}Gy)	Peak Area D-leu	Peak Area L-leu	Radiolysis % D-leu	Radiolysis % L-leu	G Value (x10^4) D-leu	G Value (x10^4) L-leu	D/L
0	21905	21894	---	---	---	---	1.00
2.02	20416	20430	6.8	6.7	2.49	2.45	0.999
3.18	20588	20575	6.0	6.0	1.40	1.40	1.00
5.91	12212	12046	7.0	7.2	8.75	9.02	1.01
8.58	13061	12132	7.6	8.1	6.56	6.99	1.06

Merwitz reports G values ranging from 10^4 to 30, indicating a chain reaction, and our G values lie in just this interval. But we found no obvious difference in the cleavage rates even between the seperately irradiated D- and L-leucine. We deduce that in their experiments some impurities or unknown factors caused a cascade process or quenched some path of the reaction.

Norden observed that the large difference of radiolysis rate between the isomers of sublimed samples did not appear between those of crystallized samples. We reported that the difference in the o-Ps intensity in sublimed samples vanished in crystallized samples. This implys the crystal effect on the interaction of elementary particles with enantiomers of amino acids, which needs to be deciphered.

TABLE 6. THE RELATIVE YIELDS OF RADIOLYTIC PRODUCTS (CO_2, NH_3 and H_2) AFTER ^{60}Co γ-IRRADIATION

Sample	Weight mg	Peak Area CO_2	Peak Area NH_3	Peak Area H_2	D / L CO_2	D / L NH_3	D / L H_2
D-Leu	8.78	80030	28266	1177035			
L-leu	8.88	82589	30063	1093120	0.98	0.95	1.09
D-leu	9.23	84131	9378	781752			
L-leu	9.25	78069	8366	910283	1.08	1.12	1.07
D-leu	10.34	94249	18311	951069			
L-leu	10.75	95132	20964	1029927	1.03	0.96	0.96
D-leu	10.39	101522	17046	1553560			
L-leu	10.59	97525	18507	1680485	1.06	0.98	0.94
D-leu	10.80	105528	18669	----			
L-leu	10.57	96524	19146	----	1.06	0.95	---

3. CONCLUSION

Of all of the proposed determinate mechanisms for the origin of chirality, the V-U hypothesis and the Yamagata hypothesis seems to be not very sound and plausible, for no undoubted experimental results have been obtained. But we think that the effort to discover the role of elementary particles in the formation of homochirality of biosphere should not be given up. As we repeatedly stated above, we need to make a clear distinction between the primary interaction process itself of elementary particles with enantiomers of amino acids and the interaction-caused asymmetric decomposition. The elementary particles serve to be only the selective factor (maybe coupled with the intrinsic energy difference of enantiomers) which may favor the growth of one particular enantiomer from fluctuation. It is not the sufficient condition that will ensure the origin of homochirality. A nonlinear kinetic reaction mechanism should be nevertheless included. It is true that no obvious stereoselective decomposition induced by elementary particle irradiation has been definitely proved. Considering that the previous experiments involved the reaction system with pure amino acids, the purer the better, in such a system an autocatalystic process or other nonlinear kinetic processs would not possibly occur. The results so far obtained are not very surprised.

Specially, we quote the report of Akaboshi (1990) on the selective decomposition of enantiomers of aspartic acid irradiated with ^{60}Co γ-rays in the mixed aqueous solution with D- or L-alanine. Akaboshi proved that in the presence of L-alanine the radiation sensitivity of L-aspartic acid decreased selectively and vice versa. The alanine could be considered as an additive which slows down the path of the decomposition of relative isomer of aspartic acid. Of course it does not mean it was exactly what happened in the history of the origin of chirality. It, however, sheds light on the effect of additives. The primordial soup is whatsoever not pure.

The above discussion supports our view that elementary particles do differentially interact with enantiomers of amino acids. The difference, though very small, could be detected by sensitive methods when the system is far from equilibrium, but it could not cause macroscopic excess of one particular enantiomer without an additive to serve as the catalyst of an amplification or to quench some path of the reaction. For further investigation, the search for the possible additives which may present in the primordial soup and complete a nonlinear kinetic reaction is highly worthwhile.

ACKNOWLEDGMENTS

We are indebted to Mr. J. Zhao, Prof. Y. Y. Wang and Asso. Prof. Y. R. Zhou for their involvement in this work. This paper was supported by Grants of State Science Commission and Grants for Doctoral Superadvisors.

REFERENCES

Akaboshi M., Kawai K., Maki H. et al, 1990: Selective decomposition of either enantiomer of aspartic acid irradiated with ^{60}Co γ-rays in the mixed aqueous solution with D- and L-alanine, Origin of Life and Evolution of the Biosphere, 20, 111-119.

Bonner, W. A., Van Dort, M. A.,Yearian, M. R., Zeman , H. D., and Li, G. C., 1976/ 1977 : Polarized Electrons and the Origin of Optical Activity , Israel J. Chem., 15 , 89-95.

Conte, E.,Fanfani,G.,Pieralice,M.,Amerotti,R.and Daddabbo, A., 1986 : Investigation on the Asymmetrical Induced Yields in ^{90}Sr-^{90}Y-beta-Irradiated D- and L-Alanines, Origins of Life, 17, 51-57.

Conte,E., Pieralice, M., 1987 : Investigation on the Chirality of Positrons from ^{22}Na

Decay and their Asymmetrical Interactions with D-, L- and DL-Alanines, IL NUOVO CIMENTO, 9, 283-288.

Garay, A.S., 1968: Origin and Role of Optical Isomery in Life , Nature, 219, 338-339.

Garay, A.S., Keszthelyi, L., Demeter, I. and Hrasko, P., 1973: Differences in the Annihilation of Positrons in Optical Isomers, Chemical Physics Letters, 23 , 549-552.

Garay, A.S., 1974: Origin of Asymmetry in Biomolecules, Nature, 250 , 332-333.

Merwitz, O., 1976: Selective Radiolysis of Enantiomers , Rad. and Environm. Biophys., 13, 63-69.

Norden, B., Liljenzin J. O. and Tokay, R. K., 1985 : Stereoselective Decarboxylation of Amino Acids in the Solid State, With Special Reference to Chiral Discrimination in Prebiotic Evolution , J. Mol. Evol., 21 , 364-370.

EXPERIMENTAL SEARCHES FOR THE ORIGIN OF BIOMOLECULAR ASYMMETRY

L. Keszthelyi

Institute of Biophysics, Biological Research Center, SZEGED, H-6701, Hungary

ABSTRACT

The problem of the origin of the asymmetry of biomolecules attracted many theoretical and experimental investigations. In spite of these studies the problem is still open. Recent experiments, experiments which need careful repetitions are reviewed and some new experiments are suggested. The experiments involve studies related to different types of physical effects possibly causing asymmetry: weak interaction, Earth rotation, magnetic and electric fields, circularly polarized light.

1. INTRODUCTION

A recent comprehensive review deals with theoretical and experimental aspects of the origin of biomolecular handedness (Bonner,1991). The conclusion of this review is to "transfer the problem of the origin of chirality out of the realm of terrestrial science and into that of astrophysics".

Similar idea appeared in a paper by Salam (1991). According to his theory quantum mechanical cooperative and condensation phenomena may give rise to second order phase transition (D to L transformation of amino acids) below a critical temperature T_c. This T_c may be a very low temperature existing only in the space. He notes that in the early atmosphere large quantity of organic material fell to the Earth ($\sim 10^6$-10^7 kg/year declining with a half life of $\sim 10^8$ years (Chiba et al., 1990).

We think, however, that we do not have yet compelling arguments to give up experimentation and theoretical considerations that the chirality of the biosphere originated on the Earth. We mention only that according to the estimation of Sagan and Khare (1971) the rate of amino acid synthesis (and not in general "organics") was so high that $\sim 10^{10}$ amino acids per second fell to one cm^2 of the surface of the primordial Earth. That means $\sim 2.5 \times 10^{11}$ kg/year of amino acids on the Earth

ISBN 0-937194-31-X

surface which is far more than the weight of organics of possible cometary origin.

In this paper we investigate 1) selected experiments from the past to see whether a new study with important extension is worth performing, 2) recent experiments (1990-1991) and 3) we suggest new experiments which seem to be important in finding a physical agent to cause and amplify asymmetry.

The "trivial" solution i.e. the generation of biomolecular asymmetry due to fluctuation should always be kept in mind whether it happened during chemical or biological evolution. This process, however, is not discussed in this paper.

2. EARLY EXPERIMENTS

One of the intriguing experiments performed more than 10 years ago is the crystallization experiment of Kovács (1979). Adding the β^- emitter ^{32}P nuclei to racemic sodium ammonium tartrate solutions and allowing them to crystallize he found that the mass of the crystallized material and also the optical activity of the dissolved crystals increased with the logarithm of the strength of radioactivity in the solution. The fact that a function was found is very important and may corroborate the results. (In most experiments of this kind the resulting asymmetries are not measured in function of radiation intensity.) This careful experiment should be repeated with β^- and β^+ radioactive sources.

Also the control data - crystallization without radioactivity - are very valuable. As pointed out (Keszthelyi, 1977), the distribution of the relative optical activities and the limit of the shift of its maximum from zero can be used to set an upper limit for the relative energy difference ε (in kT units) between L- and D- molecules. From Kovács's control data two different upper limits $\varepsilon_2 < 7\times10^{-13}$ and $\varepsilon_3 < 3\times10^{-10}$ were calculated. The argument to explain the two different values is taken from Yamagata who states that polymers of one enantiomer with advantage of ε and a length of n units are formed with $n\varepsilon$-times higher probability than the opposite enantiomer (Yamagata,1966). If a crystal containing n molecules grows mainly in planes then $n^{1/2}$ molecules are found in one direction (ε_2) and $n^{1/3}$ if the crystals grow in three dimension (ε_3). The real upper limit may be somewhere between ε_2 and ε_3. I call this procedure the linear growth model.

The estimated upper limit is much larger than the calculated energy difference between the enantiomers of amino acids ~ 10^{-17}kT (Tranter, 1985a). Our estimate based on the linear growth model may be wrong: Tranter (1985b) used the Boltzmann distribution valid for equilibrium systems to estimate the ratio of the number of L-crystals to D-crystals for quartz:

$$N(L)/N(D) = \exp(-\Delta En(L)/kT) \tag{1}$$

where $\Delta En(L)$ is the energy difference between an L- and a D-crystal containing the same number of molecules n. If this equation is used we obtain from Kovács's control experiment $\varepsilon_1 \sim 10^{-19}$ far below the theoretical limit. I call this procedure the equilibrium model.

Probably crystal growth is not an equilibrium process therefore Eq.1 may not be applicable. Other experiments should be planned to find the value of ε. One possibility is to crystallize such molecules which have heavy elements in their asymmetry center because ε is proportional to Z^6, where Z is the atomic number. Using for example platinum (Z=78) $\varepsilon_{Pt} \sim 5\times10^{-11}$. It seems quite possible to measure it if crystals with 10^{17}-10^{18} molecules can be grown. As candidates the following chiral compounds may be mentioned: tris-ethylendiamine Ir(III) tribromide or cis-dichlorobisethylenediamine Pt(II) (Burger, 1973).

3. RECENT EXPERIMENTS

The crystallization of sodium chlorate performed by Kondepudi et al., (1990) shows very interesting features that may be relevant. The crystallites grown from solution were large and transparent enough to measure their optical activity. If the solution was stirred the crystals in a single vessel had almost the same chirality: from 32 beakers 18 contained ~ 99% D- and 14 ~ 99% L-crystals. Without stirring this spontaneous resolution did not occur. The authors attribute the resolution at this macroscopic level to autocatalysis and competition between L- and D-crystals. This experiments demonstrate that the processes postulated to explain spontaneous resolution (Kondepudi and Nelson, 1985, Morozov et al., 1984) exist.

The question arises, however, why the authors found ~ 18% more D-crystals than L-crystals in the experiment. Could it be due to the effect of stirring? If it always was in the same direction then the combined effect of rotation and gravitation may have created this asymmetry. A new experiment with stirring in the opposite directions is needed. (We note that this idea was previously tried in amino acid polymerization and sodium-amonium-tartrate crystallization with negative results by Kovács et al., 1981.)

Cerenkov radiation generated by β-particles from ^{32}P in R- and S-phenolbutyric acid was measured by Garay and Ahlgren-Beckendorf (1990). In this carefully performed study the peak amplitude of the photomultiplier impulses due to Cerenkov light was smaller in R- than in S-, and both were smaller than in racemic phenylbutyric acid. The result is interesting, it involves, however, very little relation to our problem. First there are no data whether the sign of the shift changes if positrons move in the liquids, second the Cerenkov radiation has nothing in common with the destruction of molecules. It is an optical phenomenon characterized by the refractive index of the liquid. The theory of Cerenkov-radiation should be extended for chiral β-particles moving in optically active materials to understand the data.

4. NEW EXPERIMENTS

Calculations show that L-amino acids have lower energy than D-amino acids (Tranter, 1985), therefore - as A. Salam argues (Salam, 1991) - in quantum mechanical cooperative and condensation phenomena racemic mixtures should condensate in the L-state below a critical temeperature T_c. The value of T_c cannot be predicted theoretically but probably is very low. The experiments performed until now (papers by R.K. Khanna and by A. Figureau in this volume) could not yet find a transition temperature T_C testing at 4K or even at 0.6 K. The experiments should be continued probably with compounds having high Z elements at the asymmetry center as suggested above.

The asymmetries produced by polarized β-particles or in general by the weak interactions are very small. Therefore it is worthwhile to look for other asymmetries. We already pointed out that combination of Earth rotation and gravitation and also electric and magnetic fields in lightnings may produce asymmetries (Kovács et al., 1984). For the rotation experiments (Kondepudi et al., 1990), repeat and extension are needed as we noted above. In such cases the shearing forces between the non uniformly rotating layes of the liquid may play a role much larger than the gravitational effects. No experiment is known to us which imitates the circumstances of lightning. We suggest a Miller-type experiment in high external magnetic fields parallel and antiparallel to the electric field causing the electric discharge.

We do not know the size and sense of circular polarization in the short wave length U.V. light emitted by the Sun. This light does not reach the surface of the Earth through the present atmosphere. It could have acted and produced asymmetry in the atmosphere of the primitive Earth by the known asymmetric photochemical reactions. Measurements of the circular polarization of short wave U.V. light outside the atmosphere are needed.

We do not know whether the asymmetry appeared during chemical or biological evolution. A theoretical work concludes that life could have started only if the biomolecules were already chirally pure (Avetisov and Goldanskii, 1991). Theory should always be confirmed by experiments. We now suggest two studies which focus on the biological origin of the asymmetry.

The early microfossils are skeletons of the first cells filled with quartz crystallites. It is possible that the organic material served as seeds for nucleation and a preference for homochirality may be detected in the quartz microcrystallites of the fossils. This problem deserves a study.

The living systems contain D-amino acids in very small quantity. The quantity of D-amino acids in the different species should be correlated with the evolution tree. Extrapolation of these data to the origin of the tree may carry information.

5. CONCLUSION

This paper is written in the beleive that experimental work to search the origin of asymmetry of biomolecules is still important. It should be done, however, with great care: firm theoretical and experimental limits already exist. New experiments must be planned so that these limits can be overcome. If a new idea is tried and effects are found theoretical support is very important.

6. ACKNOWLEDGEMENT

The author is indebted to Prof. Stoeckenius for advices to improve the manuscript.

7. REFERENCES

Avetisov, V.A. and Goldanskii, V.I. 1991: Homochirality and stereospecific activity: evolutionary aspects. Biosystems, 25, 141-149.

Bonner, W.A. 1991: The origin and amplification of biomolecular chirality. Origins of Life Evol. Biosphere, 21, 59-111.

Burger, K. 1973: Coordination Chemistry: Experimental Methods. Butterworth, London.

Chiba, C.F., Thomas, P.J., Brookshow, L. and Sagan, C. 1990: Cometary delivery of organic molecules to the early Earth. Science, 249, 366-373.

Garay, A.S. and Ahlgren-Beckendorf, J.A., 1990: Differential interaction of chiral β-particles with enantiomers. Nature, 346, 451-453.

Keszthelyi, L. 1977: Contribution of parity violating effects to intramolecular interactions. Physics Letter, 64A, 287-288.

Kondepudi, D.P., Kaufmann, R.J. and Singh, N. 1990: Chiral symmetry breaking in sodium chlorate crystallization. Science, 250, 975-976

Kondepudi, D.K. and Nelson G.W., 1985: Weak neutral currents and the origin of biomolecular chirality. Nature, 314, 438-441.

Kovács, K.L. 1979: On the physical origin of biological handedness. Origins of Life, 9, 219-233.

Kovács, K.L., Keszthelyi, L. and Goldanskii, V.J. 1981: Unconsidered sources of chirality in Nature. Origins of Life, 11, 93-103.

Morozov, L.L., Kuz'min, V.V. and Goldanskii, V.J. 1984: Can neutral current be detected in chemical systems. J.E.T.P. Lett., 39, 414-416.

Sagan, C. and Khare, B. 1971: Long wavelength ultraviolet photoproduction of amino acids on the primitive Earth. Science, 173, 417-421.

Salam, A. 1991: The role of chirality in the origin of life. J. Mol. Evol., 33, 105-113

Tranter, G.E. 1985a: The parity violating energy differences between the enantiomers of α-amino acids. Chem. Phys. Letters, 120, 93-96.

Tranter, G.E. 1985b: Parity violating energy differences of chiral minerals and the origin of biomolecular homochirality. Nature, 318, 172-173.

Yamagata, Y. 1966: A hypothesis for the asymmetric appearance of biomolecules on Earth. J. Theor. Biol. 11, 495-498.

TRUE AND FALSE CHIRALITY

L. D. Barron
Chemistry Department, The University, Glasgow G12 8QQ, U.K.

ABSTRACT

The concept of chirality is extended to cover systems that exhibit spatial enantiomorphism on account of motion. This is achieved by applying time reversal in addition to space inversion and leads to a more precise definition of a chiral system. Although spatial enantiomorphism is sufficient to guarantee chirality in a stationary system such as a finite helix, enantiomorphous systems are not necessarily chiral when motion is involved, leading to the concept of true and false chirality associated with time-invariant and time-noninvariant enantiomorphism, respectively. Although only a truly chiral influence can induce an enantiomeric excess in a reaction that has reached true thermodynamic equilibrium, false chirality can suffice in a reaction under kinetic control due to a breakdown in microscopic reversibility analogous to that which arises in particle-antiparticle processes involving the neutral K-meson as a result of *CP* violation with the apparantly contradictory kinetic and thermodynamic aspects being reconciled by an appeal to unitarity. Consideration of falsely chiral influences greatly enlarges the range of possible chiral advantage factors in prebiotic chemical processes if far from equilibrium.

1. INTRODUCTION

A central theme in discussions of chemical evolution and the origin of life is the problem of the origin of biomolecular homochirality. Why is this homochirality based on L-amino acids and D-sugars rather than on the mirror-image versions? Any suspicion that, because of fundamental forces in the fabric of the universe (such as parity violation in the weak interactions, or something deeper), mirror-image life would not work at all appears to have been laid to rest by the recent synthesis of an enzyme (HIV-1 protease) exclusively from D-amino acids which was shown to have the same catalytic activity (within experimental error limits) as the original with respect to the corresponding mirror-image substrates (Milton et al., 1992). It has been argued that life, based on self-replication of organic homochiral polymers, could have originated only if the prebiotic organic medium had a high degree of enantiomeric purity, possibly achieved *via* a bifurcation-type transition to a chirally-pure state (Avetisov et al., 1991). Attention therefore becomes focused on *chiral advantage factors* which might influence the build up of an

ISBN 0-937194-31-X

enantiomeric excess in the products of pre-biotic chemical reactions, and on whether or not any such factors could be sufficiently large to compete with the amplification of chance fluctuations at the bifurcation point.

The word *chiral* was first coined by Lord Kelvin (1904), Professor of Natural Philosophy in the University of Glasgow, in his Baltimore Lectures delivered in 1884, to describe a figure "if its image in a plane mirror, ideally realized, cannot be brought into coincidence with itself". This definition is essentially the same as that introduced earlier by Pasteur to describe objects he called *dissymmetric*. There is no disagreement when the term chiral is applied to a stationary handed object such as a finite cylindrical helix or an asymmetric molecule (Fig. 1); but when applied to less tangible systems the concept becomes less clear.

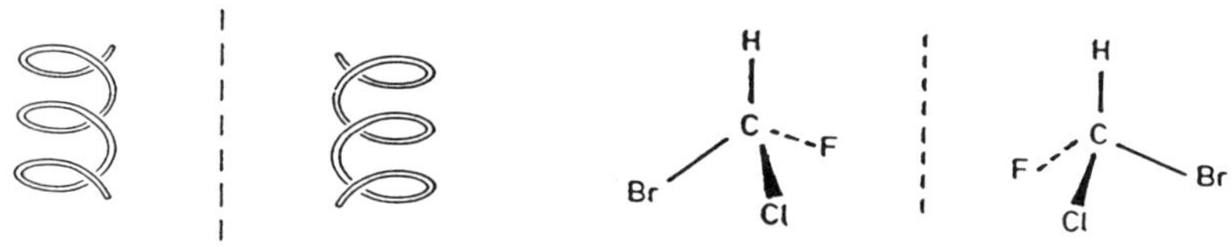

FIGURE 1. Dissymmetric (chiral) objects and their distinguishable mirror-image forms (enantiomers).

Pasteur (1848) was the first to extend the concept of dissymmetry (chirality) to other aspects of the physical world in his search for universal forces which might be connected with optical activity in nature. For example, he thought that a magnetic field, since it can induce optical rotation (the Faraday effect), generates the same type of dissymmetry as that possessed by an optically active molecule. As we shall see, this idea is quite wrong and has been the source of much confusion. However, Pasteur was correct in thinking that the combination of linear motion with a rotation does generate the same type of dissymmetry as an optically active molecule.

The application of fundamental symmetry arguments provides a deeper description of chirality than that usually encountered in the literature of stereochemistry and facilitates a proper understanding of the structure and properties of chiral molecules and the factors involved in their synthesis and transformations. It is found that, *although dissymmetry is sufficient to guarantee chirality in a stationary object such as a finite helix, dissymmetric systems are not necessarily chiral when motion is involved.* I have introduced the concept of 'true' and 'false' chirality to draw attention to this distinction (Barron, 1986a; 1986b) and have suggested that the word 'chiral' be reserved for systems that I call truly chiral. It will be appreciated from what follows that true and false chirality correspond to time-invariant and time-noninvariant enantiomorphism, respectively. This new classification has important implications for the discussion of chiral advantage factors.

2. TRUE CHIRALITY

2.1 NATURAL AND MAGNETIC OPTICAL ACTIVITY

A careful analysis of the natural and magnetic optical rotation experiments shows that the symmetry classifications of the two are quite different: the natural optical rotation observable associated with an isotropic collection of chiral molecules is a time-even pseudoscalar, whereas the magnetic optical rotation observable associated with a collection of achiral molecules is a time-odd axial vector (Barron, 1982). In fact Lord Kelvin (1904) was aware of this distinction, for his Baltimore Lectures contain the following pronouncement:

> The magnetic rotation has neither left-handed nor right-handed quality (that is to say, no chirality). This was perfectly understood by Faraday and made clear in his writings, yet even to the present day we frequently find the chiral rotation and the magnetic rotation of the plane of polarized light classed together in a manner against which Faraday's original description of his discovery of the magnetic polarization contains ample warning.

This analysis tells us that the nature of the quantum states of molecules that can support natural optical rotation is quite different from that of the quantum states that can support magnetic optical rotation. Since natural optical rotation is an odd-parity, time-even observable, the quantum states must have mixed spatial parity but definite 'time parity'; conversely, since magnetic optical rotation is an even-parity, time-odd observable, the quantum states must have definite spatial parity but mixed 'time parity' (Barron, 1982).

2.2 A NEW DEFINITION OF CHIRALITY

It should now be clear that the hallmark of a chiral system is that it can generate time-even pseudoscalar observables. This leads to the following definition (Barron, 1981; 1982) that enables true chirality to be distinguished from other types of dissymmetry:

> *True chirality is exhibited by systems that exist in two distinct enantiomeric states that are interconverted by space inversion, but not by time reversal combined with any proper spatial rotation.*

This means that the enantiomorphism shown by truly chiral systems is time-invariant. Enantiomorphism that is time-noninvariant has different characteristics that I call false chirality in order to emphasise the distinction.

Notice that a static uniform magnetic field on its own is not even falsely chiral.

It is easy to see that a stationary object such as a finite helix that is chiral according to the traditional stereochemical definition is accommodated by the first part of this definition: space inversion, associated with the parity operator *P*, corresponds to inverting the positions of all the particles in a system through an arbitrary space-fixed origin, and is a more fundamental operation than the mirror reflection traditionally invoked but provides an equivalent result. Classical time reversal, associated with the operator *T*, corresponds to reversing the motions of all the particles in the system: although time reversal is irrelevant for a stationary object, the full definition is required to identify more subtle sources of chirality in which motion is an essential ingredient. A few examples will make this clear.

2.3 TRANSLATING SPINNING CONES, SPHERES, PHOTONS AND ELECTRONS

Consider a cone spinning about its symmetry axis. Since space inversion *P* generates a version that is not superposable on the original (Fig. 2a), it might be thought that this is a chiral system. However, the chirality is false because time reversal *T* followed by a rotation R_π through 180° about an axis perpendicular to the symmetry axis generates the same system as space inversion (Fig. 2a). But if the spinning cone is also translating along the axis of spin, time reversal followed by the 180° rotation now generates a different system to that generated by space inversion (Fig. 2b). Hence a *translating* spinning cone possesses true chirality. A sphere translating along

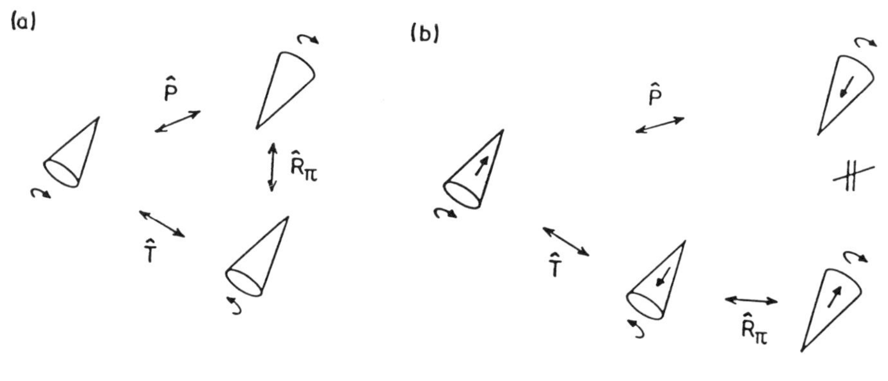

FIGURE 2. The effect of *P*, *T* and R_π on (a) a stationary spinning cone, and (b) a translating spinning cone.

the axis of spin also shows true chirality, which can be appreciated by looking at just the pattern of arrows in Fig. 2b and ignoring the cone.

The photons in a circularly polarized light beam propagating as a plane wave are in spin angular momentum eigenstates characterized by a spin quantum number s = 1 with quantum numbers m_s = +1 and −1 corresponding to projections of the spin angular momentum vector parallel or antiparallel, respectively, to the propagation direction. The absence of states with m_s = 0 is connected with the fact that photons, being massless, have no rest frame and so always move at the velocity of light. Considerations analogous to those given above for a translating spinning sphere then show that a circularly polarized photon exhibits true chirality.

The case of a spinning electron (s = 1/2, m_s = ± 1/2) is rather different to that of a circularly polarized photon because an electron has rest mass. It is clear that, whereas a stationary spinning electron is not a chiral object, an electron translating with its spin projection parallel or antiparallel to the propagation direction exhibits true chirality. These considerations expose a link between chirality and relativity, because the chirality that an observer perceives in a spinning electron depends on the relative motions of the observer and the electron. Indeed, this relativistic aspect of chirality is a central feature of elementary particle theory, especially in relation to the weak interactions. The velocity-dependence of the parity-violating aspects of the charge-changing weak interactions is particularly interesting: for relativistic electrons, only the left-handed states participate; but for nonrelativistic electrons the weak interaction still violates parity but with amplitude reduced to order v/c (Gottfried and Weisskopf, 1984).

2.4 ELECTRIC, MAGNETIC AND GRAVITATIONAL FIELDS

From the foregoing, neither a static uniform electric or magnetic field on its own constitutes a chiral influence, true or false (a static electric field **E** transforms as a time-even polar vector so that its direction is reversed by *P* but not by *T*; a static magnetic field **B** is a time-odd axial vector so that its direction is reversed by *T* but not by *P*). Likewise time-dependent uniform electric and magnetic fields. Furthermore, contrary to a suggestion first made by Curie (1894), no combination of a static uniform electric and a static uniform magnetic field can constitute a chiral system. Collinear electric and magnetic fields do indeed generate enantiomorphism, but it is time-noninvariant and so corresponds to false chirality. Thus parallel and antiparallel arrangements are interconverted by space inversion and are not superposable:

E →		**E** ←
	$\overset{P}{\longleftrightarrow}$	
B →		**B** →

But they are also intercoverted by time reversal combined with a rotation through 180°:

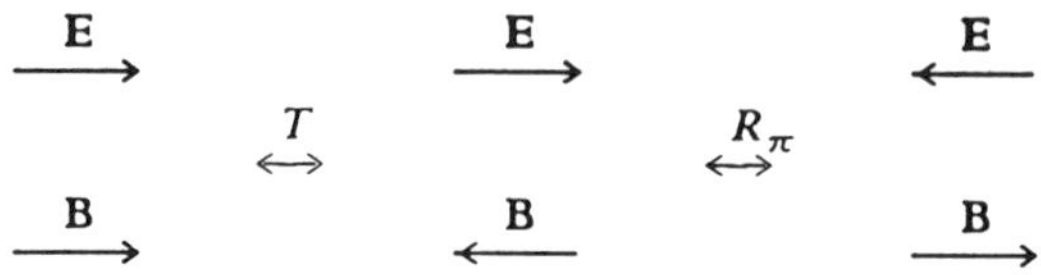

In fact the basic requirement for two collinear vectorial influences to generate chirality is that one transforms as a polar vector and the other as an axial vector, with both either time-even or time-odd. The second case is exemplified by *magneto-chiral* phenomena such as a birefringence and a dichroism induced in a chiral sample by a uniform magnetic field collinear with the propagation vector **k** of a light beam of arbitrary polarization (Wagnière and Meier, 1982; Barron and Vrbancich, 1984). Thus parallel and antiparallel arrangements of **B** and **k** are true chiral enantiomers because they cannot be interconverted by time reversal since **k**, like **B**, is time-odd.

Analogous to collinear electric and magnetic fields is the case of a rapidly rotating vessel with the axis of rotation perpendicular to the earth's surface (Dougherty, 1980). Here we have the time-odd axial angular momentum vector of the spinning vessel either parallel or antiparallel to the earth's gravitational field, itself a time-even polar vector. The physical influence in this case therefore exhibits false chirality.

2.5 ANYON SUPERCONDUCTIVITY

There are tantalizing similarities between the anyon theory of high-temperature superconductivity and the concept of false chirality. Anyons are spinning particles (or quasiparticle excitations) obeying quantum statistics that can vary continuously between those of fermions and bosons and which have been suggested to inhabit the two-dimensional world of copper oxide planes (Wilczek, 1990). Parity has a different effect in two dimensions than in three since it becomes simply a reflection in only one axis (Halperin et al., 1989). This has the consequence that, unlike a spinning system in three dimensions, the sense of a spinning system in two dimensions is invariant under PT together but not under P and T separately. Hence time-noninvariant enantiomorphism (i.e. false chirality), albeit in two dimensions, is a central feature of the anyon theory of high-temperature superconductivity.

3. ABSOLUTE ASYMMETRIC SYNTHESIS

The use of an external physical influence to produce an enantiomeric excess in what would otherwise be a racemic product of a prochiral chemical reaction is known as an absolute asymmetric synthesis (Morrison and Mosher, 1976; Mason, 1982). The subject still attracts much interest and controversy (Mason, 1983; Barron, 1986a), not least because it is an important ingredient in considerations of the prebiotic origins of biological molecules (Mason, 1988; Bonner, 1988; de Min et al., 1988).

3.1 TRULY CHIRAL INFLUENCES

If an influence can be classified as truly chiral we can be confident that it has the correct symmetry characteristics to induce absolute asymmetric synthesis, or some associated process such as preferential asymmetric decomposition, in any conceivable situation, although of course the influence might be too weak to produce an observable effect. Consider a unimolecular process in which an achiral molecule R generates a chiral molecule M or its enantiomer M^*:

$$M \underset{k_b}{\overset{k_f}{\leftrightarrows}} R \underset{k_f^*}{\overset{k_b^*}{\leftrightarrows}} M^* \qquad (1)$$

In the absence of a chiral influence, M and M^* have the same energy so no enantiomeric excess can exist if the reaction is allowed to reach thermodynamic equilibrium. However, in the presence of such an influence M and M^* will have different energies so an enantiomeric excess can now exist at equilibrium. There are also kinetic effects because the enantiomeric transition states will have different energies.

Circularly polarized photons or longitudinal spin-polarized electrons are the obvious choice, and several examples in asymmetric synthesis or preferential asymmetric decomposition are known (Mason, 1982; Bonner, 1988). Chiral effects induced by spin-polarized electrons should increase with increasing electron velocity because electron chirality is velocity-dependent.

Less obvious is the use of an *unpolarized* light beam collinear with a static magnetic field. This system exhibits true chirality so we can be confident it can induce asymmetric synthesis. A mechanism based on magneto-chiral dichroism would appear to be the most favourable (Wagnière and Meier, 1983).

3.2 FALSELY CHIRAL INFLUENCES

When considering the possibility or otherwise of absolute asymmetric synthesis being induced by a falsely chiral influence, a distinction must be made between reactions that have been allowed to reach thermodynamic equilibrium (*thermodynamic control*) and reactions that have not attained equilibrium (*kinetic control*).

The case of thermodynamic control is quite clear. Because M and M* are isoenergetic in the presence of, say, collinear electric and magnetic fields, or a spinning vessel with its axis perpendicular to the earth's surface (neglecting the very small difference due to parity violation), such falsely chiral influences cannot induce absolute asymmetric synthesis in a reaction mixture which is isotropic in the absence of the influence and which has been allowed to reach thermodynamic equilibrium (Mead et al., 1977). (The energy equivalence of M and M* follows from a consideration of the invariance of the Hamiltonian in the presence of the influence under the combined operations of space inversion and time reversal.)

The situation is less straightforward for reactions under kinetic control for, as discussed below, if microscopic reversibility and detailed balancing were to break down an enantiomeric excess could develop.

3.2.1 The Breakdown of Microscopic Reversibility: Enantiomeric Detailed Balancing

I have suggested that conventional detailed balancing, and the associated kinetic principles, might not be valid for reactions involving chiral molecules in a time-noninvariant enantiomorphous influence (Barron, 1987; 1991). This can be seen from a quantum-mechanical description of the microscopic reaction event (Sakurai, 1985). The amplitude for a transition from some initial linear momentum state **p** to some final state **p'** is written $\langle \mathbf{p'} | \mathcal{T} | \mathbf{p} \rangle$, where $\mathcal{T}$ is the operator responsible for the transition. If $\mathcal{T}$ involves purely electromagnetic interactions it will be invariant under both parity and time reversal, which enables us to write

$$\langle \mathbf{p'} | \mathcal{T} | \mathbf{p} \rangle \overset{\text{under } T}{=} \langle -\mathbf{p} | \mathcal{T} | -\mathbf{p'} \rangle \overset{\text{under } P}{=} \langle \mathbf{p}^* | \mathcal{T} | \mathbf{p}^{*\prime} \rangle \qquad (2)$$

where we have allowed the particles to be chiral, the star denoting the *P*-enantiomer. The first equality in Eq. (2), obtained from time reversal alone, is the basis of the conventional principle of microscopic reversibility and, when averaged over the complete system of reacting particles at equilibrium, of the principle of detailed balancing (Tolman, 1938). The second equality, obtained by applying space inversion to the time-reversed transition amplitude, describes the inverse process involving the *enantiomeric* particles.

Conventional detailed balancing is usually adequate for the kinetic analysis of reactions, even those involving chiral molecules, because conventional microscopic reversibility based on the assumption of *T* invariance is usually valid. This is conceptualized in terms of a potential energy profile that is the same in the forward and backward directions for a given reaction (Fig. 3).

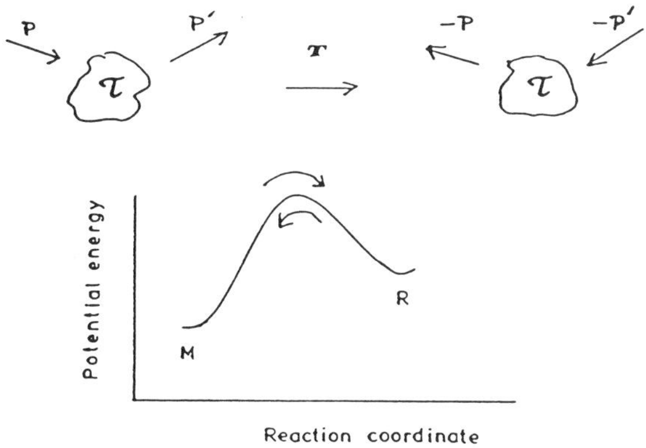

FIGURE 3. Microscopic reversibility conceptualized as the same potential energy barrier for the forward and backward processes.

However, in the presence of a time-noninvariant enantiomorphous influence such as collinear electric and magnetic fields, time reversal alone is not a symmetry operation since a different influence would be generated: space inversion must also be applied to recover the original relative orientations of **E** and **B**. The first equality in Eq. (2) is therefore no longer valid, and we must base any kinetic analysis on the relationship

$$\langle \mathbf{p'} | \mathcal{T} | \mathbf{p} \rangle \overset{\text{under } TP}{=} \langle \mathbf{p}^* | \mathcal{T} | \mathbf{p}^{*\prime} \rangle \tag{3}$$

which implies *enantiomeric* microscopic reversibility. Hence, as illustrated in Fig. 4, for reactions of chiral molecules in situations where only the

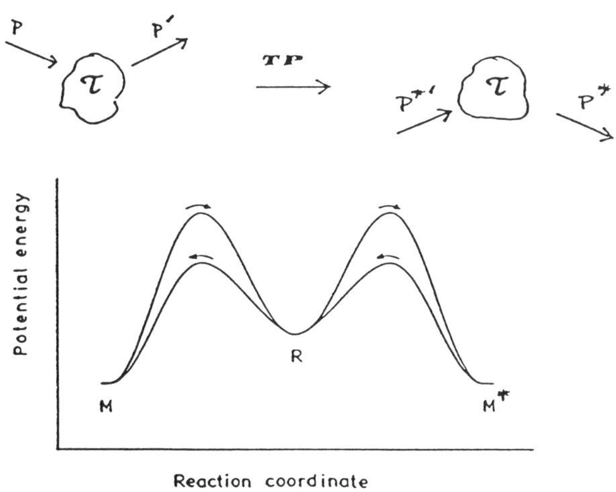

FIGURE 4. Potential energy surfaces for enantiomeric reactions in the presence of a time-noninvariant enantiomorphous influence.

combined *TP* invariance holds, the potential energy profiles for the forward and backward *enantiomeric* reactions are the same but the forward and backward profiles for the reaction of a given enantiomer are in general different.

3.2.2 Unitarity and Thermodynamic Equilibrium

It therefore appears that enantiomeric microscopic reversibility and the associated enantiomeric detailed balancing allows $k_f \neq k_f^*$ and $k_b \neq k_b^*$ if the reaction (1) takes place in the presence of a time-noninvariant enantiomorphous influence such as collinear **E**, **B**. However, this would generate an enantiomeric excess at thermodynamic equilibrium, which conflicts with the requirement that the concentrations of M and M* must be the same. This conflict between the kinetic and thermodynamic requirements can be resolved by including other pathways by which M and M* can be interconverted. Then at *true* thermodynamic equilibrium (i.e. when all of the possible infinite number of interconversion pathways have equilibrated), equal concentrations of M and M* obtain because the different enantiomeric excesses associated with each separate pathway sum to zero. The proof of this assertion (Barron, 1987; 1991) follows from the unitarity of the scattering matrix, which corresponds to the fact that the sum of the transition probabilities from a given initial state to all final states is unity. The argument is similar to that used to demonstrate the existence of equal numbers of particles and antiparticles at thermodynamic equilibrium, even when *CP* violation (see below) destroys the equality between the rates for specific particle → antiparticle and antiparticle → particle transitions, in the big bang model of the early universe (Kolb and Wolfram, 1980). All this is consistent with the demonstration that it is *unitarity*, rather than microscopic reversibility, that lies behind the validity of Boltzmann's H-theorem and hence the second law (Aharony, 1973).

4. CHIRALITY AND SYMMETRY VIOLATION

A symmetry violation (often called symmetry nonconservation) arises when what was thought to be a 'nonobservable' is actually observed (Lee, 1984). The nonobservables of relevance here are absolute chirality (i.e. absolute left- or right-handedness), absolute sense of motion and absolute sign of electric charge, being associated with the symmetry operations of parity *P*, time reversal *T* and charge conjugation (particle-antiparticle exchange) *C*. However, even if one or more of these three symmetries is violated, the combined symmetry of *CPT* is always conserved (Lee, 1984). A consideration of symmetry violation, and how it differs from spontaneous symmetry breaking, provides considerable insight into the phenomenon of molecular chirality.

4.1 PARITY VIOLATION AND TRUE ENANTIOMERS

Following the unification of the weak and electromagnetic interactions by Weinberg, Salam and Glashow in the late 1960s (Gottfried and Weisskopf, 1984), it was shown that the absolute parity violation associated with the weak interaction infiltrates to a tiny extent into all electromagnetic processes by means of a *weak neutral current* interaction mediated by the massive neutral Z^o particle. One manifestation is a small energy difference between enantiomeric chiral molecules which is susceptible to quantum-mechanical calculations giving values of the order 10^{-20} a.u. (Hegstrom, et al., 1980; Mason and Tranter, 1985; MacDermott and Tranter, 1989).

It is easy to prove that only the space-inverted enantiomers of truly chiral systems show a parity-violating energy difference (Barron, 1986b): this follows from the fact that the weak neutral current Hamiltonian transforms as a time-even pseudoscalar and so is odd under P but is invariant under T and R_π. Space-inverted enantiomers of falsely chiral systems such as a stationary rotating cone are strictly degenerate. Hence parity violation provides a cornerstone for the identification of true chirality.

Since the space-inverted enantiomers of a chiral molecule are not strictly degenerate, they are not true enantiomers (the concept of enantiomers implies the *exact* opposite). So where is the true enantiomer of a chiral molecule to be found? In the antiworld, of course! The molecule with the opposite absolute configuration but composed of antiparticles will have exactly the same energy as the original (Barron, 1981, 1982; Jungwirth et al., 1989). This follows from the *CPT* theorem and the assumption that T is not violated. So true enantiomers are interconverted by *CP*, which means that a chiral molecule is associated with two distinct pairs of true enantiomers (Fig. 5). Since P violation automatically implies C violation here,

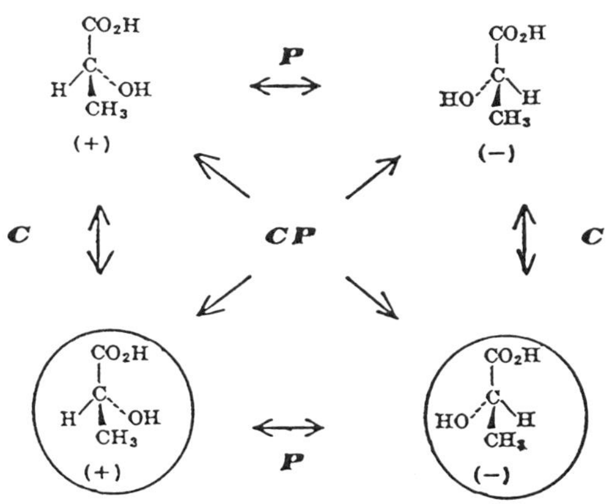

FIGURE 5. The two pairs of true enantiomers (i.e. strictly degenerate) of lactic acid that are interconverted by *CP*.

it also follows that there is a small energy difference between a chiral molecule in the real world and the corresponding chiral molecule with the same absolute configuration in the antiworld.

This more general definition of the enantiomers of a truly chiral system is consistent with the chirality that free atoms display on account of parity violation (Hegstrom et al., 1988). The weak neutral current generates only one type of chiral atom in the real world: the conventional enantiomer of a chiral atom obtained by space inversion alone does not exist. Clearly the enantiomer of a chiral atom is generated by the combined *CP* operation. Thus the corresponding atom composed of antiparticles will of necessity have the opposite 'absolute configuration' and will show an opposite sense of parity-violating optical rotation (Barron, 1982).

The space-inverted enantiomers of objects such as translating spinning cones or spheres that only exhibit chirality on account of their motion also show parity-violating differences. One manifestion is that, as mentioned above, left- and right-handed particles (or antiparticles) have different weak interactions. Again, *true* enantiomers are interconverted by *CP*: for example, a left-handed electron and a right-handed positron. Notice that right- and left-handed circularly polarized photons are automatically true enantiomers since a photon is its own antiparticle.

4.2 PARITY VIOLATION AND SPONTANEOUS PARITY BREAKING

The appearance of parity-violating phenomena is interpreted in quantum mechanics by saying that, contrary to what had previously been supposed, the Hamiltonian lacks inversion symmetry (the weak interaction potential being a pseudoscalar). This means that P and H no longer commute, so the law of conservation of parity no longer holds. Such *symmetry violation* (nonconservation) must be distinguished from *spontaneous symmetry breaking* which arises when a system displays a lower symmetry than its Hamiltonian. Natural optical activity is therefore a phenomenon arising from spontaneous parity breaking since a resolved chiral molecule displays a lower symmetry than its associated Hamiltonian: if the small parity-violating term in the Hamiltonian is neglected, the symmetry operation that the Hamiltonian possesses but the chiral molecule lacks is parity, and it is this parity operation that interconverts the two enantiomeric parity-broken (mixed parity) states.

The conventional view is that the stability of chiral enantiomers arises from the existence of a high potential energy barrier separating the enantiomeric structures (Barron, 1982; 1991). Quantum-mechanical tunnelling through this barrier leads to a finite lifetime for a resolved enantiomer, so the corresponding quantum states are time-dependent. The optical activity therefore remains observable only so long as the observation time is short compared with the interconversion time between enantiomers, which is inversely proportional to the area of the barrier. The law of parity is saved in systems displaying spontaneous parity breaking because their pseudoscalar properties average to zero over a sufficiently long observation time, or, equivalently, the space-inverted experiment is realizable. In either case absolute chirality is not observable.

This exposes a crucial distinction between spontaneous parity-breaking and parity-violating natural optical activity phenomena. The former are time-dependent and average to zero; the latter are constant in time. It is still an open question as to whether the primary source of the high stability of resolved chiral molecules such as amino acids is parity violation (since the handed quantum states become the true stationary states if the parity-violating energy difference between the enantiomers is much larger than the tunnelling splitting), spontaneous parity breaking with very long tunnelling times, or spontaneous parity breaking stabilized by the environment. There has also been much discussion on the relationship between the microscopic and macroscopic aspects of molecular chirality (Woolley, 1980; Amann, 1991), including the possibility of a phase transition to a chirally pure state induced by parity violation (Salam, 1991).

4.3 VIOLATION OF TIME REVERSAL

Violation of time reversal was first observed by Cronin and Fitch in 1964 in certain decay modes of the neutral K-meson (Gottfried and Weisskopf, 1984). Although unequivocal, the effects are very small and have not been observed in any other system. In fact *T* violation itself is not observed directly: rather, the observations show *CP* violation, from which *T* violation is implied from the *CPT* theorem.

For example, the decay mode of the long-lived neutral K-meson $K_L \rightarrow \pi^- e_r^+ \nu_l$ into negative antipions π^- plus right helical positrons e_r^+ plus left helical neutrinos ν_l occurs 1.00648 times faster than the decay mode $K_L \rightarrow \pi^+ e_l^- \tilde{\nu}_r$ into positive pions π^+ plus left helical electrons e_l^- plus right helical antineutrinos $\tilde{\nu}_r$. Since these two sets of decay products are interconverted by *CP*, this decay rate asymmetry indicates that *CP* is violated. If we naively represent this decay process in the form of 'chemical equilibria' as in Eq. (1),

$$\pi^+ + e_l^- + \tilde{\nu}_r \quad \underset{k_b}{\overset{k_f}{\leftrightarrows}} \quad K_L \quad \underset{k_f^*}{\overset{k_b^*}{\leftrightarrows}} \quad \pi^- + e_r^+ + \nu_l \tag{4}$$

a parallel is established with absolute asymmetric synthesis associated with a breakdown in microscopic reversibility discussed above since in both cases $k_f \neq k_f^*$. Thus the K_L and the two sets of decay products are the equivalents, with respect to *CP*, of R, M and M* with respect to *P*. We can therefore conceptualize the decay rate asymmetry here as arising from a breakdown in microscopic reversibility due to a time-noninvariant *CP* enantiomorphism in the forces of nature (Barron, 1987): the *CPT* theorem guarantees that the two distinct *CP*-enantiomorphous influences are interconverted by *T*. The analogy is completed by the fact that, as mentioned above, the asymmetries cancel out when summed over all possible channels at true thermodynamic equilibrium.

5. CONCLUDING REMARKS

The use of a falsely chiral influence to induce an enantiomeric excess in a prochiral chemical reaction under kinetic control *via* a breakdown in microscopic reversibility has yet to be demonstrated experimentally. However, the close parallel with the neutral K-meson system where a breakdown in microscopic reversibility has been observed gives confidence in the prediction. It is intriguing that a falsely chiral influence (including the analogous influence responsible for *CP* violation) acts as a *catalyst* in the chemical sense since it modifies potential energy barriers to change the relative rates of formation of enantiomeric products without affecting the relative energies of reactants and products so that the position of the equilibrium is not changed.

Consideration of a particular mechanism involving the conrotatory ring closure of a substituted butadiene in collinear electric and magnetic fields (Barron, 1987) shows that the function of the electric field in this case is to partially align the molecules in the fluid. This means that the electric field is not required if the molecules are already oriented, so a particularly favourable situation could be reaction and transport processes involving chiral molecules aligned in a crystal or on a surface in the presence of a magnetic field (Gilat, 1990, has suggested that a magnetic field will disturb the equilibrium of enantiomers aligned at the surface of a bulk racemic solution so that one particular enantiomer will tend to concentrate at the surface: however, this appears to conflict with the present analysis which requires aligned enantiomers to remain strictly degenerate at equilibrium in a magnetic field). In any event, since the processes in prebiotic chemical evolution were likely to have been far from equilibrium, consideration of falsely chiral influences greatly enlarges the range of possible chiral advantage factors.

There have been some interesting speculations about the possible role of T violation in processes involving chiral molecules which might be significant for biology (Garay, 1979). However, CP and hence T violation has only ever been observed in particle $\leftrightarrow$ antiparticle processes in the neutral K-meson system which is peculiar in that an observable species, the K_L, bridges the world of matter and antimatter. We can deduce from this that CP violation and the associated breakdown in microscopic reversibility will occur in principle in currently unrealizable processes involving transformations between a chiral molecule made of matter and its mirror image made of antimatter (i.e. between CP enantiomers). But there is no apparent connection with rates of processes involving mirror-image enantiomers in the real world, although from our incomplete understanding of CP violation at this time we cannot completely rule out such a possibility.

Finally, I have introduced the concept of 'true' and 'false' chirality to draw attention to the distinction between time-invariant and time-noninvariant enantiomorphism, but do not intend this to become standard nomenclature. Rather, I suggest that that the word *chiral* be reserved in future for systems classified here as truly chiral.

REFERENCES

Aharony, A., 1973: Microscopic Irreversibility, Unitarity and the H-theorem. In B. Gal-Or (Ed.), *Modern Developments in Thermodynamics*, Wiley, New York, 95-114.

Amann, A., 1991: Theories of Molecular Chirality. In W. Gans, et al. (Eds.), *Large-Scale Molecular Systems*, Plenum Press, New York, 23-32.

Avetisov, V. A., Goldanskii, V. I., and Kuz'min, V. V., 1991: Handedness, Origin of Life and Evolution, *Physics Today*, July 1991, 33-41.

Barron, L. D., 1981: Optical Activity and Time Reversal, *Mol. Phys.*, **43**, 1395-1406.

Barron, L. D., 1982: *Molecular Light Scattering and Optical Activity*, Cambridge University Press, Cambridge, 408 pp.

Barron, L. D., and Vrbancich, J., 1984: Magneto-Chiral Birefringence and Dichroism, *Mol. Phys.*, **51**, 715-730.

Barron, L. D., 1986a: Symmetry and Molecular Chirality, *Chem. Soc. Rev.*, **17**, 189-223.

Barron, L. D., 1986b: True and False Chirality and Parity Violation, *Chem. Phys. Lett.*, **123**, 423-427.

Barron, L. D., 1987: Reactions of Chiral Molecules in the Presence of a Time-Noninvariant Enantiomorphous Influence: A New Kinetic Principle Based on the Breakdown of Microscopic Reversibility, *Chem. Phys. Lett.*, **135**, 1-8.

Barron, L. D., 1991: Fundamental Symmetry Aspects of Molecular Chirality. In P. G. Mezey (Ed.), *New Developments in Molecular Chirality*, Kluwer Academic Publishers, Dordrecht, 1-55.

Bonner, W. A., 1988: Origins of Chiral Homogeneity in Nature, *Topics in Stereochemistry*, **18**, 1-96.

Curie, P., 1894: *J. Phys. (Paris)* (3), **3**, 393.

Dougherty, R. C., 1980: Chemical Geometrodynamics: Gravitational Fields Can Influence the Course of Prochiral Chemical Reactions, *J. Am. Chem. Soc.*, **102**, 380-381.

Garay, A. S., 1979: Broken Symmetries in Physics and their Relevance in Chemistry and Biology. In D. C. Walker (Ed.), *Origins of Optical Activity in Nature*, Elsevier, Amsterdam, 245-257.

Gilat, G., 1990: A Critical Experiment to Test the Chiral Interaction Hypothesis, *Chem. Phys.*, **140**, 195-198.

Gottfried, K., and Weisskopf, V. F., 1984: *Concepts of Particle Physics*, Vol. 1, Clarendon Press, Oxford, 189 pp.

Hegstrom, R. A., Rein, D. W., and Sandars, P. G. H., 1980: Calculation of the Parity Nonconserving Energy Difference Between Mirror-Image Molecules, *J. Chem. Phys.*, **73**, 2329-2341.

Hegstrom, R. A., Chamberlain, J. P., Seto, K., and Watson, R. G., 1988: Mapping the Weak Chirality of Atoms, *Am. J. Phys.*, **56**, 1086-1092.

Jungwirth, P., Skála, L., and Zahradnick, R., 1989: The Parity Non-Conserving Energy Difference Between Enantiomers and a Consequence of the CPT Theorem for Molecule-Antimolecule Pairs, *Chem. Phys. Lett.*, **161**, 502-506.

Kolb, E. W., and Wolfram, S., 1980: Baryon Number Generation in the Early Universe, *Nucl. Phys.*, **B172**, 224-284.

Lee, T. D., 1984: *Particle Physics and Introduction to Field Theory*, Harwood, Chur, 865 pp.

Lord Kelvin, 1904: *Baltimore Lectures*, C. J. Clay and Sons, London.

MacDermott, A. J., and Tranter, G. E., 1989: Electroweak Bioenantioselection, *Croat. Chem. Acta*, **62** (2A), 165-187.

Halperin, B. I., March-Russell, J., and Wilczek, F., 1989: Consequences of Time-Reversal Symmetry Violation in Models of High-T_c Superconductors, *Phys. Rev.* **B40**, 8726-8744.

Mason, S. F., 1982: *Molecular Optical Activity and the Chiral Discriminations*, Cambridge University Press, Cambridge, 273 pp.

Mason, S. F., 1983: Molecular Handedness and the Origins of Chiral Discrimination, *Int. Rev. Phys. Chem.*, **3**, 217-241.

Mason, S. F., and Tranter, G. E., 1985: The Electroweak Origin of Biomolecular Handedness, *Proc. Roy. Soc.*, **A397**, 45-65.

Mason, S. F., 1988: Biomolecular Homochirality, *Chem. Soc. Rev.*, **17**, 347-359.

Mead, C. A., Moscowitz, A., Wynberg, H., and Meuwese, F., 1977: The Influence of an Electric and Magnetic Field in Chemical Reactions, *Tetrahedron Lett.*, pp. 1063-1064.

Milton, R. C. deL., Milton, S. C. F., and Kent, S. B. H., 1992: Total Chemical Synthesis of a D-Enzyme: The Enantiomers of HIV-1 Protease Show Demonstration of Reciprocal Chiral Substrate Specificity, *Science*, **256**, 1445-1448.

de Min, M., Levy, G., and Micheau, J. C., 1988: Chiral Resolutions, Asymmetric Synthesis and Amplification of Enantiomeric Excess, *J. Chim. Physique*, **85**, 603-619.

Morrison, J. D., and Mosher, H. S., 1976: *Asymmetric Organic Reactions*, American Chemical Society, Washington D.C., 465 pp.

Sakurai, J. J., 1985: *Modern Quantum Mechanics*, Benjamin/Cummings, Menlo Park, California, 474 pp.

Salam, A., 1991: The Role of Chirality in the Origin of Life, *J. Mol. Evol.*, **33**, 105-113.

Tolman, R. C., 1938: *The Principles of Statistical Mechanics*, Oxford University Press, Oxford, 660 pp.

Wagnière, G., and Meier, A., 1982: The Influence of a Static Magnetic Field on the Absorption Coefficient of a Chiral Molecule, *Chem. Phys. Letts.*, **93**, 78-81.

Wagnière, G., and Meier, A., 1983: Difference in the Absorption Coefficient of Enantiomers for Arbitrarily Polarized Light in a Magnetic Field: A Possible source of Chirality in Molecular Evolution, *Experientia*, **39**, 1090-1091.

Wilczek, F., 1990: *Fractional Statistics and Anyon Superconductivity*, World Scientific, Singapore, 447 pp.

Woolley, R. G., 1980: Quantum Mechanical Aspects of the Molecular Structure Hypothesis, *Isr. J. Chem.*, **19**, 30-46.

CHIRAL INTERACTION AND BIOMOLECULAR EVOLUTION

G. Gilat
Department of Physics, Technion - Israel Institute of Technology,
Haifa 32000, Israel

ABSTRACT

Recent developments in the concept of chiral interaction open now new options and dynamical possibilities for biomolecules which have so far been overlooked. A few of these possibilities are mentioned, such as the control mechanism of enzymatic activity and the role played by non-ergodicity in evolutionary processes. It is shown that chiral interaction, being a surface phenomenon, does not obey Barron's symmetry constraints, which are suitable for force fields present in bulk interactions. In particular, the situation at the ocean-air surface in the prebiotic era is described, as well as the possible role played by chiral interaction in conjunction with the terrestrial magnetic field normal to the ocean surface, which could have lead to a process of deracemization at the ocean-air interface.

1. INTRODUCTION

In recent articles by Goldanskii, Kuzmin and Avetisov (1989,1991), the various approaches that could have determined the natural selection of the chiral monomers that are present in terrestial biology are thoroughly reviewed and analyzed. The advantage factor hypothesis is playing a decisive role in these considerations, since it can lead to a natural excess of one enantiomer, say L, with respect to another (D). Such an advantage factor η causes an energy gap ΔE between the energy levels $E_L \neq E_D$ which otherwise should be equal. The lower energy level usually wins in the evolutinary competition between the two enantiomers. The advantage factor may originate from existing terrestial or cosmic conditions that may favor one chirality upon another. These conditions, however, are subject to a rather severe set of symmetry constraints, formulated by Barron (1986,1987), that involve time and space transformations that limit the availability of fields combinations that can affect the advantage factors for generating enantiomeric excesses in the terrestial bio-sphere. Among the field combinations that satisfy these constraints exists the weak nuclear current (WNC), which is of cosmic origin and which has attracted much attention recently. (Kondepudi and Nelson 1983,1985, Mason and Trantor, 1984, Salam, 1991). This is so in spite of the smallness of its advantage factor being in favour of the L-enantiomer of amino acids.

ISBN 0-937194-31-X

Another aspect discussed widely in the articles by Goldanskii *et al.* (1989, 1991) concerns the possible process of deracemization of the natural mixture of both enantiomers that were likely to coexist in prebiotic era. In order to produce sufficiently long chains of the homochiral composites that exist in biomolecules, it is necessary to obtain a highly pure advantage factor, i.e. $\eta \rightarrow 1$, for the relative concentration of the L and D enantiomers. In order to accomplish such a complete deracemization it has been proposed to apply the process of spontaneous symmetry breaking of mirror symmetry. (Morozov, 1978,1979). The mechnism of such a process has been discussed, but there is no obvious phy-sical factor that exists in the biosphere that seems to be able to lead to such an extreme preference of one enantiomer with respect to another.

2. CHIRAL INTERACTIONS

In a recent series of articles a new mode of inelastic interactions between chiral systems and various media has been proposed (Gilat, 1985,1986,1991 ; Gilat and Schulman, 1985), for which chirality is a necessary condition. Such an interaction is to be labeled as "Chiral Interaction". As yet, chiral interaction has not been observed directly on molecular level, and hence it becomes necessary to look for analogous interactions in macroscopic devices. A few such examples have been presented and dicussed and they include the windmill, the Crookes' radiometer, as well as the electric cell. *In all these examples there is a certain asymmetry built into the device that enables it to choose one preferred direction of rotational motion, or flow of current, out of two possible ones.* These examples, helpful as they may be, are to be used mainly as analogies for possible similar interactions in molecular systems such as enzymes and soluble proteins, where the main analogy is meant for electric rather than for mechanical chiral interaction. A closer inspection of chiral interaction (Gilat, 1991) reveals that the interaction between the devices and the various media, (e.g. air flow, light radiation and electrolyte), occurs at the interface separating between the device and the relevant medium. It seems to be the case that surface interaction is a major feature of chiral interaction, where the discontinuity between the device and the medium at the interface establishes a contact mechanical force, or electrochemical potential, that enables the transfer of energy from the medium to the device which results in a mechanical or electric motion in a prefered direction. *The chiral asymmetry must, therefore, be present at the interface separating between the medium and the device, where chiral interaction is taking place.* Another significant feature of the chirality of these devices is that it is of physical rather than of pure geometric nature. For instance, in the Crookes' radiometer, chirality is in the colours distribution on the wings, that is, in the distribution of the optical absorption coefficient on the surface of contact, rather than in its shape, which may be totally achiral. Similar conclusion exists also for an electric cell, where the asymmetry involves the electrochemical potentials of the electrodes rather than their geometric shapes. Another interesting analysis (Gilat, 1991) reveals that apart from rather small entropic effects, such as friction or electric resistance, the operation of these devices is time-irreversible but conserves PT-invariance, where P and T stand for parity and time reversal transformations, respectively. This conclusion indicates that chiral interaction on molecular level may well be a micro-irreversible process that still obeys PT-invariance.

This brings us towards the molecular level which concerns chiral biomolecules, soluble proteins in particular. A model of chiral interaction between random ionic motions in a polar solvent, such as H_2O, and a chiral molecule, such as soluble protein, has been presented (Gilat, 1985,1991). This interaction gives rise to a perturbation similar to electric current that flows throughout the molecule, along the helical structure, in a preferred direction. In order for such a perturbation to persist, it is required that the stochastic time τ between two consecutive ions approaching the molecule be sufficiently small in comparison with the dissipation time τ_R of the flowing perturbation, or current. Another important requirement concerns the existence of an interface (Gilat, 1986) that separates between the interior of the molecule and the solvent, or in other words, the presence of surface interaction. This requirement resembles largely the situation in macro-chiral devices, and surprisingly enough, it is well known (Tschesche, 1983) that all soluble proteins become globular before they can function. Globular molecular structure of a molecule provides for the necessary microscopic interface separating between the solvent and the inner molecular region. In the absence of such a molecular interface, the solvent-air surface can replace it. In other words, chiral interaction can take place for monomeric amino-acids at the solvent-air surface. It ought to be mentioned at this stage that the well known effect of optical activity (Pasteur, 1848) is also a chiral effect, but being elastic in energy, as well as a bulk effect, it should be considered as a chiral scattering rather than as proper chiral interaction (Gilat, 1991).

The possibile existence of chiral interaction on a molecular level opens new options for processes which vave been overlooked in the past. For instance, it is conjectured (Gilat, 1991) that chiral interaction may play a decisive role in the control mechanism of enzymatic activity. Such a mechanism is still not well understood and chiral interaction may provide for a clue through the directional preference of the flow of perturbation, which may carry useful information as well.The reasoning behind this argument concerns the violation of time reversal invariance by chiral interaction, being a necessary condition for any machine-like operation, as well as for an organizing system. Another significant example concerns the non-ergodic nature of chiral interaction, being the outcome of a persistent selection of a single direction of flow out of two possible ones. Non-ergodicity is likely to be a necessary condition for any evolutionary process. It is highly reasonable to assume that systems that can easily reach thermal equilibrium, that is, being ergodic, are hardly susceptible to any evolutionary process. Therefore, non-ergodicity must play a crucial role in biomolecular evolution, and this may be associated with chiral interaction.

On the other hand, chiral interaction on a molecular level involves intrinsic molecular perturbation which is not easy to inspect directly by experiment. Nevertheless, two modes of experiments have been proposed to look for molecular chiral interaction. One experiment is direct (Gilat, 1990,1991), and it involves a magnetic field interaction with hydrophobic amino-acids at the sovent-air surface. The second is indirect (Gilat, 1991), and it is based on an assumption that chiral interaction is a necessary condition for enzymatic activity. In addition to these there exist many structural details of proteins that fit smoothly with the chiral interaction hypothesis. It is also obvious that any well known phenomenon that concerns chiral biomolecules and that can be attributed to chiral interaction serves as a support to this hypothsis.

The last subject of this section concerns the validity of Barron's symmetry constraints (Barron, 1986,1987) for the case of chiral interaction. It is quite obvious that Barron's constraints are suitable for bulk interactions in the presence of field forces, being largely of electromagnetic nature, which are continuous in space. Recently it has been pointed out by Barron (1991) that for oriented molecules, e.g., attached to a surface, a magnetic field alone is sufficient to induce microscopic irreversibility. This modification fits excellently with the principles of of chiral interaction and hence also supports its viability.

3. CHIRAL INTERACTION AND BIOMOLECULAR EVOLUTION

It is the purpose of the present section to discuss possible conditions that might have existed at ocean surface during prebiotic era, including effects that might have resulted from chiral interaction, which could have influenced evolutionary processes of amino-acids towards developing into peptide chains of pure L or D chirality. These arguments include well known physical and chemical factors to which new factors, related to chiral interaction, are incorporated.This situation has already been treateded in part in two separate contexts, being naturally linked together. The first (Gilat and Schulman, 1986) describes the possible statisical effect of chiral interaction on the Boltzman distribution of L&D enantimers at the ocean surface, whereas the second (Gilat, 1990,1991) is concerned with a direct experimental verification of chiral interaction. It ought also to be stressed that the present treatment is not meant to provide a complete solution to the problem of prebiotic natural selection, but rather to throw some light on additional factors that so far have been overlooked, which might have played a role in prebiotic molecular evolution.

As mentioned above, the physical situation has already been described (Gilat, 1990) and it concerns hydrophobic amino-acids, such as Tryptophan, Valine, Leucine and other, that may have aggregated at the ocean-air surface, so that their hydrophobic group R is attached to the water-air surface layer. Such a molecule can be considered to be aligned with respect to the normal of this surface. The presence of an interface, separating between water and air, provides for a discontinuity in the ionic concentration which allows for chiral interaction to take place (Gilat, 1986,1991) for the case of amino-acid molecules at the ocean surface. This interaction gives rise to an induced magnetic moment μ due to the chiral current (perturbation) on the loop $NH_3^+.COO^-.C$, which may include an H-bond connecting NH_3^+ and COO^-, as well as an additional H_2O molecule that can relax the angular strains on the bonding (Fig. 1).

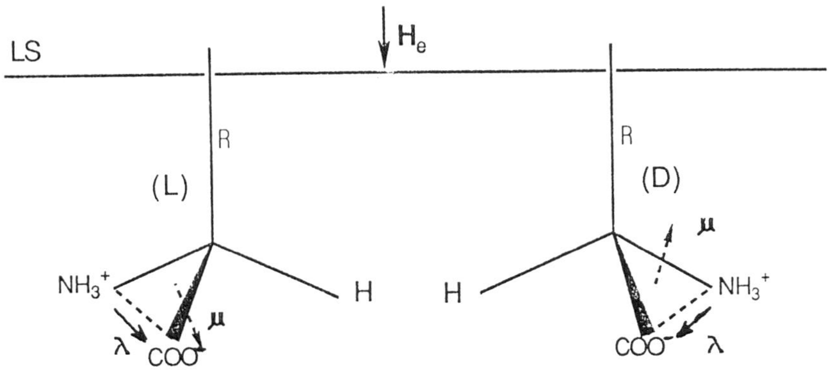

Figure 1 - Two hydrophobic amino-acid enantimers L and D are presented schematically as a tetrahedron $R.H.COO^-.NH_3^+$. The side chain R is attached to the liquid surface layer LS. H_e is the terrestial magnetic field, drawn along the normal to LS. μ is the chirality-induced magnetic moment, in opposite directions for the two enantiomers. λ is the electric dipole moment of each enantiomer.

For two aligned enantiomers L and D at the ocean surface, their respective magnetic moments μ are in opposite directions. If, in addition to this, there exists also a terrestial magnetic field H_e with a normal component to the ocean surface, then its interaction energy with μ is $E_\mu = \pm \mu.H_e$, depending on the chirality L or D. Let E_h be the hydrophobic interaction energy, then the total energy is

$$E = E_h \pm \mu.H_e \tag{1}$$

If the bulk solution is at equilibrium with the ocean-air surface, then the energy difference of $2\mu.H_e$ causes a small change of population Δn between the original racemic constituents L and D, where

$$\Delta n/n = 2\,\mu\,.H_e/kT \approx 10^{-8} - 10^{-10} \quad (2)$$

for $\mu \approx (10^{-1}\text{-}10^{-3})\,\mu_B$ and $H_e \approx 10^{-5}$ T. It is important to indicate that E is quite small in comparison to kT but still some 7 to 9 orders of magnitude larger than the WNC interaction energy (Kondepudi and Nelson, 1983,1985). Next, the possible combination of factors that could have influenced the amino-acid aggregates at the ocean-air interface in prebiotic era, is summarized.

3.1 The *concentration* of amino-acids at the ocean surface layer was *considerably larger* in comparison to that in the bulk of the ocean. This situation was a significant factor for enhancing chemical bonding at the surface in conmparison to the bulk.

3.2 The amino-acid molecules *were oriented* with respect to the ocean surface layer which could, in principle, sterically facilitate the formation of certain bondings among moleçules.

3.3 The magnetic interaction of the terrestial field $\mathbf{H_e}$ with the molecular chirality induced magnetic moment μ, was causing a slight *change in the relative population of the L and D enantiomers* at the ocean surface in the specific location where the aggregates were formed.

3.4 Due to the opposite alignments of the induced magnetic moments μ there existed a *magnetic attractive force* between opposite enantiomers L and D.

In view of these four conditions, there existed also a certain preference for the process of LD-*dimerization* and this is mainly because of the magnetic attraction between opposite enantiomers. The following estimate may be of interest. Let 10^{24} racemic monomers, i.e; of the order of 1 mole, be spread over an area of $1km^2$ of ocean surface, then the average distance between two monomers is 10A, which makes the magnetic attraction between them quite effective. On the other hand, the sign of the electrostatic force acting between two monomeric electric dipoles does not depend on their respective chirality, but rather on their relative steric position. This force is also largely reduced owing to the high dielectric constant of water. In view of this it looks plausible that LD-dimerization process did have a certain physical preference over other possible chemical reactions. It is also reasonable to assume that any LD-dimer that had been formed did not participate in any following evolutionary process. Now, if a sufficiently large fraction of the closely racemic population of L and D enentiomers could convert into LD-dimers, the relative difference Δn between the L and D populations at the ocean-air surface, being proportinal to the advantage factor, could have increased considerably. This, in fact, could well be a prescription for a process of deracemization that might have occurred at the ocean-air surface owing to the factors listed above, where chiral interaction had to play a decisive role in such a process. On the other hand, it is quite obvious that such a description is still short of a complete and convincing line of argument that can lead to a natural selection of a certain chirality of the building blocks of macro-molecules. Nevertheless, short peptide chains of pure chirality could have been formed at the ocean-air surface as a result of such a process, which might have lead to the next stage of biomolecular evolution.

4. CONCLUSIONS

The main motivation of the present article is to show that the phenomenon of chiral interaction may open new options and possibilities for molecular biology. The principal argumementation in favor of chiral interaction is based on general symmetry considerations, which are hard to refute, and this gives a firm foundation to this hypothesis, in spite of the fact that it still lacks experimental support on microscopic level. For this reason, use is being made of macroscopic chiral interactions, and this is mainly for the sake of analogy. Being, in essence, a surface phenomenon, chiral interaction conforms with Barron's symmetry constraints for prochiral reactions adjusted to molecules aligned on a surface (Barron, 1991). Few examples for such new possibilities are mentioned in the context of control mechanism of enzimatic activity and biomolecular evolution. Chiral interaction on microscopic level, being an intrinsic molecular perturbation, is not readily observable by experiment and, therefore, it must be supported by auxiliary pieces of evidence. Many structural details of proteins are in accordance with the chiral interaction hypothesis (Gilat, 1991), in particular, so is the fact that all soluble proteins become globular before they can function (Tschesche, 1983). This feature can well be associated with a molecular interface separating the protein from the solvent, which is highly required by the chiral interaction hypothesis. It is also necessary to observe that any piece of evidence that can support the assumption of chiral interaction, may simultaneously be also of significance as a biophysical fact.

In this article, the physical and chemical situations that could have existed at the ocean-air surface during prebiotic era, are described. It is shown that several simultaneous advantages had existed there, which could have enhanced the rate of deracemization at the ocean surface and so increase the advantage factor of one enantiomer of amino-acid with respect to another. In particular, the role of the magnetic field of earth together with the monomeric magnetic moment induced by chiral interaction, are described, as well as the process of LD-dimerization owing to the selective magnetic attractive force between opposite enantiomers. As a result of this process, short peptide chains of pure chirality could be generated during a sufficiently long evolutionary time. On the other hand, such a process is still inconclusive and certain problems remain unsolved, such as the formation of sufficiently long peptide chains of pure chirality; of which the well recognized enzymes are formed. In conclusion, this article is not meant to propose a complete solution to the puzzle of biomolecular evolution, but rather to offer additional and new possibilities that could have been of significance for evolutinary processes of biomolecules.

ACKNOWLEDGEMENTS

The author wishes to thank the French *Ministère des Affaires Etrangères* for its financial support as well as Dr Jean Maruani of the *Laboratoire de Chimie Physique* (UPMC and CNRS) in Paris for their kind hospitality. The author also wishes to acknowledge support from the Fund of Promotion of Research at Technion.

REFERENCES

Avetisov, V.A., V.I. Goldanskii and V.V. Kuzmin, 1991: Handedness, Origin of Life and Evolution, Physics Today, **44**, 33-41.

Barron, L.D., 1986: True and False Chirality and Parity Violation, Chem. Phys. Lett., **123**, 423-427.

Barron, L.D., 1986: Symmetry and Molecular Chirality, Chem. Soc. Rev., **15**, 189-223.

Barron, L.D., 1986: True and False Chirality and Absolute Symmetry Synthesis, J. Am. Chem. Soc., **108**, 5539-5542.

Barron, L.D., 1987: A New Kinetic Principle Based on Breakdown of Microscopic Reversibility, Chem. Phys. Lett., **135**, 1-8.

Barron, L.D., 1991: in P.G. Mezey (ed.) New Developments in Molecular Chirality, Kluwer Academic Publishers, p.22.

Gilat, G., 1985: Chiral Interaction in Biomolecules, Chem. Phys. Lett. **121**, 9-12.

Gilat, G., 1986: Interaction Between Chiral Molecules and Solvents, the Importance of Interface, Chem. Phys. Lett., **125**, 129-133.

Gilat, G., 1990: Critical Experiment to Test the Chiral Interaction Hypothesis, Chem. Phys., **140**, 195-198.

Gilat, G., 1991: Chiral Interaction in Molecular Structure, Mol. Engin. **1**, 161-178.

Gilat, G., L.S. Schulman, 1985: Chiral Interaction, Magnitude of the Effect and Natural Selection of L-Enantiomer, Chem. Phys. Lett., **121**, 13-16.

Goldanskii, V.I. and V.V. Kuzmin, 1989: Spontaneous Mirror Symmetry Breaking in Nature and the Origin of Life, Usp. Fiz. Nauk., **157**, 3-50; (English translation: Sov. Phys. Usp., **32**, 1-29, 1989).

Kondepudi, D.K. and G.W. Nelson, 1983: Chiral Symmetry Breaking in Nonequilibrium Systems, Phys. Rev. Lett., **50**, 1023-1026.

Kondepudi, D.K. and G.W. Nelson, 1985: Weak Neutral Currents and the Origin of Molecular Chirality, Nature, **314**, 438-441.

Mason, S.F. and G.E. Trantor, 1984: The Parity-Violating Energy Difference between Enantiomeric Molecules, Mol. Phys., **53**, 1091-1111.

Morozov, L.L., 1978: Spontaneous Mirror Symmetry Breaking from the Viewpoint of Biochemical, Dokl. Akad. Nauk. SSSR, **241**, 481-484; (English translation: Dokl. Biochem., **241**, 233-236, 1978).

Morozov, L.L., 1979: Spontaneous Mirror Symmetry Breaking of Biochemical-Evolution, Orig. Life, **9**, 187-195.

Pasteur, L., 1848: Ann. Chim., **24**, 457.

Salam, A., 1991: The Role of Chirality in the Origin of Life, J. Mol. Evol., **33**, 105-113.

Tschesche, H., 1983: in W. Hoppe, W. Lohmann, H. Markl and H. Ziegler (eds.), Biophysics, Springer-Verlag, Berlin, p. 37.

CHIRAL FORCES AND MOLECULAR DISSYMMETRY

Ram Mohan
M.P.I for Biophysical Chemistry
Göttingen, Germany

ABSTRACT

Pure homochiral D-sugar or L-amino acid derivatives should have been formed as a prelude to formation of homochiral polymers, which themselves "evolved" into information coded biomacromolecules, the originators for complexity and self-organised systems. Organic-metallic complexes as quintessence pre-enzymatic catalysts could have done the job of chiral discrimination in a near equilibrium "perfect reaction" process rather than by catastrophic, far from equilibrium, spontaneous symmetry breaking (Avitesov; Kondepudi and Nelson). Although, the molecular forces involved in chemical reactions are usually electromagnetic (EM) in origin, it is worthwhile to ascertain if weak neutral currents (WNC) significantly modulate these forces to play any role in chiral chemistry. The search should be for chemical compounds with large enough parity violating energy difference (PVED) rather than amino acids and sugars themselves (Mason et al.,). Higher excited states of highly charged organo-metallic type of compounds have sufficient PVED that could be experimentally detected and followed in the chiral catalytic reactions

Existence of such WNC chemistry, if proved, has important epistemological consequences: (i) life is not a rare event of chance, but could be manifested in suitable other parts of the universe and, (ii) life on this earth is a consequence of basic principles of physics and chemistry and we should know where the principle of natural selection took over, i.e., from near equilibrium to far from equilibrium.

1. INTRODUCTION

4 billion years ago, whether RNA was running the show (replication first, Freedman, 1992a; Schuster, 1993) or FeS-like oxidation-reduction autotrophic catalysts (metabolism first, Freedman, 1992b; Wächterhäuser, 1992)) were churning out a variety of organic compounds in underwater volcanic vents, chiral molecules in a pure enantiomorphic form should have been formed as a necessary requirement for the polymer takeover. According to Avetisov (1993), this is only possible by far from equilibrium , nonlinear, strong. spontaneous symmetry breaking, catastrophic bifurcation theory, where the chiral origin is a matter of chance between the two antipodes. Avetisov points out that external effects as radiolysis (Garay, 1993; Wang, 1993) experiments, even if successful could not lead to the stabilization of homochiral polymers that were formed subsequently. Similarly, one can expect that the Kondepudi-Nelson-Frank, (Kondepudi and Nelson, 1983) based studies

ISBN 0-937194-31-X

(Thiemann,1993) utilizing the intrinsic advantag factor, the parity violating energy difference (PVED), in a slow but, far from equilibrium, spontaneous symmetry breaking with positive feedback is not sufficient, even if it could be shown to produce chiral discrimination. Thus Tranter et al., calculations for thiosugar (MacDermatt, 1933) with larger PVED than for sugar at 10^{-14} kT, giving hopes of finding a molecule with at least a PVED of 10^{-12} kT that would be needed for laboratory scale experiments are of no avail. For a review on chiral evolution theories see Bonner (1988) and Bonner (1991) for a list of failed experiments. Kondepudi's recent experiment (Kondepudi et al., 1990) involving primary nucleation in $KClO_3$ in a stirred system, according to Orgel (1992) has nothing to do with chirality.

These strictures do not apply to near equilibrium asymmetric synthesis of chiral molecules or enantiomorphic conversion by organometallic catalysts. In fact, the job was so well done, it is difficult to trace back those organometallic complexes that were within the reach of nature during the prebiotic stage, and the only organometallic compounds remaining are those required functionally for energy conversion and oxygen transport.

2. FEATURES OF CHIRALITY: AN OVERVIEW

Chirality is an universal phenomenon and has two aspects: geometry and physics. For molecules, it is stereochemistry and pseudoscalar observables. The later is the scalar product of a polar vector and an axial vector as between electric and magnetic dipole moments respectively, as a result of spatial dispersion in atom or molecule. The optical strength in a coupled oscillator as in a chromophoric group is stronger than in one electron mechanism as in amino acids or sugars. This parity conserving optical activity is universally perturbed by the interaction of weak neutral currents (WNC) with spin-orbit coupling of electrons outside the nucleus, operating as polar and axial vectors respectively, creating a real energy difference between the two antipodes, however, small. Chirality is usually measured by circular dichroism, circularly polarized luminescence, optical activity, infra-red and NMR. PVED strength goes as the 6th power of neutron number (i.e., larger for the heavier atoms) but is extremely short-ranged.

In cosmology, vacuum is the interface between left and right chirality, related to the symmetry of CPT theorem (Boeyens, 1992). But with the big bang, the asymmetry of the universe is caused by CP violation. Electron is left chiral with antiparallel momentum and spin vectors, while the positron is right chiral. Photon selection rules are discussed by Steadman (1983) and can be extended to $W^{\pm}$ and Z bosons, provided the enormous mass differences are taken into account in the relativistic formulation. An observer moving towards a relativistic chiral object would see the sign of rotation reversed from when the observer is moving away from the object. Dispersion relations and sum rules, with correction to Rosenfeld-Condon dispersion relations, are discussed by Thomaz and Nussenzveig (1982). Lakhtakis (1991) deals with composite chiral media and the reciprocity relations involved in such media. Fluctuation between two enantiomers in equilibrium could enhance their optical activities. Electron, photon, K-meson etc are true chiral objects according to Barron. Barron (1993), in his trait but uncompromising search for common grounds between elementary particles and rigid isolated molecules, on the basis of basic symmetry principles, CPT theorem and a simple definition of time, distinguishes chirality between stationary and moving objects, between true and false chirality and between EM(Electromagnetic) and WNC chirality (parity conserved vs parity violated). Some of the consequences of the analyses are that the real counterpart of a L or D compound is in the other world, and that false chirality is better than true chirality in promoting a chiral reaction.

While Barron's symmetry principles apply to well defined left-right molecules that overlap each other, there exists a whole range of complex molecules from high symmetric (dissymmetric) to asymmetric chirality (Nakazaki, 1986) and factorization and sequence rules were devised by Cahn, Ingold and Prelog for them (Hirschmann and Hanson, 1983). In the same spirit, Mislow, Gilat ,Avnir and others (Buda et al., 1992) find a range of mirror asymmetric compounds, that depend on shape, that could be called "more" or "less" chiral. These molecules are poorly optically active complex molecules, rigid but connected by kinks and joints,as say, the natural phospholipids. At present, the presentation is purely geometrical like graph theory in theoretical chemistry. Chiral coefficients are quantitatively defined in terms of length, area, volume mass or electron density as a fraction between achiral and purely dissymmetric chiral. The asymmetry is larger than can be expected from WNC. The pseudoscalar observables are not yet seen in terms of the physical variables involved and their strength. Although, it may involve the one electron mechanism, as there are no chromophoric groups, it is not specifically designed for bond-stretch isomerism (Mayer, 1992) and is not related to the quantum mechanical problem of "do molecules have a shape?" (Boeyens, 1985).

With chiral coefficients fractional to diasteromers, the compounds enlisted for this formalism could neither provide a third mechanism in pseudoscalar observables for chirality, other than dipole-dipole and WNC, nor for chiral origin, as the chiral strength is diluted from a binary choice to a fractional choice. While Barron's chiral counterparts are strictly in the other world, Gilat calls even a piece of rock as chiral. The chiral coefficient is at present primarily used for pattern recognition of complex molecules and objects (Hel-Or et al., 1990).

Like Newton's apple and Kekule's serpent, Gilat (1993) draws inspiration from (spiralling) water sprinklers and wind-mills(with blades) and, like the slanted stripes on your tie, calls them as chiral objects The usual EM chirality is of no interest to him as an "equilibrium " process. He correctly points out that the Rosenfeld mechanism of EM chirality in molecules is not rigorous, although used by Mason in physical chemistry and Tinoco in biophysics. For chiral discrimination between L and D molecules, he simply orients them at a surface with oppositely directed magnetic dipole moments and lets them interact with earth's magnetic field. His chiral objects are not functional until driven by water or wind, which he calls "chiral interactions", one-sided, non-ergodic, dissipative and time symmetry breaking processes:. Enzymes are a circular string of permanent dipole moments driven by ions in solution, creating a circular current, as in benzene molecule, that is responsible for the enzyme reactions Although, his approach to enzyme reactions are unconventional and not realistic, they bear certain resemblance to time symmetry violation due to permanent dipole moment in the neutron and electron. Barron might call them as false chirality and, on the other hand,. gives a model chemical reaction tailored to resemble K-meson production where time symmetry violation occurs as a result of CP violation.

There are both quantum mechanical and quantum measurement problems involving chirality that is dealt with by algebraic quantum mechanics (Amman., 1991) and topological quantum mechanics respectively. (For a discussion of quantum mechanical problems involving chirality, knots and quasicrystals in the context of algebraic quantum mechanics, see Amman et al., 1988). Quantum measurement problems of chirality due to quantum interference and nonlocality are not yet investigated, although measurement of chirality itself has its own problems, let alone the argument whether molecules are really rigid as in the Born-Oppenheimer approximation. All this could be simplified by saying that the L and D chiral states would be localized classical observables without dispersion; the superposition principle or coherency condition being broken by superselection rules. The entanglement of the chiral objects with their environment can be severed and the chiral objects localized in the two potential wells via a Schrödinger-Langevin equation for spontaneous symmetry breaking for the case of EM chiral states which are kinetically metastable to tunneling from one chiral to another chiral state, and via Schrödinger-Kramer like equation for the parity violating case. Boeyens (1991) has given initial sketches of a gauge chemistry that he is developing for transition elements that may be useful for the organometallic compounds.

Objects with two-dimensional chirality can be identical when viewed in three dimensions. Rotation around the plane of the two-dimensional chiral object will not be superimposable, but rotation around the axis will be. Rigid symmetric two-dimensional planar ethylene, biphenyl or nucleic acid base molecules are not chiral, but on distortion ethylene or biphenyl can acquire chirality, while nucleic acid bases are rendered chiral by induced optical activity from the attached sugar. Biphenyl as a model compound has been studied on a two-dimensional lattice in a form of cellular automata by Harris and Stillinger (1990) and in more detail by Avetisov et al., (1991). Both EM and WNC chirality could be sensitive to distortion due to changes in the spatial dispersion of electronic charges.

Avnir et al., (1985) have shown by changes in fluorescence, how distortion of a molecule due to tight binding to a surface affects its dipolar properties. Andelman (1989) recently has made an interesting extension of Salem et al., (1987) work exploiting the superiority of two dimensions over three dimensions for chiral phase transition in Langmuir-Blodgett films with tripodly attached chiral molecules (natural phospholipids in this case). He studied the effects of conventional molecular forces (EM), but can be extended to take into account WNC forces and also chiral reactions. Surface—both between bulk interfaces and in constrained spaces as in pores—offer definite advantages for chiral discrimination as opposed to freely tumbling (solution) or crystallized samples. . Rigidity and closely held, compressed or twisted, crowded or confined, they offer a better chiral advantage for a molecule and for its chemical reactions. These are reasons why earlier solution-in-flask experiments could not produce chiral discrimination. However, organometallic catalysts can bind strongly to the precursors stereospecifically in solution, until the products are released.

Anyons (Wilczek, 1991) involving fractional statistics and Barry's geometric phases (topological quantum mechanics) are found in superfluid He^3. Time symmetry violation is directly exhibited. Garay's (1993) experiments, based on the idea that ß-particles showing CP violation should violate time symmetry invariance and hence lead to chiral enrichment by radiolysis, were inconclusive. Although, there are possibly cleaner scattering experiments with ß-particles, no attempt has been made to study time violation. Indeed, it was from such similar scattering experiments that helium was found to be an anyon. Barron suggests that a false chiral system would have better kinetic control than true chiral systems.

Recently, time symmetry violation in K-mesons has been shown experimentally. Intrinsic time symmetry violation due to permanent dipole moments in the neutron and electron are expected. The PVED is estimated to be smaller due to these permanent dipole moments and yet there is an experimental search to find these in Xe,Cs and TlF (Hunter, 1991). An interesting experiment would be electrophoretic mobility study of a radical ion in solution as suggested by Khirplovich (1990). Heterodyne Laser— Doppler— Velocimetry with large uncharged molecules with unpaired electrons and fully liganded complexes could be tried.

While suitable organometallic compounds could convert one chiral state to another or produce a pure enantiomer from its precursor, the reaction mechanism and the kinetic control parameters in such asymmetric synthesis are not understood well enough to lead to a more systematic approach. Kinetic control by reduction of activation energy in the transition state is too simplistic. A landscape corresponding to the connectivity in the organometallic cluster through which the reaction could be optimally steered through by the chiral forces and by the geometric restrictions between the chelated ligands and the forces from the heavy central atom, somewhat similar to a Darwinian landscape, may be needed. There exist as yet no fully satisfactory theory for reaction dynamics, let alone for asymmetric synthesis. Earlier bond-selective experiments were used to understand selective reaction paths. But recent quantum interference methods to guide selectively a single reaction product (Rice, 1992) may help us understand the forces and structural parameters involved in chemical reactions, and to ultimately check if WNC can play any role in the kinetic control of chemical reactions. Nature has indeed adopted such a guided bucket-brigade mechanism, in a circular reaction scheme, involving organometallic clusters to specifically transfer electron with high efficiency after photolysis, as in the photosynthetic apparatus. The implications of the nonunitarity (Barron) and nonergodicity (Gilat) have for time reversibility is either not demonstrated or understood in terms of the principles of chemical reactions. Barron considers that it is the S-matrix and von Neumann's principle than Onsager's principle of microscopic reversibility that is relevant. However, Von Neumann's codification of quantum mechanics is a subject of criticism by Primas (1990).

600 million years ago, when the animal kingdom erupted upon this earth, the left-right body plan was laid and it is conserved even today. Could Mislow, Gilat and others' chiral coefficients be used for pattern recognition of species evolution or for the embryonic (ontogenic) development? Meinhardt (1992) tries to explain the positioning of left-right asymmetry during limb development. The chiral evolution ended long ago, but the information is preserved and operative. There have been attempts to explain the role of the chiral molecules in the physiognomy of development. Language is asymmetric. The right brain reads from left to right, while the left from right to left; and dyslexia is the achiral condition. Plants that grow left helically, refuse to grow when distorted into a right helix refuse to grow. The helical staircase of castles are left handed so that the defenders have a better advantage. There have been attempts to measure changes in the chirality of proteins with ageing. Since we are far from equilibrium, Goldanskii and Ku'zmin (1989) are apocalyptic that pollution could one day revert us all back to the original molecules.

3. PHYSICS OF CHIRAL ATOMS

All atoms are chiral due to WNC. The Z force is fixed by the standard model and is proportional to the sixth power of the atomic number. However, since the Z bosons are massive their range of interaction is short and barely extends out to the nearest electrons. The Z forces acting on the electron tend to align its direction of motion with its spin axis making the electron's circular orbit into a right-handed helix. Because of parity violation, its mirror image, the left-handed helix does not exist.

Spectroscopically, WNC actually mixes the gerade (g) and the ungerade (u) electronic states of opposite parity so that the usual g-u electronic transition is no longer valid and the g-g or u-u magnetic dipole transition acquires a small colinear electric dipole moment from the parity mixing with resultant optical activity, which could be observed in the magnetic absorption region.

Atomic gas spectroscopy with circularly polarized light is used to measure the small chiral strength. Metals as Pb, Th, Bi and Cs have been studied (Stacey,1992). It would be useful to extend these measurements to lanthanides, transition elements and rare earths as liganded metals. Further, hardly any study exists on the influence of WNC in atomic physics. Such studies would serve as forerunner and guidance towards the spectroscopy of WNC chirality of molecules and influence of Z forces on the chemical properties of molecules, if any. The dissymetry ratio, the g.factor (rotational strength / dipole strength) for the magnetic transitions of heavy atoms, is of the order of 10^{-7}, while conventional instruments for circularly polarized light measurements require sophisticated instrument to measure differential circular polarization of 10^{-4}.

4. PHYSICS OF CHIRAL MOLECULES

Spin-orbit coupling in a molecule tend to align the electron's orbital axis against its spin axis due to the EM forces operating between the nucleus and the electron. For EM right-handed chiral molecules, the spin-orbit coupling produces left-handed helical electron due to down spiralling of spin-up electrons, while the opposite is the case for left-handed chiral molecules. A molecule with a spatial dispersion of electrons would show different regions of chirality, usually the sum being observed. The WNC induced parity mixing at one atomic center produces an electric dipole moment which is parallel or antiparallel to magnetic moment (depending on right or left chirality) arising from spin-orbit coupling at another atomic centre. The signed sum of these perturbations from all the atomic centers gives raise to the PVDE with opposite energies for the right and left enantiomer. With ordinary molecules in the ground state, the PVED corresponds to less than 1 Hz.

Mason and Tranter (1985) started quantum chemical calculations of PVED for ground states of amino acids and sugars with encouraging results of advantage factor to L-amino acids and D-sugars. However, in the absence of any experimental check, it is crucial to know simply the order of magnitude for PVED, i.e., to understand the approximations involved (Schrödinger vs. Dirac representation and the Born-Oppenheimer approximation) and the additional factors that might have been omitted. Molecules as large quantum system are increasingly coming under the scrutiny of modern quantum theory and nonlinear dynamics. With experimental evidence for intramolecular energy transfer, the Born-Oppenheimer approximation should be examined. While the Z force formula is well fixed by the phenomenological Standard model, the counterpart, spin-orbit coupling is involved in the electronic structure of the molecule. The role of relativistic spin-orbit coupling (Ermler et al., 1988) in intersystem crossing, bond lengths, Jahn-Teller effect, charge transfer, free radicals and paramagnetic systems in addition to effects of non-adiabaticity provide useful information. Enhancement of spin-orbit coupling in oriented systems as lanthanides embedded in a polymer matrix needs investigation.

Atom-like molecules would show larger PVED. Quack (1989) gives a partial list of molecules suitable for spectroscopic study. In addition to the experimental problem of probing beneath the hyperfine structure, he also discusses what should be measured in view of the welter of theories for chirality. Also, the difference in spectra between collision-free molecule and in condensed media should be known. Quack(1986) uses infra-red spectroscopy and Barra et al.,(1988) nuclear magnetic resonance for these studies, although no real success has yet been achieved.

Instead of trying to measure very low energ (low frequencies), it would be worth while to use laser spectroscopy and organometallic chelates that can be expected to show a 'g' value (rotation strength/dipole strength) of 10^{-7} or with the higher excited states of highly charged organometal complexes that may show a PVED of about 10^{-4}.

Rapid advances in laser spectroscopy and in the spectroscopy of transition, rare earth and lanthanides and their complexes could be exploited. Some of the experimental possibilities are laser cooling, nonlinear optics, two photon beating, inverse optical Faraday effect, total internal reflection, rough surfaces, localization of circularly polarized light, photochemistry and improvements in the measurement of circularly polarized light by fluorescence detected circular dichroism and Fabry-Perot to increase the sensitivity.

In addition to these measurements, it is equally important to select the molecules to be studied on the basis of their superior chiral strength (g value). From experimental nuclear physics, it is already known that one can enhance parity violation observable by orders of magnitude by nuclear amplification of symmetry violation by mixing nearly degenerate levels and the competition of weak parity allowed multipole with a strong parity forbidden one. There are also other methods for parity mixing, as by a high electric field (about 100 KV/cm) on the atom or the molecule (Demkin, 1990).

Similarly, with molecules such as highly excited highly charged organometallic complexes with a metal atomic number of about 30, a spectral g value of 10^{-4} can be obtained by looking for ΔE energy levels of opposite parity which are extremely close and by searching for maximally forbidden magnetic states with a low transition probability, W_{EM} ; upon parity mixing by WNC, this gives a very high transition probability, W_{WNC} as:

$$g = 2 \cdot \frac{\langle V \rangle}{\Delta E} \sqrt{\frac{W_{WNC}}{W_{EM}}}$$

where V is the usual parity mixing matrix. Melibaev (1979) gives some calculations for two electron ions with atomic number between 24 and 32. He also suggests beam-foil spectroscopy for production of these compounds.

5. CHEMISTRY OF CHIRAL MOLECULES: SYNTHESIS

As mentioned before, two dimensional or restricted geometries are more conducive for chiral transformations for selective synthesis or conversion than in solution or by crystallization, and external organometallic catalysts could facilitate these reactions. Chiral synthesis on layers of metal or metal compounds is a rapidly developing field in industrial chemistry. Similarly, recent progress in producing chiral chemicals directly using metals liganded or chelated to chiral inducing organic compounds has made the pharmaceutical industry aware of its potential applications (Amato, 1992). At present, these are trial and error experiments as the reaction mechanisms are not understood. The choice of metal, chiral inducing chelators, solvent and temperature are crucial. A partial rationalization by computer simulation for one of these successful "perfect reactions", the Sharpless asymmetric epoxidation using $Ti(O_i\text{-}P_r)_a$/tartarate/t-butyl peroxide is given by Sharpless et al., (1983). Transition metals are the most popular. Reviews of transition metal clusters with organic ligands (Vahrenkamp,1977) and asymmetric synthesis induced by them (Bosnich, 1981) and there chiral and optical activity properties (Sokolov, 1990) exist. Further, chiral metal complexes are used for probing local variations in DNA which may play an important role in gene expression (Barton, 1986).

Alternatively, intrinsically intramolecular reactions involving kinetically controlled minimization of steric interference between substituents in the transition state, like in the synthesis of ribose 2,4-diphosphate as main product (Eschenmoser and Lowenthal, 1992) involving aldolization of glyceraldehyde phosphate in the presence of formaldehyde is also possible. But these reactions are rare.

6. FROM MONOMERS TO OLIGOMERS

The chiral molecules produced cannot be freelancing in the solution too long, for if it wants to evolve, it has to multiply and polymerize or polymerize and multiply.

While the origin of chiral sugars and amino acids is yet disputed and the laboratory synthesis of these molecules directly with catalysts is yet to come, greater progress is made in the synthesis and replication of multimeric chiral polymers either by starting with pieces of *Tetrahymena* Ribozyme (Szostak; Joyce) or ab initio (Rebek; Von Kiedrowski) .from chiral molecules.

In addition, by innovative chemistry, topological molecules subh as helical polymers with metal chelates and knotted helical polymers can be designed, and replication can be carried out as with lithium catalyzed Na-octanoate (Bachman et al.,1992) and polymerization of ribonucleotides on montmorillinite (Ferris and Etem,1992). Regio-specificity is often metal selective. For instance, Zn^{2+} catalyses 3'-5' oligomer link, while Pb^{2+} catalyses the 2'-5' link.

According to Rebek (Famulok et al., 1992) who has been studying the chemistry of replication, the basic principle is simple. All that is needed is a host-guest relation by a covalent bond like crown ether to NH_4^+, a cyclophane to an aromatic compound, or a cyclodextrin to an aryl group, i.e., covalent link between two complementary substrates to give self-complementary systems that could act as template for their own assembly. Von Kiedrowski (Terfort and Von Kiedrowski, 1992) studied the kinetics of autocatalysis to improve the parabolic growth observed with his minimal self-replicators to an exponential growth as in biology.

Chemists have been slow (Hoffmann, 1992), but they are bringing back evolution as a topic in chemistry. Learning from the molecular biology of Watson and Crick's DNA, the molecular recognition (Ciba Foundation, 1991) of the host-guest relation of Rebek (Famulok et al., 1992) and the kinetics of Von Kiedrowski (Terfort and Von Kiedrowski 1992), replication is chemistry in disguise. Von Kiedrowski's next project will be to couple modulation of molecular recognition with the ***generation*** of chiral template molecules in one step producing a total enantioselective, autocatalytic, oligomeric synthesis, bridging the gap between chiral production and reproduction. Perhaps that is how nature did it.

Like crystallization and fractal growth, biological evolution is treated phenomenologically because it is there , but we do not know yet how it got there.

7. LOOKING BACK FROM BIOLOGY TO CHEMISTRY

Cairn-Smith's 'hypothesis (1982) of clay as originator of life lacks chirality. The much touted prebiotic soup, that according to Cairn-Smith would have poisoned itself, is the original source for homochiral production by fluctuation and with positive feedback in the nonlinear dynamic formulation of Avtiseov(1993) and Kondepudi and Nelson (1983) respectively. This soup is circumvented in the theory of hydrothermal vents (Holm ,1992).and not needed for the organometallic catalysts described here. Avetisov and Kondepudi and Nelson should be looking for D-sugars than L-amino acids by their mechanisms.

The structure of D-sugar-phosphate-nucleic acid bases with hydrogen bonds is a good model for division of labor —catalysis, self-replication and information (coded sequences), all combined in one single ancient RNA molecule; later the RNA molecules differentiatied into double helices protecting the genetic code and the functional enzymes. Nucleic acid bases by themselves are prone to stacking just as in colloid chemistry. In the more orderly case of template directed molecular recognition, the attached sugar induces excitonic coupling between the stacked bases. Chiral evolution was fully completed in the chemical prebiotic stage. Homochirality is responsible for the secondary and higher order structures of proteins and DNA—Pattern formation and pattern recognition as in the second genetic code that switches the genes "on" and "off". The multipolar structure due to spatial dispersion is conducive for specific aggregations that are functional and make possible very long polymers such as chromosomes.

Eschenmoser and Lowenthal (1992) consider glyceraldehyde phosphate leading to ribose 2.4 -diphosphate as a potential prebiotic chemical. Pentoses are more flexible than hexoses and DNA formed with hexoses give rigid linear DNAs both in single and double stranded forms with higher melting points and hence more A-A, T-T pairing than A-T pairing.

Orgel's (1992) challenge to Rebek and Von Kiedrowski is to make replicating informatio coded polymers, i.e., heteropolymers with sequence specificity, efficiency, fidelity and regio-specificity. Rebek is synthesizing artificial proteins on a nucleic acid matrix to model a simple genetic code.(See also Ponnamperuma, 1993). There have also been attempts to make artificial selection like Joyce's (Beaudry and Joyce.,1992) directed evolution. If these artificial systems can eventually mimic biomacromolecules in specificity, fidelity and mutation, i.e., to self-organized complexity, when do these near equilibrium processes crossover into the Darwinian far from equilibrium phenomena?

8. CONCLUSION

The extremely small intrinsic advantage factor due to parity violation in organic molecules, both in the ground state and in the excited state, amplified either by positive feedback or by radiative damage, does not have sufficient strength to form stable homochiral polymers. The choice is left between a catastrophic process that does not care between right and left or by specific catalytic processes as with organometal clusters for chiral discrimination. To verify that indeed an advantage factor due to electroweak currents plays a part in chiral chemistry, a search should be made in the arsenal of chemistry for compounds with the largest PVED that could be tested experimentally,both spectroscopically and in chemical reactions. Organometal complexes that are enantioselective are a suitable choice. Once this accomplished, then the search can go on to establish what was available to nature in its prebiotic stage. If no such weak current chemistry exists, then we have to concede with Monod, that life is a matter of chance.

REFERENCES

Amman, A., ,L.Cederbaum and W, Gans, (Eds.,), 1988; Fractals, quasicrystals, chaos, knots and algebraic quantum mechanics, Dodrecht,Kluwer, NATO 235C.

Amman, A., 1991; Theories of molecular chirality: A short review. In Large scale molecular systems , quantum and stochastic aspects- beyond the simple molecular picture (Eds.,Gans, W., A. Blumen and A. Amman,Plenum Press, NATO 255B, pp. 23-32.

Amato, I., 1992; Looking glass chemistry (Research news), Science, 256, 964-966.

Andelman, D., 1989; Chiral discrimination and phase transitions in Langmuir monolayers. J. Am. Chem. Soc., 111, 6536-.

Avetisov, A., 1993; Bifurcation with symmetry breaking as the cause of biological big bang, In: chemical evolution and origin of life(Eds., C.Pnnamperuma and J.Chela-Flore), Deepak Press, U.S.A

Avetisov. A., V.I. Goldanskii. S.N. Grechukha and V.V. Kuz'min, 1991; Nonequilibrium generation of optical activity in solid 1,1'-biphenyl , Chem. Phys. Lett., 184, 526-530.

Avnir, D., E. E. Wellner and M. Ottolenghi, 1989; Photophysical recognition of chiral surfaces, J. Am. Chem. Soc., 111, 2001-2003.

Bachman. P.A., P.L. Luisi and J. Lang, 1992; Autocatalytic self-replicating micelles as models for prebiotic structures, Nature, 357, 57-59.

Barra, A.L., J.B. Robert and L. Wiesenfeld, 1988; Possible observation of parity nonconservation by high resolution NMR, Europhys. Lett. 5, 217-212.

Barron, L., 1993; True and false chirality. In: Chemical evolution and origin of life, (Eds., C. Ponnamperuma and J. Chela-Flores), Deepak Press ,U.S.A

Barton, J.K., 1986; Metals and DNA: Molecular left-handed complements, Science, 233, 727-734.

Beaudry, A.A., and G.F. Joyce, 1992; Directed evolution of an RNA enzyme , Science, 257, 635-641.

Boeyens J.C.A., 1985; Molecular mechanism and the structural hypothesis. In: Structure and Bonding, Springer-Verlag, 63, 67-101.

Boeyens, J.C.A., 1991; Gauge chemistry. In: Large scale molecular systems(Eds., Gans,W., A. Blumen and A. Amman), Plenum Press, NATO 258B.

Boeyens, J.C.A., 1992; The geometry of quantum events, Speculat. Sci. Technol., 15, 192-210.

Bosnich, B., 1981; Asymmetric synthesis mediated by transition metal complexes . In: Topics in stereochemistry (Ed., Geoffrey, G.L.,), Academic Press, 12, 119-154.

Bonner, W.A., 1988; Origins of chiral homogeneity in nature, In: Topics in stereochemistry (Eds., Eliel, E.l., and S.W.Wilen), Interscience, 18, 1-95.

Bonner, W.A., 1991; The origin and amplification of biomolecular chirality, Origins of life, 21, 59-111.

Buda., A.B., T. Auf der Hyde and K. Mislow, 1992; On quantifying chirality, Angew. Chem. Int. Ed. Engl., 31, 989-1007.

Cairns-smith, A.G., 1982; Genetic takeover and the mineral origin of life, Cambridge Univ. Press.

Ciba Foundation. 1991; Host-guest molecular recognition from chemistry to biology , Ciba Foundation Symposium, Wiley, 158.

Demkin, V.P., N.L. Kupchinski and O.G. Ravinskaya, 1990, Angular distribution of probabilities from excited Stark states of He atoms, Opt. Spectrosc. U.S.S.R, 68, 724-726.

Ermler, W.C., R.B. Ross and P.A. Christiensen, 1988; Spin-orbit coupling and other relativistic effects in atoms and molecules, In: Advances in quantum chemistry(Ed., P.O.Lowdin), Academic Press, 19, 139-182.

Eschenmoser, A., and E. Lowenthal, 1992; Chemistry of potentially prebiological natural products, Chem. Soc. Revs., 21, 1-16.

Famulok, M., J.S.Novak and Rebek.1992; Self-replicating systems, Acta. Chem. Scand., 46, 315-524.

Ferris. J.P., and G. Etem, 1992; Oligomerization of ribonucleotides on montmorillonite: Reaction of the 5'-phosphorimiazolides of adenosine, Science, 257, 1587-1389.

Freedman. D.H., 1992a., The handmade cell, Discover, 13, no.8, 46-52.

Freedman, D.H., 1992b; Molding the metabolism, Discover, 13, no.8, 36-45.

Garay, A.S., 1993; Experimental studies on the possibility of chirality dependent time reversal, In: Chemical evolution and origin of life(Eds., C. Ponnamperuma and J.Chela-Flores), Deepak Press, U.S.A.

Gilat, G., 1993; Chiral interaction and biomolecular evolution, In: Chemical evolution and origin of life (Eds., C. Ponnamperuma and J. Chela-Flores), Deepak Press, U.S.A.

Goldanskii, V.I., and V.V.Kuz'min, 1989; Spontaneous breaking of mirror symmetry in nature and the origin of life, Sov. Phys. Usp.,. 32. 1-29.

Harris, J.G., and F.H. Stillinger, 1990; Spontaneous development of optical activity during freezing: A kinetic lattice model , J. Chem. Phys., 92. 6140-6151.

Hel-Or, Y., S. Peleg and Avnir, 1990, Two dimensional rotational dynamic chirality and a chiral scale. Langmuir, 6, 1691-1695.

Hirschmann H., and K.R. Hanson, 1983; On factorizing chirality and stereoisomerism, In: Topics in stereochemistry (Eds., Allinger, N.L., E.L . Eliel and S.H. Wilen), Academic Press,14, 183-229.

Hoffmann, S., 1992; Artificial replication systems, Angew. Chem. Int. Ed. Engl., 31, 1013-1016.

Holm,N.G., (Ed.,), 1992, Marine hydrothermal systems and origin of life, Origins of life, 22, nos., 1-4.

Hunter, L.R., 1991; Test of time reversal invariance in atoms, molecules and the neutron, Science, 252, 73-79.

Khriplovich, I.E., 1990. Annapole moment of chiral molecules, Z.Phys., 17D, 31.-

Kondepudi, D.K., and R.W.,Nelson, 1983; Chiral symmetry breaking in nonequilibrium systems, Phys. Rev. Lett., 50. 1023-1026.

Kondepudi, D.K., R.J. Kaufman and W. Singh, 1990, Chiral symmetry breaking in sodium chlorate crystallization, Science`, 250, 975-979.

Lakhtakis, A.I., 1991,Recent contributions to classical electromagnetic theory of chiral media. What next? Speculat. Sci. Technol., 14, 2-17.

MacDermott, A.J., 1993; The weak forces and the origin of life, In:Chemical evolution and the origin of life (Eds., C. Ponnamperuma and J. Chela-Flores), Deepak Press, U.S.A.

Mayer, J.M., 1992; Bond stretch isomers: Fact or artifact? Angew. Chem. Int. Ed. Engl., 31, 286-287.

Mason, S.F., and G.E. Tranter, 1985; The electroweak origin of molecular handedness, Proc. R. Soc., London, Ser. A 317. 45-65.

Meinhardt, H., 1992; Pattern formation in biology: A comparison of models and experiments, Rep. Prog.Phys., 55, 797-849.

Melibaev.M., 1979;; Parity nonconservation effects for highly excited states of multiply charged ions, Opt. Spectrosc. U.S.S.R, 47, 236-239.

Nakazaki, H., 1984; Synthesis and sterochemistry of chiral organic molecules with high symmetry, In: Topics in stereochemistry, Academic Press, 15, 199-251.

Orgel. L.E., 1992; Molecular replication , Science, 358, 203-209.

Ponnamperuma, C., 1993; Physicochemical bases for the origin of genetic code, In: Chemical evolution and origin of life(Eds., C.Ponnamperuma and J. Chela-Flores), Deepak Press, U.S.A.

Primas, H., 1990,; Induced nonlinear time evolution of open quantum objects , In: Sixty-two years of uncertainty: historical, philosophical, physics enquiries into the foundation of quantum physics,(Ed., A.Miller), Plenum Press, NATO 226B.

Quack, M., 1986; On the measurements of the parity violating energy difference between enantiomers, Chem. Phys. Lett., 132, 147-153.

Quack, M., 1989; Structure and dynamics of chiral molecules, Angew. Chem. Int. Ed. Engl, 28, 571-586.

Rice, S.A., 1992; New ideas for guiding the evolution of a quantum system, Science,

258, 412-413.

Salem.L., X.Chapuisat, G.Segal, C.Hiberly, C.Minot. C.Laforestier and P. Sautet, 1987; Chirality forces, J. Am. Chem. Soc., 109, 2887-2894.

Schuster, P., 1993; Evolution in the RNA world, In: Chemical evolution and origin of life(Eds. C. Ponnamperuma and J. Chela-Flores),Deepak Press, U.S.A.

Sharpless, K.B., C.H. Behrens, T. Katsuki, A.W.M. Lee, V.S. Martin, N.Takatani, S.M. Viti, P.S. Walker and S.S. Wooodard, 1983; Stereo and regioselective opening of chiral 2,3-epoxy alcohols. Versatile routes to optically pure natural products and drugs. Unusual kinetic resolution, Pure & Appl. Chem., 55, 588-604.

Sokolov, V.I.,(Ed.,)1990, Chirality and optical activity in organometallic compounds, Gordon & Breach.

Stacey, D.N., 1992; Experiments on the electro-weak interaction in atoms, Physica. Scripta., T40,, 15-22.

Steadman, G.E., 1983; Space-time symmetries and photon selection rules, Am. J. Phys., 51,750-755.

Terfort, A., and G. Von Kiedrowski, 1992, Self-replication by condensation of 3-aminobenzamidine and 2- formylphenoxy acteic acids, Angew. Chem. Int. Ed. Engl.,31, 654-656.

Thomaz, H.T., and H. H. Nussenzweig, 1982; Dispersion relation and sum rules to natural optical activity, Ann. Phys., 139,14-35.

Thiemann, W ., 1993; In: Chemical evolution and origin of life,{Eds., C. Ponnamperuma and J. Chela-Flores),Deepak Press,U.S.A

Vahrenkamp, H., 1977; Recent results in the chemistry of transition metal clusters with organic ligands . In: Novel chemical effects of electronic behaviour, Structure and Bonding, Springer-Verlag, Vol.32.

Wächterhäuser, G., 1992; Groundworks for an evolutionary biochemistry: The iron-sulphide world, Progr.biophys, molec.biol, 53, 65-201.

Wang, W., W. Jilan and J. Jiang, 1993; Differences in radiolysis behaviour of D,L amino acid in primary stage and thermodynamic equilibrium state, In: Chemical evolution and origin of life(Eds., C.Ponnamperuma and J. Chela-Flpres),Deepak Press, U.S.A.

Wilczek, F., 1991; Anyons, Sci.Am., 264, ,no.5, pp.24-31.

VIROIDS AND VIRUSES AT THE ORIGIN OF ORGANIZED LIFE

Luis J.BOYA
Departamento de Física Teórica, Facultad de Ciencias
Universidad de Zaragoza. E-50009 ZARAGOZA, Spain

Patricia BOYA
Departamento de Microbiología, Facultad de Ciencias
Universidad de Navarra. E-31080 PAMPLONA, Spain

A B S T R A C T

We argue for the necessity of an intermediate step in the primordial biochemical world, linking the longest random chains of about one hundred nucleotides formed spontaneously, to the next complexity level, this being perhaps similar to the actual small RNA viruses, with a few thousand nucleotides. We subscribe to the *viroid* hypothesis of DIENER and MOYA; the main reasons for this primordial character of viroids, beside general concordance with the "RNA world" idea, are: 1) intermediate length, around 300 nucleotides; 2) they are made up of pure RNA, no proteins; 3) "selfish character", carrying no coding capability; 4) non-specific "rolling on" replication mechanism, probably with help of host RNA polymerase II; and 5) possible monophyletic origin.

We propose also an initial RNA-amino acid relation starting as viroidlike entities in a pool of abiotically originated organic molecules, develop, evolve, mutate and expand. The genetic code was perhaps created then. This allowed these small RNA fragments to evolve into primitive *viruses* (probably like the actual RNA plant or phage viruses), a few thousand nucleotides long, with associated proteins including coat and perhaps replicases, characteristic of RNA viruses.

Finally some comments are offered about the transition to the first DNA-carrying cells (DNA viruses and eventually primitive prokaryotes).

CHEMICAL EVOLUTION: ORIGIN OF LIFE
C. Ponnamperuma and J. Chela-Flores (Eds.)

ISBN 0-937194-31-X

1. THE PROBLEM.

In the RNA world, it is difficult to assemble and conserve more than one hundred nucleotides spontaneously, because of the high level of mutation in RNA chains and the lack of an error-correcting mechanism (Eigen, 1971). But the simplest (Eu- and Archaeo-) Bacteria have over a million nucleotides (nt); how has Nature bridged over a factor of 10^4 ?

Obviously, *viruses* come immediately to mind, because of their intermediate nucleotide length, 3.10^3 to 2.10^5 nt; it is surprising how little attention viruses have attracted as primordial genetic material; most authors tend to consider viruses as degenerate, extreme parasites, an aside in the evolution of life (Stanier,1988; Davis *et al.*, 1976). In this paper we shall entertain the idea that *viroids* and later some kind of RNA viruses were intermediate milestones in the evolution of organized life, following the fundamental pioneer works of Diener (1989) and Moya (Elena *et al.*, 1991; Elena and Moya, 1991); although these authors refer only to viroids, the transition to viruses seems to be the next natural step. An obvious difficulty stands in the way: today apparently only *solanaceae* and other flowering plants carry viroids, and these are very recent in geologic time; the answer is of course that viroids got refuge in these plants for some reasons, after living quasi-free in the primordial soup and being for ages guests in other hosts. The fact that virus and viroids tend to concentrate in chloroplasts and nuclei (Riesner and Gross, 1985), suggests a bacteriophage-like period, and this by itself will already extend the age of viroids back at the beginning of the proterozoic; in fact, single-stranded (ss) RNA viruses are common as phages today.

2. VIROIDS.

Viroids are a modern addition to simple reproducing "organisms". (Diener 1971). They have the smallest known size for self-replication, around 350 nucleotides, and are not very specific; for example, the first identified viroid, the Potato Spindle Tuber Viroid (PSTVd), grows in tomatoes even better that in the original potato plant. Today several dozen viroids are known.

So the hypothesis runs:

Viroids represent the oldest molecular fossils, capable of reproduction and propagation, as relics from the primitive RNA world (Diener, 1989)

We now review and reinforce this statement, extend it further, put it in relation with the next evolutionary step, namely the genetic code through the RNA-aminoacid relation, set up a scenario for appearance of viruses, and approach the RNA --->> DNA transition (Lazcano *et al.*, 1988).

3. ASSSUMPTIONS IN THE RNA WORLD.

We contemplate an RNA world (Gilbert, 1986), with self-assembled or more likely ribozyme-catalyzed chains of nucleotides (nt), up to about one hundred of them, taken form a pool with the P-activated nucleotides A, G, C and U. These chains undergo an incipient alteration, elongation and duplication, by Watson-Crick bonds, by opening up the stems, by cytosine-uracil degradation, or by other means; synthesis of complementary chains has been already obtained, even *in vitro* . Errors in the reproducing "genomes" are multifold, perhaps around 10^{-3} (per nt); also "wrong" G-U pairs should be present.

All this occurs *in absence of proteins;* but aminoacids are present and abundant in the pool. Energy is coming from phosphates, beside external sources like lightning, unscreened ultraviolet radiation, some meteorite bombardment, vulcanism, etc. For the origin of active pyrophosphates see (Yamagata *et al.*, 1991).

Formation of viroids. In some circumstances, perhaps *not generically* three or four chains of random RNA may join, making up a viroid-like object, with ca. 300 nt. It would be preserved by some unknown mechanism: isolation, autocatalytic self-cleavage, differential advantages, etc. Eventually, these viroids-to-be propagate by selfreproduction, and mutate. Mutations in the unpaired "loops" of the RNA structure are harmless ("neutral"), and should be more frequent. Much later, they should evolve as small, single-stranded, plant RNA-like viruses (see below).

4. FEATURES OF ACTUAL VIROIDS AS PRIMITIVE INTERMEDIARIES.

These have been mainly pointed out by DIENER and MOYA, (Diener, 1989; Elena *et al.*, 1991). Let us review them quickly:

1) *Viroids have a small length, 250-400 nucleotides,* intermediate between "spontaneous" RNA polymers, 100 nt long, and small RNA plant viruses (e.g. a typical plant virus, Tobacco Mosaic Virus (TMV), has an RNA of about 6000 nt).

2) *Viroids are made of naked, pure RNA, with no protein coat;* this befits perfectly with the RNA world idea, as previous to the polypeptide synthesis and therefore they are self-consistent RNA objects.

3) *They have no coding capability; they seem to be purely "selfish" RNA;* this fits again with the RNA-previous-to-protein idea, and also with the primordial role of introns (Doolittle, 1978). Nevertheless, viroids with some autocatalytic capability will survive better; catalytic properties of introns were discovered by Cech (Cech *et al.*, 1981). The first "viroid" might be made of pure random chains, later viroids with some ribozyme-like activity will take over.

4) *High C+G content.* This produces a more robust union than A-U, hence more likely to be conserved. For example, we have a (G+C)/(A+U) ratio of 65:35 in PSTVd (Gross *et al.*, 1978).

5) *Parasites of plants.* Viroids tend to gather in the nucleus of plant cells, whereas plant viruses usually thrive in chloroplasts (Zaitlin and Hull, 1987). Plant parasitism is the only way for viroids to propagate, because to invade animal cells will require adhesion proteins, which viroids lack; on the other hand, plasmodesmata will allow cell to cell passage in plants. Also there are some reasons to suppose that eukaryote algae are more primitive that metazoa (Knoll, 1992). If it could be proven that actual viroids got into higher plants via algae or more primitive eukaryotes, the hypothesis will be reinforced. Some similarities between viroids and the RNA of *snurps*, the nuclear spliceosome components, have been remarked (Riesner and Gross, 1985). Also, as stated, this will ease the next step, the transition to viruses, which eventually got hold of plants...

6) *Stem and loop structures.* Viroids exhibit the usual double- and single-stranded structure of most RNAs. This make them similar to primitive forms of RNA: viruses, tRNA, rRNA, etc. These structures tend to be stable by themselves, presenting a mixture of ss and ds RNA parts.

7) *Rolling on mechanism of duplication,* probably by host RNA polymerase II, which is non-specific. It is well known that usual plant viruses and phages require specific replicases coded in the genome of the virus. Viroids, as more primitive, probably duplicated riboenzymatically; today

remnants, however, have taken advantage of plant cell protein polymerases.

8) Related phylogeny; indeed, the claim has been made (Diener,1989; Elena *et al.*, 1991) that viroids can be monophyletic. In their family it is included also a satellite hepatitis virus, so viroid-like entities are already not confined to be plant invaders.

9) Self-cleaving; conserved "hammerhead" structures, at least in some cases. This will reinforce the intron-like character (Diener, 1993).

5. SOME FURTHER CONSIDERATIONS

So far no proteins are needed; now it is the moment to invoke them. A fairly hypothetical sequence of events, with a view towards increasingly complexity , can be the followings:

1) The first stable relations between the RNAs and particular amino acids (aa) are established; affinity among both species is already well studied. This is to be seen as a step towards transfer RNA, which are of small size, about 85 nt and probably in old times carried other functions (Weiner and Maziels, 1991).

2) Ribozymes, already present, start to bind aa's to make simple polypeptides; today it is becoming increasingly clear that the RNA moiety is the active center in the protein synthesis apparatus of the ribosomes (Noller *et al.* 1992; Piccirilli *et al.*, 1992).

3) Elongation and fusion of polypeptides; they will be selected if they cover and protect the elongating RNA (one could envisage, of course, many other scenarios). This would suggest to study virus coat proteins as being about the oldest existing proteins; primitive replicases might be other possibility (Lazcano, 1986).

4) These protected RNAs are now allowed to enlarge, perhaps again by fusion, up to ten times in length. That we take as the decisive step towards small ss RNA viruses. Today, they are the smallest true viruses (e.g. the phage MS2 has 3559 nt, Fiers *et al.*, (1976)).

5) The relationship between RNA and aa´s increases; polypeptide synthesis is concentrated in some points of the genome; aa are transported by these small RNA fragments; this is how we envision the incipient *genetic code* (see below).

6. THE GENETIC CODE.

The genetic code is not random; it is best understood if developed in three stages; because G-C is the strongest bond, we suggest the following serires of events:

1) Accept first that the third letter in the codon is irrelevant; in fact, two letters are already enough for intron excision. Then starting by using only C and G nucleotides, we obtain just four important amino acids:

GGN: codes for glycine, the simplest amino acid.
GCN: codes for alanine, the second simplest
CGN: codes for arginine, which is basic (NH_2), to neutralise acidity in RNA; very abundant in histones.
CCN: codes for proline, the only aa allowing chain twisting.

2) As a second step, bring in adenine and uracil; they are already present in the primitive soup . Keeping still the irrelevance of the third letter, we obtain something like this:

UUN:	Phe.	UCN:	Ser	UAN:	Tyr
CUN:	Leu	ACN:	Thr	CAN:	His
AUN:	*Start*	UGN:	Cys	AAN:	Lys
GUN:	Val	AGN:	*Stop*	GAN:	Asp

3) Separate the third letter in Pu and Py in some cases; e.g.

GAPy = Asp, GAPu = Glu

4) Finally, reshuffle Start/Stop signals; and code for Trp, Met.

Of course, the whole of the genetic code problem requires much more that our very elementary sketch; see e.g. (Fox, 1987). We did only hypothesize, to provide continuity with the viroid idea. For example, some aa´s are present in meteorites, and are abundant in the soup; others, like arginine, are not. Also, selenium-mediated amino acids would perhaps be present. There is again the issue of the modified genetic code in mitochondriae, ciliates, etc; the impression one gets is that they are "converted" to more primitive code; e.g. Trp is coded by two codons, not just one, in agreement with step three above. For an study of specific the affinity between amino acids and (anti)codons, see Ponnamperuma (1993).

7. THE REALM OF VIRUSES.

Protected by proteins, forming a kind of spores which can propagate and invade new ecological niches, viruses might now thrive and diversify; the environment is still rich to allow them to incorporate directly activated nucleotides from the primitive soup. Purine coenzymes are ready, as well as amino acids and sulphur and phosphorus compounds, all from the pool.

Today we have a large paraphernalia of small RNA viruses: about 400 in plants only (Zaitlin and Hull, 1987), but also abundant as phages and even as animal viruses. It is hard to imagine the fate and evolution of these "replicators"; eventually most of them should "starve", probably by deplenishment or exhaustion of the primitive organic soup. Also the mutation rate should be very high, 10^{-3} or 10^{-4} per nt per cycle. At the same time, proteins diversify and start to take up catalytic functions, for which they are much better prepared that ribozymes, and more specific. Also the coding and noncoding parts of the genome are ready to compete; introns, in the modern sense of the world, are born.

But RNA viruses still do not have correcting mechanisms: hence the length of the viruses is still small, up 10-15 thousand nt; still *no lipid membrane* exists, but there are many coat proteins, capsids, replicases, etc.

8. TRANSITION TO THE DNA.

Two significant discoveries led to the DNA stage of evolution: first, uracil methylates to thymine, and is stabilised: sometimes cytosine spontaneously deaminates to uracil, adding to the natural errors in duplication. Methylation is a well used resort of Nature to "tag" nucleotides. Secondly, and more important, ribose becomes deoxyribose by deoxidation in carbon 2′; this takes away all the catalytic properties of RNA, and also allows, by steric reasons, the much more robust DNA double helix (beta form). That both crucial advances are performed by a simple loss of an oxygen in position 2′ we take as a complete quiz of Nature. Henceforth the biological work gets much better organized:

1) RNA keeps the functions associated to the t-, r- and m-RNA forms and functions. It also keeps a small but non-specific, crucial catalytic activity, today present in splicing and protein synthesis, as primer in DNA replication, etc.

2) DNA is preserved for *storage* and *back-up* functions only; it can now grow in principle indefinite, because of the much bigger stability of the double helix.

However, two more achievements are necessary for the definitive success of the biosphere, i.e. of the biological world. 1) The enzyme reverse trascriptase (RT), to produce DNA from RNA. Today, the number of RTs is increasing, and its functions are better understood; there seem to be no major obstacles to a sort of primitive reverse trascriptase, made of protein subunits, perhaps a single one. 2) Error-correcting enzymes, as to preserve the information for such long chains, which reach a million nucleotide in the simplest nondegenerate bacteria (i.e., excluding extreme parasitic forms like Mycoplasmas). From the way the actual duplication of the DNA is self-correcting, one can argue that the error correcting mechanism developed probably at the same time as reverse transcriptase or a bit later, when DNA replication was fully established with the help of a multimeric-protein enzymatic apparatus.

As fossils of this period, one can suggest the study of some small DNA viruses, like ØX 174, phages of the T-series, etc.

9. FIRST CELLS.

The time window in which these developments occurred is rather narrow. The planet earth formed 4.54 Ga. ago, but was probably not supportive of life at all until 3.9 or 3.8 Ga. ago, because absence of continents, heavy meteorite bombardment, and perhaps excessive heat. On the other hand, fully developed cyanobacteria did exist around 3.55 Ga., which leaves "only" over two hundred million years for the stated steps, from random short RNA chains of nucleotides to evolved photosynthetic and/or methanogenic archaebacteria of 1-2 million nucleotides. Moreover, if the carbon isotope anomaly in the Isua supracrustal formation in Greenland is confirmed to be biogenic, that would put back the biological clock just around the time the earth could support life. In other words, we come to the striking conclusion that life started at the very moment that it was at all possible!

The following table summarises the progression towards complexity

Table I: LEVELS OF COMPLEXITY

ENTITIES	LENGTH(nt)	ORIGIN
Random RNA chains	100	spontaneous
Viroids	350	casual fusion of former
RNA Viruses	3.550-10.000	polypepeptide helped
DNA Viruses	up to 150.000	fully enzymatic
Mycoplasmas	800.000	degenerate prokaryote
Archaebacteria	1.5 to 2 Mb	3.6 Ga?
(Eu)bacteria	2- 4 Mb	3.5 Ga.(Photosynthesis?)
First Eukaryote	10-100 Mb (?)	2.1 Ga. [1]
Earliest metaphyta	??	1-1.5 Ga. [2]
First metazoan (Ediacaran fauna)	??	700 Ma.
H. sapiens	3 10^9 nt	200.000 y.[3]

Ref [1]:(Han and Runnegar, 1992); Ref [2]:(Knoll and Walter, 1992); Ref [3]:(Cann, 1987)

We do not pretend to convey a too optimistic impression that we understand thoroughly the first stages of organized life on earth. The following is a short list of problems which we judge as the most serious ones:

1) Build and sustain pieces 100 nt long of RNA with stem and loop structures, which will contain ONLY the four selected nucleotides and ONLY the sugar ribose!

2) Still unsolved is the origin of the biological "tree": which came first, the primitive archaebacteria, perhaps of the Thermo-Sulfo class (Woese-Zillig; see e.g. Woese, 1990; Zaitling *et al.*, 1990), or the primitive photosynthetic cyanobacteria, as argued by Cavalier-Smith (Cavalier-Smith, 1991) ?

REFERENCES

Cann,R.L. et al.,1987:"Mitichondrial DNA and human evolution",Nature *325*,31-36

Cavalier-Smith,T.,1991:"The evolution of cells", in Evolution of life, S.Osawaa and T.Honjo eds., Springer, Tokyo,pp 271-304

Cech,T.R., 1981:"In vitro splicing of the ribosomal RNA precursor of *Tetrahymena*", Cell *27*,487-496

Davis *et al.*, 1976: Microbiology, Harper and Row, Hagestown, Maryland

Diener, T.O., 1971:"Viroids", Virology, *45*, 411-428

Diener, T.O., 1989: "Circular RNAs: Relics of precellular evolution? ", Proc.Natl.Acad.Sci.USA, *86*, 9370-9374

Diener,T.O.,(1993):"Small Pathogenic RNAs of Plants:Living Fossils of the RNA World?",(these Proceedings)

Doolittle,F.W., 1978:"Genes in pieces: were they ever together?",Nature *272*,581-582

Eigen, M., 1971: "Self-organization of matter and the evolution of biological macromolecules", Naturwiss. *58,* 465-523

Elena,S.F. and A.Moya, 1991:"Algunos parametros evolutivos de los RNA de viroides y de tipo viroide", Rev. Cien. (Ins. D'Estud.Balear.), *8,* 9-15

Elena,S.F. *et al.*, 1991: "Phylogeny of viroids, viroidlike satellite RNAs, and the viroidlike domain of hepatitis Δ virus RNA", Proc.Natl.Acad.Sci.USA, *88,* 5631-5634

Fiers,W. *et al.*,1976:"Complete nucleotide sequence of bacteriophage MS2 RNA",Nature *260,*500-507

Fox,T.D.,1987:"Alterations of the genetic code",Ann.Rev.Genet. *21,*67-133

Gilbert,W., 1986:"The RNA World", Nature,*319,* 618

Gross H.J. et al.,1978:"Nucleotide sequence and secondary structure of potato spindle tuber viroid",Nature 273,203-208

Han,T.M. and B.Runnegar,1992:"Megascopic Eukaryotic Algae from a 2.1 Billion-Year-Old Formation",Science 257, 232-235

Knoll,A.H. and M.R.Walter,1992:"Latest Proterozoic stratigraphy and Earth history", Nature *356,*673-678

Knoll,A.H.,1992:"The Early Evolution of Eukaryotes:A Geological Perspective", Science *256,*622-627

Lazcano et al.,1988:"The Evolutionary Transition from RNA to DNA in Early Cells", J.Mol.Evol *27,* 2283-290

Lazcano,A.,1986:"Prebiotic Evolution and the origin of cells",Treb.Soc.Catal.Biol. *39,*73-103

Noller,H.F. *et al.*,1992:"Unusual Resistence of Peptidyl Transferase to Protein Extraction Procedures", Science *256,*1416-1419

Piccirilli,J.A. *et al.*,1992:"Aminoacyl Esterase Activity of the *Tetrahymena* Ribozyme, Science *256,*1420-1424

Ponnamperuma,C.,1993):"Physicochemical Basis for the Origin of the Genetic Code" (these Proceedings)

Riesner,D. and H.J.Gross,1985:"Viroids", Ann.Rev.Biochem. *54,* 531-564

Stanier,R.Y., 1988: Microbiology, Prentice-Hall, Englewood, New Jersey

Weiner,A.M. and N.Maziels,:"The genomic tag model for the origin of protein synthesis" in Evolution of life, S.Osawa and T.Hong eds., Springen, Tokyo, pp 51-66

Woese,C.R. 1990:"The progenote", Science *247,*789-794

Yamagata *et al.*, 1991:"Volcanic production of polyphosphates and its relevance to prebiotic evolution", Nature *352,* 516-519

Zaitling,M. and R.Hull, 1987:"Plant Virus-Host Interactions", Ann.Rev.Plant.Physiol.*38,* 291-390

THE ROLE OF NEOTENY AND SOCIOGENESIS IN THE EVOLUTION OF CELL STRUCTURE

Vladimír J.A.Novák
Laboratory of Evolutionary Biology
Praha, ČS-128 00, ČSFR

ABSTRACT

In the research of questions concerning the origin of life, the utilisation of regularities of biological evolution is very important next to chemical and physical (biophysical) notions, beginning with the prebiological period. Among them the law of neoteny and the principle of sociogenesis have not been elaborated sufficiantly from this point of view till now. The origin of the DNA double helix may be viewd as the phylogenetically earlist and most general case of neoteny. The character of neoteny is also shared by the first phase of sociogenesis on each of the five grades of individuals. - The origin of the first organisms and their gradual transformation into prokaryotic cell correspond to the first grade of sociogenesis with its five phases. This has been suggested by our hypothesis of coacervate-in-coacervate. Examples of the five phases appear among extant viruses. - The theory of polyploid synbacteriogenesis has been formulated as an alternative to the generally accepted symbiogenesis theory. According to it, the eukaryotic cell organelles have originated by gradual internal divisions of the prokaryon (the ring gland molecule) of the prokaryotic cell. In the course of this very long phylogenetic process (between the past 3,5 and 1,5 thousand million years) the first several generations of the DNA molecules constituted the nucleus, a number of subsequent generations transformed into mitochondria. At about the same time the microtubules were synthetised and formed kinetosomes and mitotic figures as well as undulipodia. The last series of internal DNA generations formed plastides, first of all the chloroplasts of green plants. - The prebiotic chemical evolution towards the origin of life is assumed to have taken place in accumulations of organic matter in the primary water basins along the estuaries of the original dry land streams. This material concentrated in the form of coacervates in which the synthesis took place of the first polypeptides (proteinoids). Their structure has determined the bases sequences of the primary polynucleotides which grew to the beginnings of the genetic code. Round the replicating DNA molecules within the primary coacervate the formation of the first coacervate-in-coacervate individuals started.

The first conclusions and notions on the origin of life on the surface of this planet were based on chemical data and experiments. (Oparin 1922, Haldane 1929, Miller 1953 etc.). An important further step was dome by

ISBN 0-937194-31-X

taking into account physical and especially biophysical regularities up to the biological level, as it was done systematically by Segal and his group (1983). Little attention has been paid, however, to the action of regularities on the early Earth surface so far. Among the law of neoteny and the principle of sociogenesis have been of special importance since the beginnings of polymerisation of the first peptides and nucleotides, for the understanding of the chemical evolution.

The neoteny as a case of shifting of morphogenetic processes results in a change in form and function due to a reduction of some of the end stages of the ancestral ontogenetic differentiation. In this way various juvenile features appear in adults (Fig. 1). Neoteny is a contradiction of the epigenesis which consists in an addition of new morphogenetic stages at the end of ontogenesis which did not occur in the ancestors. It does not interfere with the law of recapitulation as some authors thought (cf.Novák 1992) except that some of the recapitulating stages may be reduced.

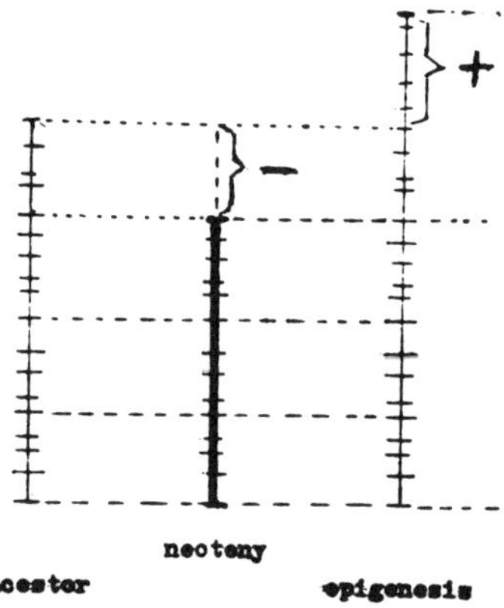

Figure 1. Scheme of neoteny and epigenesis. The endstages of ancestral morphogenesis are reduced in neoteny (-), whereas new ones have evolved in epigenesis (+). (From Novák 1982).

From the point of view of genetics, three types of neoteny may be distinguished :1) The non-hereditary individual neoteny, only concerning one generation, a given individual, often being of teratological character, produced by some exceptional, unrepeated action of environment. 2) The non-hereditary phylogenetic neoteny, produced by various environmental factors working for a number of generations (cf.Novák 1988). 3) Hereditary neoteny produced by mutation of a feature (gene) of an organism resulting in a species specific change of the neotenic character.

Neoteny is of a very common occurence, not only in plants and animals as well as in protists (both protozoans and unicellular algae) and in prokaryots (Novák 1982), but also in subcellular organisms.

As a most common example of the above case, the origin of the double helix of the nucleic acid molecules may be mentioned (Novák 1982). The duplication of the chain is an obligatory stage in the replication of the monohelix DNA (as in some viruses) after which the both chains separate to form two daughter molecules. A reduction of this last stage results in a permanent double helix common to all cellular organisms. Here new bonds have developed. The same happened with the RNA helix where the doubled form has been restrained, however, to a few virus groups. (Figure 2).

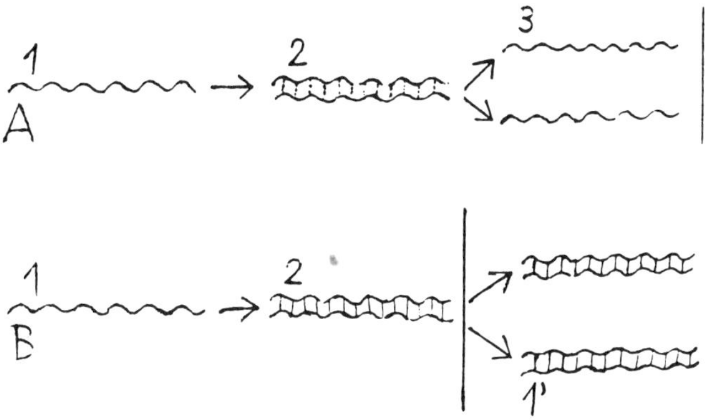

Figure 2. <u>The origin of the DNA doublehelix</u> as a case <u>of neoteny (From Novák 1982)</u>

A - monohelix DNA replication process (1,2,3)
B - doublehelix DNA replication(1,2,2'), the stage 3 is lacking here.

The status of viruses is to be mentioned in this connection. The accepting of viruses for indepenaent organisms has been generally refused long time. Their exclusive occurence inside the host cells and their dependance on the host enzymes viewd as an indisputable evidence of their having been derived from a preceding cell structure. This is no more so. Their way of reproduction, their hereditary features and a survey of the whole system of animal, plant and prokaryot viruses shows sufficiantly their organismic character. As to the phylogenetic origin of viruses, three main theories have been suggested : 1) They are stages in the evolution of the living matter from monomolecular to the prokaryotic level with secondary simplifications due to their parasitic way of life (C.f.Blaškovič 1978; Novák 1982). 2) They originated from earlier bacteria reduced up to a single nucleic acid molecule. 3) They are remnants of various DNA containing cell structures (plasmids, parts of genome, mitochondria atc.) which became independant of the host cell and penetrated into other cells changing their host metabolism to reproduce themselves. The theories do not exclude each other, however, and the three different ways may be occured parallely in different groups of viruses.

A very general role of neoteny is that which is in connection with the sociogenesis. The principle of sociogenesis (Novák 1967,1975,1982) is one of the main laws of biological evolution (on about the same level of generality as the principle of natural selection). It may be the most briefly characterized by the following three theses : 1) All known organisms, both extant and extinct ones, belong to one of the five grades (structural levels) of individuals, the zero grade being individual nucleic acid molecules (like viroids) replicating in a suitable medium. They begin with all various viruses and prokaryotic cells (grade I), eukaryotic protists, i.e. unicellular plants, fungi and animals form the grade II, lower multicellulars (lower plants and non-metameric animals) are the grade III, higher multicellulars (bryophytes and tracheophytes so as metameric animals) the grade IV and tufts of higher plants and societies and metameric animals including the human society the grade V (Figure 3A). 2) Every higher grade organism has passed through all the lower grades gradually during its phylogenesis. 3) Each of the higher grades of individuals has passed through the same five basic phases of sociogenesis in its evolution from the preceding lower grade : a) non-separation, b) differentiation, c) formation of an internal medium, d) evolution of correlation mechanisms, e) integration to a higher grade individual. The sequence and duration of all these phases is more or less the same in principle in all the five grades of sociogenesis. (Figure 3B).

The neoteny concerns first phases of sociogenesis -the non-separation in all five grades. It consists in that the second step in the reproduction process, the separation of the segmented daughter individuals, does not take place, so that they remain together and form the beginning of a colony (society). The first step towards sociogenesis consists thus in the preservation of an early ontogenetic (juvenile) stage into the adult period. This explains, together with the high selection value of any association and cooperation, the very general occurence of all sociogenesis phenomena. The neotenic character of the colony (society) formation is clearly evident from the comparison of the formation of the multicellular state in the sporozoan Plasmodium where it is just a transitory stage in the ontogenesis towards the unicellular infecting stage. In colonial protozoans like the volvocid Eudorina, on the other hand, the multicellular state is the permanent adult stage, whereas the unicellular state is just a temporary reproduction stage like in multicellulars (Figure 4).

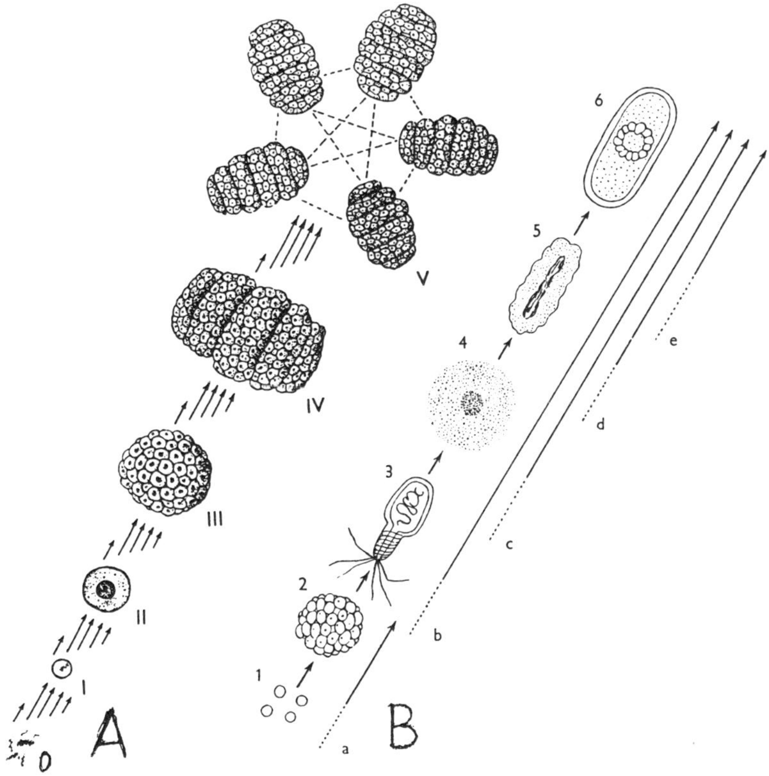

Figure 3. Schema of the principle of sociogenesis. A - Five grades of sociogenesis. I to V (grade 0 are monomolecular replicating nucleic acids in their life medium (primary coacervate or protoplasm of a cell). Grade V individuals are interconnected not by physical (morphological) ties, but by behaviour ones (instincts, nervous activity). The arrows between the various grades denote the phases of sociogenesis. B - Phases of sociogenesis in the evolution of the grade I individuals. a - non-separation, b - differentiation, c - formation of internal environment, d - correlation, e - integration into a prokaryotic cell. (After Novák 1982).

As suggested by the coacervate-in-coacervate hypothesis (Novák,Liebl 1971, Novák 1974 etc.), findings resulting from a comparative study of the ultrastructure of all groups, viruses are the main source of evidence for the character and successive sourse of the evolution of prokaryotic cell (protocell or "die Prozyte" according to Sc˙wemmler(1979) structure. Thre is a practically unbroken sequence of intermediate forms between the lowest (monomolecular) viruses of the plasmid or viroid type and the primitive, phylogenetically lowest prokaryotic cell like those of mycoplasms or rickettsiae. The main steps in this evolution are the following ones : The coacervation of proteinoids (the original abiogenetic polypeptides, cf.Fox 1969) (1). The polymerization within this primary coacervate of abiotic nucleotides with a bases sequence determined by the peptide sequence of the proteinoids. (Cf.Segal et al. 1983). (2). Replication of the first determined nucleic acids within the coacervate. (3) Formation of a secondary coacervate inside the primary one originating from the new coded (sphaero) proteins round the replicating molecule (4).

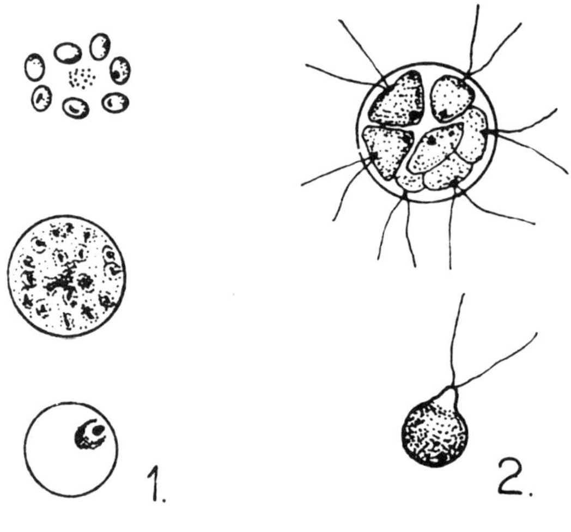

Figure 4. Neotenic character of the non separation as the first phase of sociogenesis in the colonial protozoans (From Novák 1983).
A) Plasmodium sp., (Sporozoa)-monocellular and multicellular stages in the life cycle
B) Eudorina sp. (Volvocidae) - reproduction stage and multicellular adult stage.

In the above way more and more complicated protobiont bodies evolved gradually. This phylogenetic process can be reconstructed on the basis of the structure of various extant viruses, so more that it is simplified due to their parasitic way of life (being often also of the character of neoteny).

Whereas in lower viruses the protein molecules round the central DNA were freely exposed to the environment (the primary coacervate), in phylogenetically higher groups a surface biomembrane has developed thanks to which the evolution of a specific internal environment has been enabled, a more suitable for survival of the original individuals than the external one.

Little if anything has been known on the correlation mechanisms in the evolution of the prokaryotic cell so far. The spreading of specific substances inside the internal fluid of the virion of the highest viruses seem to may be viewd as the beginnings of this phase. In thi s way the protocells of the phylogenetically lowest bacteria like mycoplasms, chlamidiae or rickettsiae have gradually developed.

As an alternative conception of the eukaryotic cell evolution to the symbiogenesis theory (Margulis 1981) being nearly generally accepted now, the hypothesis of polyploid synbacteriogenesis has been formulated (Studitsky 1975, 1981, Novák 1983). (It is different but parallel to Genom-Duplication-Invagination hypothesis = Successive hypothesis, cf.Schwemmler 1979). According to it, the extant eukaryotic cell originated by several divisions of the prokaryon (the ring gland molecule) of a prokaryotic cell and their gradual differentiation into individual cell organelles like nucleus, mitochondria, microtubules, plastides etc. The whole phylogenetic process from the prokaryotic to the first eukaryotic organisms needed ca. two thousand million years. (Figure 5).

The first generations of the polyploid DNA molecules constituted the components of the nucleus. A number of subsequent generations of the molecules (inside the mother cells)in new conditions (increased O_2 content) transformed mitochondria. Microtubules in the form of their various adaptations like kinetosomes, mitotic figures and undulipodia and other motility mechanisms etc. have developed at about the same time. The last series of internal DNA generations have evolved to plastids which became chloroplasts in the necessary conditions of sufficiant light intensity and the presence of green pigments (chlophils). Similar structures have evolved in free-living organisms like aerophilic oxydizing bacteria, spirochetes and cyanophytes at about the same time or earlier.

Much may be objected against the hypothetical sequence of these events. I think, however, that the difficulties here are less serious than those of the symbiogenesis conception. Few decisive criteria may be found between the both hypotheses at present state of our notions. Nevertheless , they are less cases of mere chance increasing inprobability to be assumed in synbacteriogenesis than in symbiogenesis. Take e.g. the repeated phagocyting and survival inside the strange conditions of the host cytoplasm, to say nothing of the lack of evidence for phagocytosis in Prokaryota. On the other hand, even if a consistent study of ecology of nucleic acid molecules (both free-living in coacervate-like media and in the cytoplasm and nucleus of cells of most various organisms (did not yet start, it is fully imaginable that the DNA daughter molecules can constitute new bases sequences (gen systems) among them such ones similar to free-living organisms groups (like archeobacteria). They can be equally well imagined to loose parts of some of their helix occuring in previous conditions inside the forming nucleus (may be some eubacterial sequences ?) (See the foot-note on the last page).

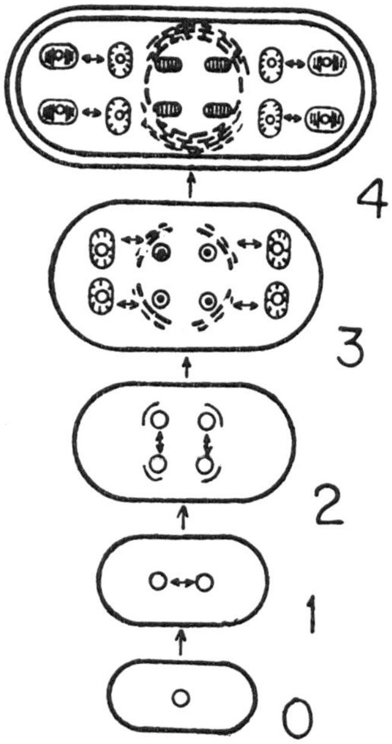

Figure 5.
<u>The polyploid synbacteriogenesis theory</u>.
Four main stages in the evolution of prokaryotic cell.
0- bacterial cell with the DNA ring chromosome.
1 - diploid protocell
2 - Increased tetraploid cell with the beginnings of nucleus
3 - Origin of mitochondria by a further series of DNA molecules
4 - primary eukaryotic cell with chloroplasts formed by the last series of internal nucleic acid molecules (From Novák 1983).

In this way the whole story of phylogenetic origin of both prokaryotic and eukaryotic cell structure may be told in the main traits with no much speculation. The chemical evolution in the direction of origin of life is assumed to have started in the accumulations of organic matter in the primary water basins along the estuaries of the original dry land streams. The material containing aminoacids and peptides concentrated in the form of coacervated in which the first abiotic polypeptides polymerized (cf.proteinoids) Fox 1969, Oparin 1968, Segal et al. 1983). The structure of these polypeptides polymerized has determined the sequence of bases of the first polynucleotides, the components of which were brought to the place by reinwater from the volcanic areas the most probably. The primary nucleic acids originating in this way have constituted the beginnings of the genetic code gradually. Both DNA and RNA originating in this way were replicated by proteinoids of the primary coacervate. Both of them have also formed the both types of double helix molecules as we know them in some of the extant viruses. From the four types, however, only the DNA double helix has survived to the first (prokaryotic) cell stage.

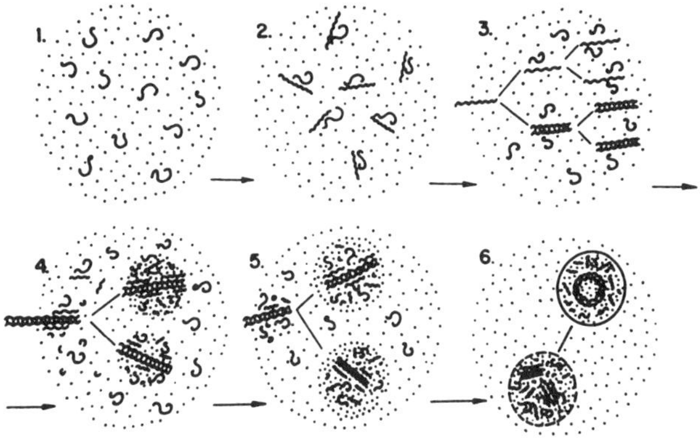

Figure 6. The coacervate-in-coacervate theory. Six consecutive stages in evolution from the first polypeptides and nucleic acid molecules to the prokaryotic cells.

a - primary coacervate with the first polypeptides, b - first nucleic acid molecules are formed by protoenzymatic activity of proteinoids, c- initial replication of monohelix DNA and formation of double helix molecules, d - formation of the secondary coacervate round the replicating DNA molecules, e - differentiation of nucleic acid molecules, initial RNA, f - formation of surface membrane and origin of prokaryotic cell. (From Novák 1982).

Round the nucleic acid molecules in the coacervate started new, coded proteins to be produced. They were, however, not much different from the previous abiotic polylypeptides which gave origin to the genetic code. They accumulated round the replicating DNA molecules forming thus a secondary coacervate inside the primary one. This was the beginning of first organisms which reached their life charactristics gradually. The first of them was the reproduction (replication capacity) brought from the prebiotic state, with gradually improving heredity mechanisms. With the reproduction the action of natural selection started favouring each improvements

Table 1. Differences between the theories of symbiogenesis (SG) (Margulis 1981) and polyploid synbacteriogenesis (PSB) (Studitsky 1975, Novák 1983) of eukaryotic cell evolution

SG theory	PSB theory
The eukaryotic cell organelles originated from prokaryotic endosymbionts surviving their phagocytosis.	The DNA containing organells of eukaryots originated by gradual differentiation of polyploid DNA molecules.
Nucleus evolved from prokaryotic nucleoid differentiating into chromosomes, nucleolus, nuclear membrane with various RNAs + proteins	Nucleus with its organells evolved by association of daughter DNA molecules of first series of generations of DNA.
Photosynthetic prokaryots became the source of atmospheric oxygen. Symbionts originated from which chloroplast evolved later.	Photosynthetic prokaryots produced the atmospheric oxygen and the mutagenic energy to produce the autogenic chloroplasts.
Mitochondria evolved from aerobic oxidizing bacteria which became the first endosymbionts	Mitochondria evolved in increasing oxygen content by later generations of DNA daughter molecules paralelly to the free living bacteria.
Microtubules were produced by endoparasitic spirochetes differentiating into kinetosomes, mitotic figures and undulipodia.	Microtubules were produced by the differentiating DNA molecules and differentiated into kinetosomes and undulipodia at about the same time in spirochetes.
Chloroplasts evolved from autotroph endosymbionts of external origin from autotroph prokaryots.	Chloroplasts originated by internal differentiation of the mentioned DNA molecules as the last series of daughter DNA molecules.

The replication function of both nucleic acids and proteins soon differentiated into the reproduction of the new individuals and the growth of their bodies (virions). First the proteins were produced and accumulated separately drom the DNA molecules. This is demonstrated by the capsids of various lower viruses. The DNA molecules increased just by growth of their helix in length originally, by which new groups of genes originated. These structures then determined the transcription of RNA molecules which differentiated soon to form the transfer RNA, ribosome RNA, messenger RNA gradually producing proteins with new properties and functions. Metabolism formed by anabolism and katabolism was the first of them, being quite indispensable for all organisms since the very beginnings. They soon differentiated into a series of special functions each like nutrition, respiration, hydratation etc. With the later differentiation of organismus into autotrophic sessile plants and heterotrophic motile animals the intermediate position of fungi, a series of further functions developed specific for each of the groups. The formation of biomembranes, both external and internal ones, has meant a further important step. It resulted in compartmentalisation and specialization of the enzymes action. In this way the protoplasm (cytoplasm and nucleoplasm) gradually developed and the internal medium evolved to the recent state.

The sense and usefulness of a scientific theory do not depend, according to my opinion, so much on its possible later confirmation, but on its synthetizing the recent state of research and orienting the attention of research workers on questions deserving deeper concern and interest. The most of new data and conclusions of this paper have been based on the excellent book by Lynn Margulis (1981) the great contribution of which to the question would not be lessened if the present approach would prove to be more probable by the further research.

REFERENCES

Blaškovič, D., 1978: The Bases of Medical Virology (in Slovakian), SAV Publisher, Bratislava, 522 pp.

Fields, B.N., and D.M.Knipe, 1986: Fundamental Virology, Raven Press Book Ltd., part I, 494 pp.

Fox, S.W., 1969: Self-ordered polymers and propagative cell-like systems, Naturwissenschaften, 56, 1-9.

Haldane, J.B.S., 1928: The origin of life, Rationalist Annual 148.

Klinkowski, M., 1967: Pflanzliche Virologie, Akad.Verlag, Berlin, Bd.I, 388 pp.

Margulis, L., 1978: Symbiosis in Cell Evolution, W.H.Freeman and Comp., San Francisco, IInd ed. 1981 419 pp.

Miller, S.L., 1953: A production of amino acids under possible primitive earth conditions, Science 117, 528-529.

Novák, V.J.A., 1967: Sociability or association of individuals of the same species (in Russian), Zhurn.obshch. biol., Moscow, 28: 387-396.

Novák, V.J.A., 1975: Das Prinzip der Sociabilität (in German), Nova Acta Leopoldina, Halle, 42 :475-488.

Novák, V.J.A., 1982: The Principle of Sociogenesis, Academia Praha, 214 pp.

Novák, V.J.A., 1983: Neoteny as one of the general laws of evolution, General Questions of Evolution, Proc. Int.Coll. Liblice, ČSAV Praha, 329-345.

Novák, V.J.A., 1988: The dialectics of the relation between hereditary and non-hereditary changes, Evolution and Morphogenesis II, Proc.Int.Symp.Plzeň 1984, CSAV Praha, 89-95.

Novák, V.J.A., 1974: The coacervate-in-coacervate theory of the origin of life, in Dose, K. et al.(Eds.) The Origin of Life and Evolutionary Biology, Plenum Press, N.York, 355-368.

Novák, V.J.A., 1992: The Darwin-Haeckel law of the recapitulation and present-day evolutionary biology, Acta Med.Fac.UK Plzeň, 60:11-24.

Oparin, A.I., 1922: Origin of Life (Reports of the Meeting of the Russ.Botanic Society), Moscow. Appeared in Moscovskii rabochii 1924.

Oparin, A.I., 1968: Genesis and Evolutionary Development of Life, Acad.Press N.York and London, 203 pp.

Schwemmler, W., 1979: Mechanismen der Zellevolution, W.de Gruyter, Berlin, N.York, 275 pp.

Segal, J., U.Körner and K.P.Leiterer, 1983: Die Entstehung des Lebens aus biophysikalischer Sicht, G.Fischer, Jena, 454 pp.

Studitsky, A.N., 1975: The Darwin-Severtzov's rule and its significance for the theory of evolutionary origin of the cell-structure, Evolutionary Biology, Proc.Int. Conf.Liblice, 309-324.

Studitsky, A.N., 1981: The Evolutionary Morphology of the Cell, Nauka, Moscow, 280 pp.

(Note to the p. 8) The observed codding of about 50 % of features of chloroplasts, ca. 65 % of mitochondria and 80 % of microtubule bodies in the cell nucleus (cf.Schwemler, 1979) is another argument in favour of the polyploid synbacteriogenesis conception. The parallel evolution of the structures inside the cell plasm and in free-living organisms seems to have been made possible by the conditions in the cell similar to those in the primary coacervate.

CONFERENCE ON
CHEMICAL EVOLUTION AND THE ORIGIN OF LIFE
Trieste, Italy 26–30 October 1993

PRELIMINARY PROGRAMME

MONDAY, 26 OCTOBER 1992

Morning Session

Chairperson: ***H. Baltscheffsky,*** University of Stockholm

9:00 - 10:30 Opening Remarks
Professor Abdus Salam

General Overview
C. Ponnamperuma, University of Maryland

10:30 - 11:00 **BREAK**

11:00 - 12:30 Earliest Records of Life on Earth
J.W. Schopf, University of California, Los Angeles

12:30 - 14:30 **LUNCH**

Afternoon Session

Chairperson: ***W. Thiemann,*** University of Bremen

14:30 - 16:00 Phosphate in Models for Molecular Evolution
G. Arrhenius, University of California, San Diego

16:00 - 16:30 **BREAK**

Discussion and Short Contributions

Discussion Leaders: ***I.G. Draganic,*** Institute of Nuclear Sciences, Belgrade
W. Wang, Peking University

16:30 - 18:00 **Contributions:**

Biological Markers in Precambrian Sediments - Indian Subcontinent
M.S. Chadha, Bhabha Atomic Research Centre, Bombay

Could Comets be Carriers of Intact Homochiral Biomolecules from Interstellar Space?
R.K. Khanna, University of Maryland

The Need to Study the Origin of Life
K. Jamil, Indian Institute of Chemical Technology, Hyderabad

TUESDAY, 27 OCTOBER 1992

Morning Session

Chairperson: ***G. Arrhenius,*** University of California, San Diego

9:00 - 10:30	Prebiotic Chemistry ***C. Ponnamperuma,*** University of Maryland
10:30 - 11:00	**BREAK**
11:00 - 12:30	Evolution in an RNA World ***P. Schuster,*** University of Vienna
12:30 - 14:30	**LUNCH**

Afternoon Session

Chairperson: ***N. Rahman,*** ICS, Trieste

14:30 - 16:00	From Amino Acids to Chemically Active Peptides ***A. Brack,*** Centre de Biophysique Moleculaire, Orleans
16:00 - 16:30	**BREAK**

Discussion and Short Contributions

Discussion Leaders: ***M.S. Chadha,*** Bhabha Atomic Research Centrc, Bombay
G. Gilat, Israel Institute of Physics, Haifa

16:30 - 18:00	**Contributions:**
	Chromoproteinoids and their Ability to Form Boundary Structures ***B. Heinz,*** Palomar College, San Marcos, California
	Ionizing Radiation and Water in Chemical Processing in Space and on Early Earth ***I.G. Draganic,*** Institute of Nuclear Sciences, Belgrade
	Chemical Effects of Ionizing Radiation and Sonic Energy in the Context of Chemical Evolution ***A. Negron-Mendoza,*** Institute of Nuclear Science, UNAM
	The Role of Neoteny and Sociogenesis in the Evolution of Cell Structure ***V.J.A. Novak,*** Laboratory of Evolutionary Biology, Prague
18:15	**RECEPTION** *(Cafeteria of the Main Building)*

WEDNESDAY, 28 OCTOBER 1992

Morning Session

Chairperson: ***B. Heinz,*** Palomar College, San Marcos, California

9:00 - 10:30	Chemical Origin and Early Evolution of Biological Energy Conversion ***H. Baltscheffsky,*** University of Stockholm
10:30 - 11:00	**BREAK**
11:00 - 12:30	Small Pathogenic RNAs of Plants: Living Fossils of the RNA World? ***T.O. Diener,*** University of Maryland
12:30 - 14:30	**LUNCH**

Afternoon Session

Chairperson: ***A.S. Garay,*** Texas A&M University

14:30 - 16:00	The Weak Force and the Origin of Life ***A. MacDermott,*** University of Oxford
16:00 - 16:30	**BREAK**

Discussion and Short Contributions

Discussion Leaders: ***K. Jamil,*** Indian Institute of Chemical Technology, Hyderabad
R. Mohan, Max Planck Institute, Gottingen

16:30 - 18:00	**Contributions:**
	Viruses and Viroids at the Origin of Organized Life ***L.J. Boya,*** University of Zaragoza, Spain
	True and False Chirality ***L.D. Barron,*** University of Glasgow
	Experiments in Search of the Origin of Biomolecular Asymmetry ***L. Keszthelyi,*** Biological Research Centre, Szeged, Hungary

THURSDAY, 29 OCTOBER 1992

Morning Session

Chairperson: ***L.D. Barron,*** University of Glasgow

9:00 - 10:30	Bifurcation with Symmetry Breaking as the Cause of Biologial Big Bang ***V.A. Avetisov,*** Semenov Institute of Chemical Physics, Moscow
10:30 - 11:00	**BREAK**
11:00 - 12:30	The Origin of Chirality, the Role of Phase Transitions, and their Induction in Amino Acids ***Abdus Salam,*** ICTP, Trieste
12:30 - 14:30	**LUNCH**

Afternoon Session

Chairperson: ***L. Keszthelyi,*** Biological Research Centre, Szeged, Hungary

14:30 - 16:00	Chirality: Hypothesis and New Experiments Theoretical Considerations: Spontaneous Regulating Mechanisms That May Have Led to the Origin of Life ***J. Chela-Flores,*** ICTP, Trieste and IDEA, Caracas An Attempted Experiment ***R.K. Khanna,*** University of Maryland
16:00 - 16:30	**BREAK**

Discussion and Short Contributions

Discussion Leaders: ***A. Negron-Mendoza,*** Institute of Nuclear Science, UNAM, Mexico
V.J.A. Novak, Laboratory of Evolutionary Biology, Prague

16:30 - 18:00	**Contributions:** Search for Phase Transitions Changing Molecular Chirality ***A. Figureau,*** Institute of Nuclear Physics, Lyon Chiral Interaction and Biomolecular Evolution ***G. Gilat,*** Israel Institute of Technology, Haifol Chiral Forces and Molecular Dissymmetry ***R. Mohan,*** Max Planck Institute, Gottingen Differences in Radiolysis Behaviour of D, L Amino Acid in Primary Stage and Thermodynamic State ***W. Wang,*** Peking University

FRIDAY, 30 OCTOBER 1992

Morning Session

Chairperson: ***A. Falaschi,*** ICGEB, Trieste

9:00 - 10:30	Experimental Studies on the Possibility of Chirality Dependent Time Reversal ***A.S. Garay,*** Texas A&M University
10:30 - 11:00	**BREAK**
11:00 - 12:30	Physicochemical Basis for the Origin of the Genetic Code ***C. Ponnamperuma,*** University of Maryland
12:30 - 14:30	**LUNCH**

Afternoon Session

Chairperson: ***J.W. Schopf,*** University of California, Los Angeles

14:30 - 16:00	The Search for Extraterrestrial Intelligence ***J. Heidmann,*** Observatoire de Paris
16:00 - 16:30	**BREAK**
16:30 - 18:00	Summary: Origin of Life 1992

- END OF CONFERENCE -

****

SATURDAY, 31 OCTOBER 1992

Venue: Room 237 – Main Building

Special Session

Chairperson: ***C. Ponnamperuma,*** University of Maryland

9:00 - 10:30	Discussion of New Experiments on the Origin of Amino Acid Chirality *(By special invitation)*
10:30 - 11:00	**BREAK**
11:00 - 12:30	Continuation of the Morning Discussion
12:30 - 14:30	**LUNCH**

LIST OF PARTICIPANTS

P. Aich, Saha Institute of Nuclear Physics, Biophysics Division, 37 Belgachia Road, 700 037 Calcutta, India

I. Arif, Bandung Institute of Technology, Department of Physics, Facility of Science and Mathematics, Jl Ganesha 10, P.O. Box 273, 40132 Bandung, Indonesia

G.O. Arrhenius, Scripps Institution of Oceanography, University of California, San Diego, Mail Code A-025, La Jolla, California 92093, USA

V. Avetisov, Nn Semenov Institute Biochemical Physics, 4 Ulitsa Kosygina, 117334 Moscow, Russia

F.L. Babani, Biological Research Centre, Rr. Irfan Tomini, Tirana, Albania

H. Baltscheffsky, University of Stockholm, Arrhenius Laboratory, Department of Biochemistry S-106 91 Stockholm, Sweden

L. Barron, University of Glasgow, Department of Chemistry, G12 8QO Glasgow, United Kingdom

N.Y.K. Baryeh, University of Science and Technology, Department of Biological Sciences, Kumasi, Ghana

L.J. Boya, Universidad de Zaragoza, Departamento de Fisica Teorica, Facultad de Ciencias, E-50009 Zaragoza, Spain

A. Brack, Centre National de La Recherche Scientifique, Centre de Biophysique-Moleculaire, 1A Avenue de La Recherche Scientifique, 45071 Cedex 2 Orleans, France

J.C. Cecchini, Universidad Nacional de Tucuman, Instituto de Fisica, Fac. Ciencias Exact. Technologia, Av. Roca 1900, RA-4000 San Miguel de Tucuman, Argentina

M.S. Chadha, Bhabha Atomic Research Centre, Bio-Organic Division, 400 085 Bombay, India

J. Chela-Flores, Instituto Internacional de Estudios Avanzados, Apartado 17606, Parque Central, 1015-A Caracas, Venezuela

S. Chen, Fudan University, Institute of Mathematics, 220 Handan Road, 200433 Shanghai, People's Republic of China

C. Cosmovici, Consiglio Nazionale Delle Ricerche, Instituto di Fisica Dello Spazio Interplanetario, Via G. Galilei, Casella Postale 27, 00044 Frascati, Italy

S. Dhanuskodi, Bharathidasan University, Department of Physics, Faculty of Science, Palkalaiperur Campus, Tiruchirapalli, 620 024 Tamil Nadu, India

J.M. Diaz, Universidad Nacional de Colombia, Departamento de Fisica, Facultad de Ciencias, P.O. Box 53912, Cra 30 Calle 45, Santa Fe de Bogota, Colombia

T.O. Diener, University of Maryland, Department of Botany, College Park, Maryland 20742, USA

B. Dorjsuren, Academy of Science, Institute of Biotechnology, 51 Ulan Bator, Mongolia

I. Draganic, The Insitute of Nuclear Sciences Vinca, P.O. Box 522, 11001 Belgrade, Serbia

A. Falaschi, Area di Ricerca, International Centre for Genetic Engineering and Biotechnology, Padriciano 99, 34100 Trieste, Italy

P. Fernandes Machado, Universidade Federal Do Rio de Janeiro, Instituto de Quimica, Department de Fisico-Quimica, Ct-Bloco A. Sala 406, Ilha Do Fundao, Cidade Universitaria, 21910 Rio de Janeiro, Brazil

A. Fernandez, The Medical School, Department of Biochemical and Molecular Biology, P.O. Box 016129, Miami, Florida 33101–6129, USA

A.R.J. Figureau, Institut de Physique Nucleaire de Lyon, 43 Blvd Du 11 Novembre 1918, 69622 Villeurbanne, France

C.M. Gallo Lopez-Aliaga, Universidad Peruana Cayetano Heredia, Laboratory of Biochemistry, Department of Physiological Sciences, Honorio Delgado 430, P.O. Box 5045, San Martin de Porres, Lima, Peru

A. Garay, Texas A&M University, Department of Biochemistry/Biophysics, College Station, Texas 77843–2128, USA

G. Gilat, Israel Institute of Technology Technion, Department of Physics, Technion City, 32 000 Haifa, Israel

W. Hebel, Directorate General for Science, Research and Development, Commission of the European Communities, Rue de La Loi 200, B-1049 Brussels, Belgium

J. Heidmann, Observatoire de Paris, 92195 Principal Cedex, Meudon, France

B. Heinz, Palomar College, 1140 West Mission Road, San Marcos, California 92069, USA

K. Jamil, Indian Institute of Chemical Technology, Biology Division, 500 007 Hyderabad, India

L. Keszthelyi, Institute of Biophysics, Biological Research Center, P.O. Box 521, Temesvari Kert. 62, H-6701 Szeged, Hungary

M.A. Khan, University of Karachi, Department of Botany, 75270 Karachi, Pakistan

R.K. Khanna, University of Maryland at College Park, Laboratory of Chemical Evolution, Department of Chemistry, College Park, Maryland 20742, USA

G. Lenaz, Universita degli Studi di Bologna, Dipartimento di Biochimica, Facolta di Medicina, Via Irnerio 48, 40126 Bologna, Italy

V. Lopac, University of Zagreb, Department of Physics, Faculty of Chemical Engineering and Technology, Marulicev Trg 19, 41000 Zagreb, Croatia

L.-F. Luo, Neimenggu University, Department of Physics, Hohhot, People's Republic of China

A.J. MacDermott, University of Oxford, Physical Chemistry Laboratory, South Parks Road, OX1 3QZ Oxford, United Kingdom

B. Mathew Graham, North Eastern Hill University, Institute of Fundamental Research on Complex Systems, Bijni Complex, Bhagyakul, 793 003 Shillong, India

R. Mohan, National Institute of Health, Department of Health and Human Services, Bethesda, Maryland 20892, USA

S. Nasi, Universita di Roma La Sapienza, Centro di Studio Per Gli Acidi Nucleici, C/O Dipartimento de Genetica e Biologia Molecolare, 00185 Roma, Italy

K.N. Nedev, University of Sofia, Biophysics Laboratory, Facility of Physics, 5 James Baucher Boulevard, 1126 Sofia, Bulgaria

A. Negron-Mendoza, Universidad Nacional Autonoma de Mexico, Instituto de Ciencias Nucleares, Apdo. Postal 70–543, Circuito Exterior, C.U., 04510 DF Mexico City, Mexico

V.J.A. Novak, Czechoslovak Academy of Sciences, Laboratory of Evolutionary Biology, Oldrichova 39, 12800 Prague, Czech and Slovak

O.R. Olguin, Universidad Nacional de San Luis, Escuelas de Fisica Y de Matematicas, Ejercito de Los Andes 950, 5700 San Luis, Argentina

L. Peliti, Universita di Napoli, Dip. di Scienze Fisiche, Mostra D'Oltremare, Pad. 19, 80125 Napoli, Italy

C. Ponnamperuma, University of Maryland at College Park, Laboratory of Chemical Evolution, Department of Chemistry, College Park, Maryland 20742, USA

A. Popescu, University of Bucharest, Department of Biophysics, Faculty of Physics, P.O. Box 5211, Bucharest, Romania

M.K. Rahman, University of Dhaka, Department of Biochemistry, 1000 Dhaka, Bangladesh

N.K. Rahman, Area di Ricerca I.C.C., Padriciano 99, 34012 Trieste, Italy

H.J.P. Reitboeck, Philipps-Universitat Marburg, Ag Angewandte Physik U. Biophysik, Fachbereich Physik, Renthof 7, D-3550 Marburg, Germany

T.Z. Rizvi, Pakistan Council of Scientific and Industrial Research Laboratories, Shahrah-E-Jalal-U-Din Roomi, 54600 Lahore, Pakistan

M. Rizzotti, Universita degli Studi di Padova, Dipartimento di Biologia, Via Trieste 75, 35121 Padova, Italy

A. Salam, International Centre for Theoretical Physics, P.O. Box 586, Miramare 34100 Trieste, Italy

A.A. Salama, University of Khartoum, Department of Biology, Faculty of Education, P.O. Box 406, Omdurman, Sudan

J.W. Schopf, University of California, Cseol, Geology Building, Los Angeles 90024, USA

P. Schuster, University of Vienna, Institut Fur Theoretische Chemie Und Strahlenchemie, Wahringerstrasse 38, 1090 Vienna, Austria

P.A.S. Silva, Universidade de Lisboa, Departamento de Fisica, Campo Grande, Ed.Cl-Piso 4, 1700 Lisbon, Portugal

W.H.P. Thiemann, Universitat Bremen, Fachbereich 2-Chemie, Postfach 33 04 40, Leobenerstrasse Nw 2, 2800 Bremen, Germany

N. Tuteja, Arca di Ricerca, International Centre for Genetic Engineering and Biotechnology, Padriciano 99, 34100 Trieste, Italy

C.-X. Wang, University of Science and Technology of China, Centre for Fundamental Physics, 24 Jinzhai Road, Anhui, 230026 Hefei, People's Republic of China

W. Wang, Beijing University, University of Peking, Department of Technical Physics, 100871, Beijing, People's Republic of China

R. Weerasooriya, Institute of Fundamental Studies, Hantana Road, Kandy, Sri Lanka

R. Weiner, University of Marburg, Department of Physics, Mainzergasse 33, 3550 Marburg, Germany

Y.Y.J. Yan, Peking University, Department of Physics, 100871 Beijing, People's Republic of China

T. Yohannes, Addis Ababa University, Bahir Dar Teachers' College, P.O. Box 79, Bahir Dar, Ethiopia